Equilibrium Statistical Physics

Marc Baus · Carlos F. Tejero

Equilibrium Statistical Physics

Phases, Phase Transitions, and Topological Phases

Second Edition

 Springer

Marc Baus
Faculté des Sciences
Université Libre de Bruxelles
Brussels, Belgium

Carlos F. Tejero
Departamento Física Aplicada I
Universidad Complutense de Madrid
Madrid, Spain

ISBN 978-3-030-75434-1   ISBN 978-3-030-75432-7 (eBook)
https://doi.org/10.1007/978-3-030-75432-7

This Springer imprint is published by the registered company Springer Nature Switzerland AG
The registered company address is: Gewerbestrasse 11, 6330 Cham, Switzerland

To our daughters Katia, Erika[†], Belén and Eva

Preface to the Second Edition

As mentioned in the preface to the first edition, at present Statistical Physics is concerned with much more states of matter than was the case originally. In order to catch up with the recent interests, we have completely rewritten Chap. 12, which is now devoted both to the study of topological line defects in nematic liquid crystals and to the study of the Kosterlitz-Thouless topological phase transition in a classical spin system. We also corrected a few misprints, clarified our explanation of the Monte Carlo method in Chap. 7, updated the references, defined the bi-axial nematic liquid crystals in Appendix E, added a study of the Gaussian fluctuations in Appendix B.4, as well as two more Appendices (F and G) devoted to mathematical questions of interest for the understanding of the Kosterlitz-Thouless transition.

Brussels, Belgium   Marc Baus
Madrid, Spain   Carlos F. Tejero
June 2020

Preface to the First Edition

The purpose of this textbook is to introduce the student to a basic area of macroscopic physics, namely the statistical mechanical study of the different phases of matter, as well as the phase transitions between them. Although many books on statistical physics, for both equilibrium and non-equilibrium systems, are already available, they largely differ in contents. This generally reflects not only the different interests of their authors, but also the epoch in which they were written. For instance, the early books did usually devote much space to problems of the solid state, whereas later ones do include, moreover, several aspects of the liquid state. At present, however, the main emphasis in physics is on soft matter (e.g. liquid crystals, colloids, polymers), and therefore these particular states of matter have also been included in this volume. The main purpose of this textbook will consist, hence, in providing its students with a first introduction, within the general framework of equilibrium statistical physics, to a much larger variety of phases and phase transitions than was previously the case for textbooks of statistical mechanics. Many of these novel topics do, of course, deserve a more detailed study than the one which can be provided here. Indeed, in the spirit of a first introduction, only very simple models of these phases will be given, but more detailed information can be found in the suggestions for further reading given in the References.

For pedagogical reasons, the subject matter of this book has been divided into four parts (Parts I–IV), and a series of appendices (Appendices A–D), devoted to some mathematical tools used in the main text.

In Part I (Chaps. 1–3), a summary is provided of the mechanical (Chap. 1) and thermodynamical (Chap. 2) basis of the postulates of equilibrium statistical physics (Chap. 3). Although most students of equilibrium statistical mechanics will in general have a prior knowledge of, classical and quantum, mechanics and of equilibrium thermodynamics, by including a summary of these topics the present book becomes more self-contained. A fact generally well appreciated by the students.

In Part II (Chaps. 4–6), the general principles of equilibrium statistical physics are illustrated for the simple case of the non-interacting or ideal systems. The three main Gibbs ensembles, microcanonical (Chap. 4), canonical (Chap. 5) and grand canonical (Chap. 6), are studied, together with some of their standard applications. These topics, although simple in principle, do nevertheless draw the students' attention to some of the subtleties of statistical physics.

In Part III (Chaps. 7–9), these general principles are applied to the less simple interacting or non-ideal systems. This, of course, can only be done approximately and the most current approximation methods for the study of classical interacting systems are summarized (Chap. 7). Based on these methods, some simple models for most of the current phases of matter are introduced (Chap. 8), and the phase transitions between some of these phases are subsequently studied (Chap. 9). Although the simple models of these phases considered here are only caricatures of the real systems, their study will prepare the students for the more realistic, but also more difficult, studies found in the current literature.

Finally, Part IV (Chaps. 10–13) is devoted to an introduction to some more advanced material. This includes an introduction to critical phenomena and the renormalization group calculation of critical exponents (Chap. 10), the study of interfaces and the calculation of the surface tension (Chap. 11), the study of topological defects (in nematic liquid crystals) and the resulting texture (Chap. 12), and the classical theories of the phase transformation kinetics (Chap. 13). It is hoped that exposing the students to an elementary treatment of these more advanced topics will encourage them to also study these topics in more detail.

The present textbook is aimed at students of condensed matter physics, physical chemistry or materials science, but throughout the level of rigor is one which will be most familiar to students of theoretical physics. Likewise, the references quoted at the end of each chapter mainly focus on books where supplementary material can be found with a similar degree of rigor, i.e. those texts most useful to the readers of the present one.

This volume is an enlarged version of a previous text, originally written in Spanish (C. F. Tejero y M. Baus, *Física estadística del equilibrio. Fases de la materia,* A.D.I., Madrid (2000), ISBN 84-931805-0-5). We are indebted to M. López de Haro for its translation into English and also for helpful and interesting suggestions. We also thank T. de Vos for his help with the figures. This final version owns much to the many discussions with our students at the Faculté des Sciences (Université Libre de Bruxelles) and at the Facultad de Ciencias Físicas (Universidad Complutense de Madrid) and with our colleagues, M. Fisher, R. Lovett, J. P. Ryckaert and J. M. Ortiz de Zarate, all of whom are gratefully acknowledged. Finally, we would also like to thank our wifes, Myriam and Isabel, for their support.

Brussels, Belgium   Marc Baus
Madrid, Spain   Carlos F. Tejero
June 2007

Contents

Part I
Basics

Chapter 1
Mechanics

Abstract In this first chapter, some of the basic concepts of Hamiltonian mechanics (both classical and quantum mechanical), which are required for the study of statistical physics, are summarized. For a more detailed analysis of these topics the reader is referred to the books included in the Further Reading.

1.1 Classical Mechanics

In classical mechanics, the location of a particle or material point in Euclidean space is determined by its position vector $\mathbf{r}$, whose components are the Cartesian coordinates x, y, z, where $\mathbf{r} = x\mathbf{e}_x + y\mathbf{e}_y + z\mathbf{e}_z$. Here, $\mathbf{e}_x, \mathbf{e}_y$, and $\mathbf{e}_z$ are unit vectors in the direction of the Cartesian axes. The above-mentioned three variables define univocally the position of the particle and, therefore, it is said that the material point has three degrees of freedom. Note that the position of the material point may, for instance, be likewise determined by using its spherical or cylindrical coordinates. Hence, the notation q_1, q_2, q_3 will be used generically for any set of three independent variables defining univocally the position of the particle. Such variables are referred to as generalized coordinates of the material point. In the case of a system of N material points, the number of degrees of freedom is, therefore, $3N$, and the position of the particles is completely specified by $3N$ generalized coordinates $q_1, ..., q_{3N}$ that are functions of time $q_i \equiv q_i(t)$ $(i = 1, ..., 3N)$.

Even if the generalized coordinates of a system of N material points are known at a certain instant of time, it is not possible to predict its temporal evolution since the generalized velocities, $\dot{q}_i \equiv \dot{q}_i(t) = dq_i(t)/dt$, may still be arbitrary. The fundamental principle of classical dynamics states that if the values of the coordinates $(q \equiv q_1, ..., q_{3N})$ and of the generalized velocities $(\dot{q} \equiv \dot{q}_1, ..., \dot{q}_{3N})$ are known simultaneously at a given time, then one can determine the value of these variables at any other time.

Let $L_N(q, \dot{q})$ denote the Lagrange function or Lagrangian of the system which, in what follows, will be assumed not to depend explicitly on time (conservative system). This function can be written as

M. Baus and C. F. Tejero, *Equilibrium Statistical Physics*,
https://doi.org/10.1007/978-3-030-75432-7_1

$$L_N(q, \dot{q}) = H_N^{\text{id}}(q, \dot{q}) - U_N(q), \tag{1.1}$$

where the first term in the r.h.s., $H_N^{\text{id}}(q, \dot{q})$, is the kinetic energy of the system and the second one, $U_N(q)$, is the potential energy. In Cartesian coordinates, the kinetic energy of a system of N particles of mass m is given by

$$H_N^{\text{id}}(\mathbf{v}^N) = \sum_{j=1}^{N} \frac{1}{2} m v_j^2, \tag{1.2}$$

where $\mathbf{v}_j = \mathbf{v}_j(t) = d\mathbf{r}_j/dt$ is the velocity of particle j, whose position vector is $\mathbf{r}_j \equiv \mathbf{r}_j(t)$, and $\mathbf{v}^N \equiv \mathbf{v}_1, ..., \mathbf{v}_N$. Note that although in this coordinate system the kinetic energy is only a function of the generalized velocities, when (1.2) is expressed, for instance, in cylindrical coordinates then it becomes a function of both the coordinates and the generalized velocities.

In what follows the potential energy will be written as

$$U_N(\mathbf{r}^N) = H_N^{\text{int}}(\mathbf{r}^N) + H_N^{\text{ext}}(\mathbf{r}^N), \tag{1.3}$$

with

$$H_N^{\text{int}}(\mathbf{r}^N) = \frac{1}{2} \sum_{i=1}^{N} \sum_{i \neq j=1}^{N} V(|\mathbf{r}_i - \mathbf{r}_j|), \tag{1.4}$$

and

$$H_N^{\text{ext}}(\mathbf{r}^N) = \sum_{j=1}^{N} \phi(\mathbf{r}_j), \tag{1.5}$$

where $V(|\mathbf{r}_i - \mathbf{r}_j|)$ is the pair potential interaction between particles i and j, which has been assumed to be a function of the modulus of the relative distance between them, $|\mathbf{r}_i - \mathbf{r}_j|$, and $\phi(\mathbf{r}_j)$ is the potential due to an external field acting on particle j, while $\mathbf{r}^N = \mathbf{r}_1, ..., \mathbf{r}_N$.

1.2   Hamilton's Equations

Even though the laws of classical mechanics may be formulated in terms of the generalized coordinates and velocities of the particles (Lagrangian dynamics), in this chapter only the Hamiltonian dynamics will be considered. In this case, the description of the mechanical state of the system is performed using the generalized coordinates ($q \equiv q_1, ..., q_{3N}$) and their conjugate momenta ($p \equiv p_1, ..., p_{3N}$), where

$$p_i \equiv p_i(t) = \frac{\partial L_N(q, \dot{q})}{\partial \dot{q}_i}. \tag{1.6}$$

The time evolution of these variables is given by the system of $6N$ first-order differential equations or Hamilton's equations, namely

$$\dot{q}_i = \frac{\partial H_N(q, p)}{\partial p_i}, \quad \dot{p}_i = -\frac{\partial H_N(q, p)}{\partial q_i}, \quad (i = 1, \dots, 3N), \tag{1.7}$$

where $H_N(q, p)$ is the Hamilton function or Hamiltonian of the system, defined by the equation

$$H_N(q, p) = \sum_{i=1}^{3N} p_i \dot{q}_i - L_N(q, \dot{q}), \tag{1.8}$$

which is a particular case of a Legendre transformation (see Appendix A).

From (1.1), (1.2) and (1.6) it follows that $\mathbf{p}_j = m\mathbf{v}_j$ and hence

$$H_N\left(\mathbf{r}^N, \mathbf{p}^N\right) = \sum_{j=1}^{N} \frac{1}{2m} \mathbf{p}_j^2 + U_N\left(\mathbf{r}^N\right), \tag{1.9}$$

which is the Hamiltonian of a system of N particles of mass m in Cartesian coordinates, with $\mathbf{p}^N \equiv \mathbf{p}_1, \dots, \mathbf{p}_N$.

Since in Hamilton's dynamics the mechanical state of a system of N particles is determined at each time by $3N$ generalized coordinates and their $3N$ conjugate momenta, one may represent such a state by a point in the $6N$-dimensional space generated by $q_1, \dots, q_{3N}, p_1, \dots, p_{3N}$. This space is known as the phase space or Γ-space of the system. According to (1.7), the mechanical state follows in time a trajectory in Γ-space (phase space trajectory), and one and only one trajectory passes through a point of this space.

The formal solution to Hamilton's equations can be written as

$$q_i^t = q_i\left(q^0, p^0; t\right), \quad p_i^t = p_i\left(q^0, p^0; t\right), \quad (i = 1, \dots, 3N), \tag{1.10}$$

where q_i^t and p_i^t are the generalized coordinates and their conjugate momenta at time t, which are functions of the initial mechanical state $q^0, p^0 = q_1^0, \dots, q_{3N}^0, p_1^0, \dots, p_{3N}^0$ and of time.

1.3 External Parameters

In all of the above it has been assumed that the Hamiltonian of a system of N material points is a function of the generalized coordinates and their conjugate momenta.

Throughout the text, systems are studied in which the Hamiltonian is also a function of certain parameters, referred to as external parameters because they depend on the position of bodies not belonging to the system. A first example is the case of a particle of mass m and charge e placed in a constant and uniform magnetic field $\mathbf{B}$. If $\mathbf{A}(\mathbf{r})$ denotes the vector potential

$$\mathbf{A}(\mathbf{r}) = \frac{1}{2}\mathbf{B} \times \mathbf{r}, \tag{1.11}$$

and c is the speed of light in vacuum, the Hamiltonian of the particle will be

$$H_1(\mathbf{r}, \mathbf{p}; \mathbf{B}) = \frac{1}{2m}\left(\mathbf{p} - \frac{e}{c}\mathbf{A}(\mathbf{r})\right)^2, \tag{1.12}$$

which is a function of the external parameter $\mathbf{B}$. Upon development of the square in (1.12) one finds

$$\frac{1}{2m}\left(\mathbf{p} - \frac{e}{c}\mathbf{A}(\mathbf{r})\right)^2 = \frac{1}{2m}\mathbf{p}^2 - \frac{e}{2mc}\mathbf{l} \cdot \mathbf{B} + \frac{e^2}{2mc^2}\mathbf{A}^2(\mathbf{r}), \tag{1.13}$$

where

$$\mathbf{l} = \mathbf{r} \times \mathbf{p} \tag{1.14}$$

is the angular momentum of the particle. The last two contributions in the r.h.s. of (1.13) are called, respectively, paramagnetic and diamagnetic. In most cases the diamagnetic term is much smaller than the paramagnetic one, so the Hamiltonian of a particle of mass m and charge e placed in a constant and uniform magnetic field $\mathbf{B}$ can be written in the form

$$H_1(\mathbf{r}, \mathbf{p}; \mathbf{B}) = \frac{1}{2m}\mathbf{p}^2 - \boldsymbol{\mu} \cdot \mathbf{B}, \tag{1.15}$$

where

$$\boldsymbol{\mu} = \frac{e}{2mc}\mathbf{l} \tag{1.16}$$

is the magnetic moment of the particle. Note that in an ideal system of N identical particles placed in a constant and uniform magnetic field one has

$$H_N(\mathbf{r}^N, \mathbf{p}^N; \mathbf{B}) = \sum_{j=1}^{N} H_1(\mathbf{r}_j, \mathbf{p}_j; \mathbf{B}). \tag{1.17}$$

Therefore,

$$\frac{\partial H_N\left(\mathbf{r}^N, \mathbf{p}^N; \mathbf{B}\right)}{\partial \mathbf{B}} = -\sum_{j=1}^N \mu_j \equiv -\mathbf{M}_N\left(\mathbf{r}^N, \mathbf{p}^N\right), \tag{1.18}$$

where μ_j is the magnetic moment of particle j and $\mathbf{M}_N(\mathbf{r}^N, \mathbf{p}^N)$ is the total magnetic moment dynamical function. When the magnetic field is modified, the variation of the Hamiltonian is thus given by

$$d H_N\left(\mathbf{r}^N, \mathbf{p}^N; \mathbf{B}\right) = \frac{\partial H_N\left(\mathbf{r}^N, \mathbf{p}^N; \mathbf{B}\right)}{\partial \mathbf{B}} \cdot d\mathbf{B} = -\mathbf{M}_N\left(\mathbf{r}^N, \mathbf{p}^N\right) \cdot d\mathbf{B}, \tag{1.19}$$

which is the mechanical work performed when modifying the external parameter $\mathbf{B}$.

In the majority of the problems that are analyzed in physics, the particles are placed in a container R of volume V, which is another example of an external parameter. Even if within the container the interaction of a particle with the walls may be negligible, this interaction becomes more important when the particle is closer to the wall. In contrast with (1.12), the particle–container interaction is not known exactly in classical mechanics. A possible way to take into account this interaction consists in writing the Hamiltonian of the system as

$$H_N\left(\mathbf{r}^N, \mathbf{p}^N; V\right) = \sum_{j=1}^N \frac{1}{2m} \mathbf{p}_j^2 + U_N\left(\mathbf{r}^N\right) + \sum_{j=1}^N \phi_R\left(\mathbf{r}_j\right), \tag{1.20}$$

where $\phi_R(\mathbf{r}_j)$ is the potential that keeps the particle inside the container. Such a potential vanishes when the particle is within the container and becomes (infinitely) repulsive as the particle approaches the wall. In this way the Hamiltonian of the system explicitly depends on the volume of the container. In analogy with (1.19), the variation of the Hamiltonian when the volume changes is

$$d H_N\left(\mathbf{r}^N, \mathbf{p}^N; V\right) = \frac{\partial H_N\left(\mathbf{r}^N, \mathbf{p}^N; V\right)}{\partial V} dV = -p_N\left(\mathbf{r}^N, \mathbf{p}^N; V\right) dV, \tag{1.21}$$

where

$$p_N\left(\mathbf{r}^N, \mathbf{p}^N; V\right) \equiv -\frac{\partial H_N\left(\mathbf{r}^N, \mathbf{p}^N; V\right)}{\partial V}, \tag{1.22}$$

is the pressure dynamical function. The expression (1.21) is the mechanical work performed upon changing the external parameter V.

In what follows, the Hamiltonian of a system of N particles that involves one or several external parameters α will be indicated as $H_N(q, p; \alpha)$.

1.4 Dynamical Functions

Let $a(q, p)$ be a dynamical function, i.e., a function of the mechanical state of the system. Its time evolution is induced by Hamilton's Eqs. (1.7), namely

$$
\begin{aligned}
\dot{a}(q, p) &= \sum_{i=1}^{3N} \left(\frac{\partial a(q, p)}{\partial q_i} \dot{q}_i + \frac{\partial a(q, p)}{\partial p_i} \dot{p}_i \right) \\
&= \sum_{i=1}^{3N} \left(\frac{\partial a(q, p)}{\partial q_i} \frac{\partial H_N(q, p; \alpha)}{\partial p_i} - \frac{\partial a(q, p)}{\partial p_i} \frac{\partial H_N(q, p; \alpha)}{\partial q_i} \right), \\
&= \{a, H_N(\alpha)\}
\end{aligned}
\tag{1.23}
$$

where the Poisson bracket of two dynamical functions, $a = a(q, p)$ and $b = b(q, p)$, has been defined as

$$
\{a, b\} \equiv \sum_{i=1}^{3N} \left(\frac{\partial a(q, p)}{\partial q_i} \frac{\partial b(q, p)}{\partial p_i} - \frac{\partial a(q, p)}{\partial p_i} \frac{\partial b(q, p)}{\partial q_i} \right).
\tag{1.24}
$$

In particular, since the generalized coordinates and their conjugate momenta are independent variables, from (1.24) it follows that

$$
\{q_i, q_j\} = 0, \quad \{p_i, p_j\} = 0, \quad \{q_i, p_j\} = \delta_{ij},
\tag{1.25}
$$

where δ_{ij} is Kronecker's delta ($\delta_{ij} = 1$, if $i = j$, and $\delta_{ij} = 0$, if $i \neq j$). Observe that with the notation used for the Poisson bracket $\{a, b\}$, the generalized coordinates and momenta do not appear explicitly. For this reason, when the Hamiltonian $H_N(q, p; \alpha)$ appears in a Poisson bracket, it will be written as $H_N(\alpha)$, like in (1.23).

Every dynamical function $C(q, p)$ conserved in the time evolution, i.e., $\dot{C}(q, p) = 0$, is denominated integral or constant of motion of the system. According to (1.23), every integral of motion verifies the condition

$$
\{C, H_N(\alpha)\} = 0.
\tag{1.26}
$$

An immediate consequence of (1.23) is the law of conservation of energy $\dot{H}_N(q, p; \alpha) = 0$. Note further that if $C_1(q, p)$ and $C_2(q, p)$ are two integrals of motion, one has

$$
\{H_N(\alpha), \{C_1, C_2\}\} = 0,
\tag{1.27}
$$

i.e., $\{C_1, C_2\}$ is also a constant of motion, as can be derived from Jacobi's identity

$$
\{C_1, \{C_2, C_3\}\} + \{C_2, \{C_3, C_1\}\} + \{C_3, \{C_1, C_2\}\} = 0
\tag{1.28}
$$

valid for any three dynamical functions, if $C_3(q, p) = H_N(q, p; \alpha)$.

Once an integral of motion, for instance the energy, is known, the phase space trajectory has to move on the "surface" of the Γ space, $H_N(q, p; \alpha) = E$, where E is the value of this integral of motion (such a surface is a "volume" of $6N - 1$ dimensions). If the system has another integral of motion, $C(q, p)$, the trajectory in phase space has to move also on the surface $C(q, p) = C^*$, where C^* is the value of this other integral of motion. It follows then that, if $6N - 1$ integrals of motion may be found, the mechanical state of the system can be determined for every instant of time. In this case one says that the system is integrable. It is well known that some simple mechanical systems are integrable. Thus, all the mechanical systems having only one degree of freedom are integrable in view of the principle of conservation of energy. Systems with two or three degrees of freedom subjected to a central force in the plane are also integrable. The integrability of dynamical systems is, however, the exception rather than the rule. In general, in the absence of external forces, a mechanical system only has seven additive integrals of motion (these are the integrals of motion whose value for a system formed by various subsystems is the sum of the values for each subsystem): the Hamiltonian H_N $(\mathbf{r}^N, \mathbf{p}^N; \alpha)$, the linear momentum $\mathbf{P}_N(\mathbf{p}^N)$, and the angular momentum $\mathbf{L}_N(\mathbf{r}^N, \mathbf{p}^N)$, where:

$$\mathbf{P}_N(\mathbf{p}^N) = \sum_{j=1}^N \mathbf{p}_j, \quad \mathbf{L}_N(\mathbf{r}^N, \mathbf{p}^N) = \sum_{j=1}^N \mathbf{r}_j \times \mathbf{p}_j. \tag{1.29}$$

These laws of conservation follow from the homogeneity of time and the isotropy and homogeneity of space.

The time evolution Eq. (1.23) for the dynamical function $a(q, p)$ may also be written as

$$\dot{a}(q, p) = -\mathcal{L}_N a(q, p), \tag{1.30}$$

where $\mathcal{L}_N$ is the linear differential operator:

$$\mathcal{L}_N \equiv \{H_N(\alpha),\} = \sum_{i=1}^{3N} \left(\frac{\partial H_N(q, p; \alpha)}{\partial q_i} \frac{\partial}{\partial p_i} - \frac{\partial H_N(q, p; \alpha)}{\partial p_i} \frac{\partial}{\partial q_i} \right) \tag{1.31}$$

is called the Liouville operator of the system. The formal solution of (1.30) is

$$a(q^t, p^t) = a(q^0, p^0; t) = e^{-\mathcal{L}_N^0 t} a(q^0, p^0; t = 0), \tag{1.32}$$

where $q^t, p^t = q_1^t, \ldots, q_{3N}^t, p_1^t, \ldots, p_{3N}^t$ and $\mathcal{L}_N^0$ acts on the variables q^0, p^0.

1.5 Quantum Mechanics

Due to Heisenberg's uncertainty principle, in quantum mechanics the position and momentum of a material point may not be determined simultaneously with arbitrary precision. If $q_1, \ldots, q_{3N}$ and $p_1, \ldots, p_{3N}$ denote the Cartesian coordinates and their conjugate momenta of a classical system of N material points, in quantum mechanics these variables are replaced by operators $\hat{q}_1, \ldots, \hat{q}_{3N}$ and $\hat{p}_1, \ldots, \hat{p}_{3N}$ that satisfy the commutation rules, which are the quantum mechanical equivalent to (1.25):

$$\left[\hat{q}_i, \hat{q}_j\right] = 0, \quad \left[\hat{p}_i, \hat{p}_j\right] = 0, \quad \left[\hat{q}_i, \hat{p}_j\right] = i\hbar\delta_{ij}\hat{I}, \tag{1.33}$$

where $\hat{I}$ is the identity operator and

$$[\hat{a}, \hat{b}] \equiv \hat{a}\hat{b} - \hat{b}\hat{a} \tag{1.34}$$

is the commutator of the operators $\hat{a}$ and $\hat{b}$. When $[\hat{a}, \hat{b}] = 0$ the operators are said to commute. In the representation of coordinates, the prescription for the operators $\hat{q}_i$ and $\hat{p}_i$ is obtained by associating to the Cartesian coordinate q_i the operator $\hat{q}_i$, multiplication by q_i, and to its conjugate momentum p_i the operator $\hat{p}_i = -i\hbar\partial/\partial q_i$, where h is Planck's constant and $\hbar = h/2\pi$.

Given a dynamical function $a(q_1, \ldots, q_{3N}, p_1, \ldots, p_{3N})$, the associated operator, $\hat{a}$, is obtained after substitution of the variables q_i and p_i by their associated operators $\hat{q}_i$ and $\hat{p}_i$, namely

$$\hat{a} \equiv a\left(q_1, \ldots, q_{3N}, -i\hbar\frac{\partial}{\partial q_1}, \ldots, -i\hbar\frac{\partial}{\partial q_{3N}}\right). \tag{1.35}$$

For instance, the Hamiltonian operator corresponding to the Hamiltonian (1.20) is:

$$\hat{H}_N(V) = -\frac{\hbar^2}{2m}\sum_{j=1}^{N}\nabla_j^2 + U_N(\mathbf{r}^N) + \sum_{j=1}^{N}\phi_R(\mathbf{r}_j), \tag{1.36}$$

where $\nabla_j = \partial/\partial\mathbf{r}_j$. In some cases, the correspondence (1.35) may not be well defined and hence it is necessary to recur to a method of symmetrization of the operator associated to the dynamical function (e.g., to the dynamical function qp one associates the operator $(\hat{q}\hat{p} + \hat{p}\hat{q})/2$, because the operators $\hat{q}$ and $\hat{p}$ do not commute).

Throughout the text, the Hamiltonian operator will be written in the form $\hat{H}_N(\alpha)$, where α indicates one or several external parameters.

1.6 Self-adjoint Operators

The operators that represent observable magnitudes in quantum mechanics are self-adjoint linear operators acting on the Hilbert space $\mathcal{H}$ of the system under consideration. This space consists of the square-integrable wave functions $\Psi(q) = \Psi(q_1, ..., q_{3N})$ in coordinate space, that is

$$\int dq |\Psi(q)|^2 < \infty, \tag{1.37}$$

where $\int dq = \int dq_1 ... \int dq_{3N}$, $|\Psi(q)|^2 = \Psi(q)\Psi^*(q)$ and $\Psi^*(q)$ indicates the complex conjugate of $\Psi(q)$. According to Born's probabilistic interpretation, the wave function completely defines the dynamical state of the system so that $|\Psi(q)|^2$ is the probability density to find the system in the state $q = (q_1, ..., q_{3N})$ in coordinate space, i.e., $\int dq |\Psi(q)|^2 = 1$ (see Appendix B).

The Hilbert space has the properties of a vector space. Thus, if $|\Psi_1\rangle = \Psi_1(q)$ and $|\Psi_2\rangle = \Psi_2(q)$ are two elements of $\mathcal{H}$, all the linear combinations $\lambda_1|\Psi_1\rangle + \lambda_2|\Psi_2\rangle$, where λ_1 are λ_2 are arbitrary complex numbers, also belong to $\mathcal{H}$. On the other hand, one may define a scalar product as

$$\langle\Psi_1|\Psi_2\rangle = \int dq \Psi_1^*(q)\Psi_2(q), \tag{1.38}$$

which is linear in the vector on the right and anti-linear in the one on the left, i.e., $\langle\Psi_3|\lambda_1\Psi_1 + \lambda_2\Psi_2\rangle = \langle\Psi_3|\lambda_1\Psi_1\rangle + \langle\Psi_3|\lambda_2\Psi_2\rangle$, $\langle\Psi_1|\Psi_2\rangle = \langle\Psi_2|\Psi_1\rangle^*$. If $\langle\Psi_1|\Psi_2\rangle = 0$ the two functions are said to be orthogonal.

Given an operator $\hat{a}$, the adjoint operator $\hat{a}^\dagger$ whenever it exists, is defined by

$$\langle\Psi_1|\hat{a}\Psi_2\rangle = \langle\hat{a}^\dagger\Psi_1|\Psi_2\rangle. \tag{1.39}$$

And if $\hat{a} = \hat{a}^\dagger$, it is said that $\hat{a}$ is a self-adjoint operator.

When the system is in a state $|\Psi\rangle$ the expectation value of an operator $\hat{a}$ in such a state is

$$\langle\hat{a}\rangle = \frac{\langle\Psi|\hat{a}\Psi\rangle}{\langle\Psi|\Psi\rangle}, \tag{1.40}$$

so that, if $\hat{a}$ is self-adjoint, one has $\langle\Psi|\hat{a}\Psi\rangle = \langle\hat{a}^\dagger\Psi|\Psi\rangle = \langle\hat{a}\Psi|\Psi\rangle = \langle\Psi|\hat{a}\Psi\rangle^*$ and $\langle\hat{a}\rangle$ is real.

1.7 Eigenvalue Equation

The expectation value (1.40) is interpreted in quantum mechanics as the result of the measurement of the observable magnitude corresponding to the operator $\hat{a}$ when the system is prepared in the state $|\Psi\rangle$. An important case is that of a self-adjoint operator, $\hat{a}$, whose dispersion $\langle \hat{a}^2 \rangle - \langle \hat{a} \rangle^2 = 0$, i.e., the case of an operator for which one can state with certainty that the result of the measurement in a state $|\Psi\rangle$ takes a prescribed value (see Appendix B). This condition is written as

$$\frac{\langle \Psi | \hat{a}^2 \Psi \rangle}{\langle \Psi | \Psi \rangle} - \left(\frac{\langle \Psi | \hat{a} \Psi \rangle}{\langle \Psi | \Psi \rangle} \right)^2 = 0, \tag{1.41}$$

and since $\langle \Psi | \hat{a}^2 \Psi \rangle = \langle \hat{a} \Psi | \hat{a} \Psi \rangle$, because $\hat{a}$ is self-adjoint, it follows that

$$\langle \hat{a} \Psi | \hat{a} \Psi \rangle \langle \Psi | \Psi \rangle = \langle \Psi | \hat{a} \Psi \rangle^2. \tag{1.42}$$

Note that (1.42) is a case in which Schwarz's inequality becomes an equality, thus implying that $|\hat{a} \Psi\rangle$ and $|\Psi\rangle$ are proportional, i.e.,

$$\hat{a} |\Psi\rangle \equiv |\hat{a} \Psi\rangle = a |\Psi\rangle, \tag{1.43}$$

which is known as the eigenvalue equation for the operator $\hat{a}$. Every real number a in (1.43) is called an eigenvalue of $\hat{a}$ and the nonzero solutions $|\Psi\rangle$ of this equation are called eigenvectors of the operator. Throughout the text examples will be analyzed in which the spectrum of the operator $\hat{a}$ is discrete, i.e., its eigenvalues form a finite (or infinite) numerable set so that the eigenvalue equation may be cast in the form

$$\hat{a} |\mathbf{a}_n\rangle = a_n |\mathbf{a}_n\rangle. \tag{1.44}$$

Note that if $|\mathbf{a}_n\rangle$ is an eigenvector of the operator $\hat{a}$, the same is true for the vector $\lambda |\mathbf{a}_n\rangle$, where λ is an arbitrary complex number. Therefore, on account of (1.38), it is always possible to choose λ in such a manner that the eigenvectors will be normalized, namely

$$\langle \mathbf{a}_n | \mathbf{a}_n \rangle = 1, \tag{1.45}$$

and hence they are defined up to an arbitrary constant phase. It should be noted that two eigenvectors of the eigenvalue equation yielding different eigenvalues are orthogonal. To see this, consider two solutions to (1.44):

$$\hat{a} |\mathbf{a}_1\rangle = a_1 |\mathbf{a}_1\rangle, \quad \hat{a} |\mathbf{a}_2\rangle = a_2 |\mathbf{a}_2\rangle. \tag{1.46}$$

If one takes the complex conjugate of the first equation and multiplies the result by $|\mathbf{a}_2\rangle$, this yields

$$\langle \hat{a}^\dagger \mathbf{a}_1 | \mathbf{a}_2 \rangle = \langle \mathbf{a}_1 | a_1 | \mathbf{a}_2 \rangle, \tag{1.47}$$

while, if the second equation is multiplied by $\langle \mathbf{a}_1 |$, the outcome is

$$\langle \mathbf{a}_1 | \hat{a} \mathbf{a}_2 \rangle = \langle \mathbf{a}_1 | a_2 | \mathbf{a}_2 \rangle. \tag{1.48}$$

From the subtraction of (1.47) and (1.48) it follows that:

$$0 = \langle \hat{a}^\dagger \mathbf{a}_1 | \mathbf{a}_2 \rangle - \langle \mathbf{a}_1 | \hat{a} \mathbf{a}_2 \rangle = (a_1 - a_2) \langle \mathbf{a}_1 | \mathbf{a}_2 \rangle. \tag{1.49}$$

Therefore, if $a_1 \neq a_2$, then necessarily $\langle \mathbf{a}_1 | \mathbf{a}_2 \rangle = 0$.

If there is a set of eigenvectors of $\hat{a}$ with the same eigenvalue, one refers to the latter as a degenerate eigenvalue. Although the eigenvectors corresponding to a degenerate eigenvalue are not necessarily orthogonal, it is always possible to look for linear combinations of them such that any two of the resulting vectors are orthogonal (Schmidt's orthogonalization), namely $\langle \mathbf{a}_i | \mathbf{a}_j \rangle = \delta_{ij}$, where δ_{ij} is Kronecker's delta. For instance, let a_1 be a three times degenerate eigenvalue of (1.44) and $|\mathbf{a}_1\rangle$, $|\mathbf{a}'_1\rangle$, and $|\mathbf{a}''_1\rangle$ three non-orthogonal eigenvectors corresponding to this eigenvalue. Assume that the first vector has been normalized, i.e., $\langle \mathbf{a}_1 | \mathbf{a}_1 \rangle = 1$. If the eigenvector $|\mathbf{a}_2\rangle$ is defined by the expression $c_2 |\mathbf{a}_2\rangle = |\mathbf{a}'_1\rangle - |\mathbf{a}_1\rangle\langle \mathbf{a}_1 | \mathbf{a}'_1 \rangle$, where c_2 is a constant, and one multiplies this equation by $\langle \mathbf{a}_1 |$ the result is $c_2 \langle \mathbf{a}_1 | \mathbf{a}_2 \rangle = \langle \mathbf{a}_1 | \mathbf{a}'_1 \rangle - \langle \mathbf{a}_1 | \mathbf{a}_1 \rangle\langle \mathbf{a}_1 | \mathbf{a}'_1 \rangle = 0$, i.e., $|\mathbf{a}_1\rangle$ and $|\mathbf{a}_2\rangle$ are orthogonal. Therefore, it is only necessary to adjust c_2 in order to normalize $|\mathbf{a}_2\rangle$. Similarly, the eigenvector $c_3 |\mathbf{a}_3\rangle$ is defined as $c_3 |\mathbf{a}_3\rangle = |\mathbf{a}''_1\rangle - |\mathbf{a}_1\rangle\langle \mathbf{a}_1 | \mathbf{a}''_1 \rangle - |\mathbf{a}_2\rangle\langle \mathbf{a}_2 | \mathbf{a}''_1 \rangle$, which is orthogonal to $|\mathbf{a}_1\rangle$ and $|\mathbf{a}_2\rangle$, and may again be normalized with the right choice for the constant c_3. Thus, $|\mathbf{a}_1\rangle$, $|\mathbf{a}_2\rangle$, and $|\mathbf{a}_3\rangle$ are orthonormal. Note that, by construction, this process of orthogonalization is not unique.

Given an orthonormal set of eigenvectors $|\mathbf{a}_n\rangle$ of a self-adjoint operator $\hat{a}$, the operator forms a complete set if any vector $|\Psi\rangle$ of $\mathcal{H}$ may be expanded in a unique way as

$$|\Psi\rangle = \sum_n \langle \mathbf{a}_n | \Psi \rangle |\mathbf{a}_n\rangle. \tag{1.50}$$

One then says that the eigenvectors $|\mathbf{a}_n\rangle$ constitute a basis for $\mathcal{H}$. Given the aforementioned point that any eigenvector is defined up to an arbitrary phase, by convention two bases will be considered identical when their eigenvectors only differ in their phases.

Note that when there is a degenerate eigenvalue, the set of eigenvectors of the operator $\hat{a}$ is not unique, and so the same applies to the expansion (1.50). Let $\hat{b}$ be another self-adjoint operator that commutes with $\hat{a}$, i.e., $[\hat{a}, \hat{b}] = 0$. Then $\hat{a}$ and $\hat{b}$ possess a set of common eigenvectors represented by $|\mathbf{a}_n, \mathbf{b}_m\rangle$. As a matter of fact, if $|\mathbf{a}_n, \mathbf{b}_m\rangle$ is an eigenvector of $\hat{a}$ with eigenvalue a_n, from (1.34) and (1.44) one has

$$\hat{a}\hat{b}|\mathbf{a}_n, \mathbf{b}_m\rangle = \hat{b}\hat{a}|\mathbf{a}_n, \mathbf{b}_m\rangle = \hat{b}a_n|\mathbf{a}_n, \mathbf{b}_m\rangle = a_n\hat{b}|\mathbf{a}_n, \mathbf{b}_m\rangle, \tag{1.51}$$

and from the last equality it follows that $\hat{b}|\mathbf{a}_n, \mathbf{b}_m\rangle$ is an eigenvector of $\hat{a}$, i.e., $\hat{b}|\mathbf{a}_n, \mathbf{b}_m\rangle = b_m|\mathbf{a}_n, \mathbf{b}_m\rangle$. If this system is unique, namely if the pair of numbers (a_n, b_m) is not degenerate, one says that $\hat{a}$ and $\hat{b}$ form a complete set of observables that commute. In such a case, any state $|\Psi\rangle$ of $\mathcal{H}$ has a unique expansion of the form of (1.50) in the basis $|\mathbf{a}_n, \mathbf{b}_m\rangle$. For any other case, one may look for a third self-adjoint operator $\hat{c}$ that commutes with $\hat{a}$ and $\hat{b}$, etc. In this fashion, if $\hat{a}, \hat{b}, \hat{c}, \ldots$ constitute a complete set of observables that commute, the state of the system is completely determined. In quantum mechanics, this is the process of preparation of the system, which consists in performing upon it simultaneous measurements of a complete set of commuting observables.

To close this section, some simple concepts about operators which will be needed in Chap. 3 are now introduced. Let $|\mathbf{a}_n\rangle$ be an orthonormal basis of $\mathcal{H}$. The spectral resolution of the identity operator $\hat{I}$ is

$$\hat{I} = \sum_n |\mathbf{a}_n\rangle\langle\mathbf{a}_n|. \tag{1.52}$$

The trace of an operator $\hat{b}$ is defined as:

$$\mathrm{Tr}\hat{b} = \sum_n \langle\mathbf{a}_n|\hat{b}|\mathbf{a}_n\rangle, \tag{1.53}$$

which is independent of the choice of the basis and hence an intrinsic characteristic of the operator.

An operator $\hat{b}$ is bounded if for every state $|\Psi\rangle$ one has

$$\langle\Psi|\hat{b}|\Psi\rangle \leq \lambda\langle\Psi|\Psi\rangle, \tag{1.54}$$

where λ is a positive constant independent of $|\Psi\rangle$.

An operator $\hat{b}$ is positive (indicated by $\hat{b} \geq 0$), when it is bounded and for all $|\Psi\rangle$ one has:

$$\langle\Psi|\hat{b}|\Psi\rangle \geq 0. \tag{1.55}$$

1.8   Schrödinger's Equation

Once the system has been prepared, the state evolves in time according to Schrödinger's equation:

$$i\hbar\frac{\partial|\Psi(t)\rangle}{\partial t} = \hat{H}_N(\alpha)|\Psi(t)\rangle, \tag{1.56}$$

where $|\Psi(t)\rangle$ is the wave function at time t and $\hat{H}_N(\alpha)$ is the Hamiltonian operator, which is a function of one or several external parameters represented by the label α. If $|\Psi_n\rangle = |\Psi_n(t=0)\rangle$ is an eigenvector of $\hat{H}_N(\alpha)$ with eigenvalue $E_n^{(N)}(\alpha)$, i.e.,

$$\hat{H}_N(\alpha)|\Psi_n\rangle = E_n^{(N)}(\alpha)|\Psi_n\rangle, \tag{1.57}$$

one says that the state $|\Psi_n\rangle$ is stationary and from Schrödinger's equation it follows that

$$|\Psi_n(t)\rangle = e^{-iE_n^{(N)}(\alpha)t/\hbar}|\Psi_n\rangle. \tag{1.58}$$

Thus, in the time evolution the state only changes its phase and it remains an eigenvector of $\hat{H}_N(\alpha)$ with eigenvalue $E_n^{(N)}(\alpha)$ at any instant of time.

In the next few lines, the eigenvalues of some stationary states of simple Hamiltonians that appear often in the text are considered.

1.8.1 Free Particle

The Hamiltonian of a free particle of mass m in a cubic box R of side L and volume $V = L^3$ is

$$H_1(\mathbf{r}, \mathbf{p}; V) = \frac{1}{2m}\mathbf{p}^2 + \phi_R(\mathbf{r}), \tag{1.59}$$

where $\mathbf{p} = p_x\mathbf{e}_x + p_y\mathbf{e}_y + p_z\mathbf{e}_z$ is the particle's momentum and $\phi_R(\mathbf{r})$ the potential that keeps the particle inside the box. The associated operator $\hat{H}_1(V)$ is according to (1.35):

$$\hat{H}_1(V) = -\frac{\hbar^2}{2m}\left(\frac{\partial^2}{\partial x^2} + \frac{\partial^2}{\partial y^2} + \frac{\partial^2}{\partial z^2}\right) + \phi_R(\mathbf{r}), \tag{1.60}$$

and the stationary states of this operator are

$$-\frac{\hbar^2}{2m}\left(\frac{\partial^2}{\partial x^2} + \frac{\partial^2}{\partial y^2} + \frac{\partial^2}{\partial z^2}\right)|\Psi_{n_x,n_y,n_z}\rangle = \varepsilon_{n_x,n_y,n_z}(V)|\Psi_{n_x,n_y,n_z}\rangle. \tag{1.61}$$

Note that although in (1.61) the potential $\phi_R(\mathbf{r})$ does not appear explicitly, in order to locate the particle inside the box the condition is imposed that the eigenvectors vanish at the extremes of the box. In such a case the eigenvalues of $\hat{H}_1(V)$ are quantized as given below:

$$\varepsilon_{n_x,n_y,n_z}(V) = \frac{\pi^2\hbar^2}{2mV^{2/3}}(n_x^2 + n_y^2 + n_z^2), \quad (n_x, n_y, n_z = 1, 2, \ldots). \tag{1.62}$$

The ground state or lowest energy level is non-degenerate and corresponds to the quantum state $n_x = n_y = n_z = 1$. Since all the other energy levels are degenerate, the Hamiltonian operator of a free particle does not constitute a complete set of observables.

Another form of quantization, that will be used throughout the text, consists in dividing the Euclidean space in cubic boxes of side L and volume $V = L^3$ and to impose the condition that the eigenvectors be periodic with period L in the three directions of space. In such a case, the eigenvalues of $\hat{H}_1(V)$ are quantized as follows:

$$\varepsilon_{n_x,n_y,n_z}(V) = \frac{2\pi^2\hbar^2}{mV^{2/3}}\left(n_x^2 + n_y^2 + n_z^2\right), \quad (n_x, n_y, n_z = 0, \pm 1, \ldots), \qquad (1.63)$$

and hence the energy of the ground state ($n_x = n_y = n_z = 0$) is zero.

The eigenvalues (1.62) and (1.63) are obtained by looking for plane wave solutions to (1.61) and requiring that the wave function either vanishes at the extremes of the box or be periodic with period L in the three directions of space. In the latter case, if $\mathbf{k}$ is the wave vector with Cartesian components k_x, k_y, and k_z, the condition of periodicity is expressed as

$$e^{i[k_x x + k_y y + k_z z]} = e^{i[k_x(x+L) + k_y(y+L) + k_z(z+L)]},$$

i.e.,

$$k_x L = 2\pi n_x, \quad k_y L = 2\pi n_y, \quad k_z L = 2\pi n_z,$$

where $n_x, n_y, n_z = 0, \pm 1, \ldots$ Upon substitution of these results in the expression of the energy,

$$\varepsilon_{n_x,n_y,n_z}(V) = \frac{\hbar^2}{2m}\left(k_x^2 + k_y^2 + k_z^2\right), \qquad (1.64)$$

one arrives at (1.63).

It should be noted that when the quantum numbers n_x, n_y, n_z change by one unit, the components of the wave vector change by $|\Delta k_x| = |\Delta k_y| = |\Delta k_z| = 2\pi/L$, so that in the limit $L \to \infty$ these variables may be considered as continuous. The number of quantum states of a free particle whose wave vector lies between $\mathbf{k}$ and $\mathbf{k} + d\mathbf{k}$ is

$$\rho(k_x, k_y, k_z)dk_x dk_y dk_z = \frac{dk_x}{|\Delta k_x|}\frac{dk_y}{|\Delta k_y|}\frac{dk_z}{|\Delta k_z|} = \frac{V}{8\pi^3}dk_x dk_y dk_z, \qquad (1.65)$$

i.e., the density of quantum states in $\mathbf{k}$-space is a constant whose value is $V/8\pi^3$. Therefore, the number of quantum states of a particle whose wave number $k = |\mathbf{k}|$ is between k and $k + dk$ is

$$\rho(k)dk = \frac{V}{8\pi^3}4\pi k^2 dk = \frac{V}{2\pi^2}k^2 dk, \tag{1.66}$$

and for each of them there is an energy state $\varepsilon = \hbar^2 k^2/2m$; hence the number of quantum states of a particle whose energy lies between ε and $\varepsilon + d\varepsilon$ is

$$\rho(\varepsilon)d\varepsilon = 2\pi V \left(\frac{2m}{h^2}\right)^{3/2} \sqrt{\varepsilon}d\varepsilon, \tag{1.67}$$

a result that will be used in the study of ideal quantum gases carried out in Chap. 6.

1.8.2 Harmonic Oscillator

The Hamiltonian of a one-dimensional harmonic oscillator of mass m and angular frequency ω is

$$H_1(q, p; \omega) = \frac{1}{2m}p^2 + \frac{1}{2}m\omega^2 q^2, \tag{1.68}$$

where q is the Cartesian coordinate and p its conjugate momentum. The Hamiltonian operator $\hat{H}_1(\omega)$ associated to this dynamical function is, according to (1.35),

$$\hat{H}_1(\omega) = -\frac{\hbar^2}{2m}\frac{\partial^2}{\partial q^2} + \frac{1}{2}m\omega^2 q^2. \tag{1.69}$$

The stationary states of this operator are

$$\left(-\frac{\hbar^2}{2m}\frac{\partial^2}{\partial q^2} + \frac{1}{2}m\omega^2 q^2\right)|\Psi_n\rangle = \varepsilon_n(\omega)|\Psi_n\rangle, \tag{1.70}$$

with

$$\varepsilon_n(\omega) = \left(n + \frac{1}{2}\right)\hbar\omega, \quad (n = 0, 1, 2\ldots). \tag{1.71}$$

Since there is only one eigenvector for each eigenvalue, $\hat{H}_1(\omega)$ forms by itself a complete set of observables. Note that since the energy $\varepsilon_n(\omega)$ is univocally determined by the non-negative number n, this number may be interpreted as the number of particles with energy $\hbar\omega$. By definition, the number of particles n is not a conserved variable.

1.8.3   Particle in a Magnetic Field

The angular momentum of a particle l has been defined in (1.14) so that, upon application of the correspondence rule (1.35), the angular momentum vector operator $\hat{\mathbf{l}}$ is given by

$$\hat{\mathbf{l}} = -i\hbar\mathbf{r} \times \nabla, \tag{1.72}$$

whose components

$$\begin{aligned}
\hat{l}_x &= -i\hbar\left(y\frac{\partial}{\partial z} - z\frac{\partial}{\partial y}\right) \\
\hat{l}_y &= -i\hbar\left(z\frac{\partial}{\partial x} - x\frac{\partial}{\partial z}\right) \\
\hat{l}_z &= -i\hbar\left(x\frac{\partial}{\partial y} - y\frac{\partial}{\partial x}\right),
\end{aligned} \tag{1.73}$$

verify the following commutation rules:

$$\left[\hat{l}_x, \hat{l}_y\right] = i\hbar\hat{l}_z, \quad \left[\hat{l}_y, \hat{l}_z\right] = i\hbar\hat{l}_x, \quad \left[\hat{l}_z, \hat{l}_x\right] = i\hbar\hat{l}_y. \tag{1.74}$$

If one defines the scalar operator,

$$\hat{\mathbf{l}}^2 = \hat{l}_x^2 + \hat{l}_y^2 + \hat{l}_z^2, \tag{1.75}$$

it is readily verified that it commutes with each of the components of $\hat{\mathbf{l}}$, namely

$$\left[\hat{\mathbf{l}}, \hat{\mathbf{l}}^2\right] = 0, \tag{1.76}$$

and hence with any function of these components. One may then form a basis of eigenvectors common to both $\hat{\mathbf{l}}^2$ and one of the components of $\hat{\mathbf{l}}$, say $\hat{l}_z$. The eigenvalue equations for these operators read:

$$\hat{\mathbf{l}}^2\left|\Psi_{l,m}\right\rangle = l(l+1)\hbar^2\left|\Psi_{l,m}\right\rangle, \quad \hat{l}_z\left|\Psi_{l,m}\right\rangle = m\hbar\left|\Psi_{l,m}\right\rangle. \tag{1.77}$$

If the spin angular momentum $\hat{\mathbf{s}}$ is included, one may generalize (1.73–1.77) through replacement of the angular momentum $\hat{\mathbf{l}}$ by the total angular momentum $\hat{\mathbf{j}} = \hat{\mathbf{l}} + \hat{\mathbf{s}}$, namely

$$\hat{\mathbf{j}}^2\left|\Psi_{j,m}\right\rangle = j(j+1)\hbar^2\left|\Psi_{j,m}\right\rangle, \quad \hat{j}_z\left|\Psi_{j,m}\right\rangle = m\hbar\left|\Psi_{j,m}\right\rangle, \tag{1.78}$$

and it can be shown that (1) the value of j in (1.78) is a non-negative number, either an integer or an odd half integer and (2) the eigenvalue of m in (1.78) is one of the $(2j+1)$ quantities $-j, -j+1, ..., j-1, j$.

The Hamiltonian of interaction of a particle of angular momentum **l** in a uniform and constant magnetic field **B** is, according to (1.15), given by

$$H_1(\mathbf{r}, \mathbf{p}; \mathbf{B}) = -\boldsymbol{\mu} \cdot \mathbf{B}, \tag{1.79}$$

where $\boldsymbol{\mu}$ is the magnetic moment of the particle (1.16).

If the spin is included, the Hamiltonian operator of the interaction of the particle with the field **B** is thus

$$\hat{H}_1(\mathbf{B}) = -\frac{1}{\hbar} g \mu_B \hat{\mathbf{j}} \cdot \mathbf{B}, \tag{1.80}$$

where $\mu_B = e\hbar/2mc$ is Bohr's magneton (note that in μ_B, e is the magnitude of the charge of the particle and m its mass, not to be confused with the quantum number appearing in (1.78)), and g is the Landé factor:

$$g = 1 + \frac{j(j+1) + s(s+1) - l(l+1)}{2j(j+1)}, \tag{1.81}$$

and $s(s+1)\hbar^2$ are the eigenvalues of $\mathbf{s}^2$ (the introduction of the Landé factor in the Hamiltonian operator is valid when the latter is acting on the eigenfunction $|\Psi_{j,m}\rangle$).

If the magnetic field is directed along the z-axis, $\mathbf{B} = B\mathbf{e}_z$, the eigenvectors of $\hat{H}_1(\mathbf{B})$ are $|\Psi_{j,m}\rangle$ and the eigenvalues are quantized as follows:

$$\varepsilon_m(B) = -g\mu_B m B, \quad (m = -j, -j+1, \ldots, j-1, j). \tag{1.82}$$

1.9 System of Identical Particles

It is well known that in classical mechanics identical particles are considered distinguishable. In contrast, in quantum mechanics identical particles are indistinguishable which in turn implies that the wave function of a system of N identical particles has to be either symmetric (bosons or particles with integer spin) or antisymmetric (fermions or particles with an odd half integer spin) upon exchange of any two particles. Consider an ideal system of N identical particles defined by the Hamiltonian operator $\hat{H}_N(\alpha)$, which is the sum of one-particle operators:

$$\hat{H}_N(\alpha) = \sum_{j=1}^{N} \hat{H}_1^{(j)}(\alpha). \tag{1.83}$$

Due to the identical nature of the particles, the operators $\hat{H}_1^{(j)}(\alpha) \equiv \hat{H}_1(\alpha)$ have the same eigenvalue equation which is written as

$$\hat{H}_1(\alpha)|\phi_i\rangle = \varepsilon_i(\alpha)|\phi_i\rangle, \tag{1.84}$$

where $\varepsilon_i(\alpha)$ is the energy of the quantum state i whose eigenvector is $|\phi_i\rangle$, which is assumed to be normalized, i.e., $\langle\phi_i|\phi_i\rangle = 1$.

The wave function

$$\left|\phi_{i_1}^{(1)}\right\rangle\left|\phi_{i_2}^{(2)}\right\rangle\dots\left|\phi_{i_N}^{(N)}\right\rangle, \tag{1.85}$$

in which the subscript i_j represents the quantum state of the jth particle (which has been indicated with a superscript), is an eigenfunction of $\hat{H}_N(\alpha)$ whose eigenvalue is

$$\varepsilon_{i_1}(\alpha) + \varepsilon_{i_2}(\alpha) + \dots + \varepsilon_{i_N}(\alpha).$$

Since the values i_j are not necessarily different, the energy of the system may be written as

$$E_{\{n_i\}}^{(N)}(\alpha) = \sum_i {}^*n_i\varepsilon_i(\alpha), \tag{1.86}$$

where n_i is the number of particles (occupation number) in quantum state i and the asterisk indicates that these numbers must satisfy the following condition:

$$\sum_i n_i = N. \tag{1.87}$$

The wave functions of the system are linear combinations of the form (1.85), symmetric (bosons) or antisymmetric (fermions) under the exchange of any two particles. Then one has $0 \le n_i \le N$ in the case of bosons and $n_i = 0, 1$ in the case of fermions. As an example, consider a system of two particles, 1 and 2, in two quantum states, i_1 and i_2. The possible wave functions of the system are

(a)   bosons

$$\begin{aligned}
&\left|\phi_{i_1}^{(1)}\right\rangle\left|\phi_{i_1}^{(2)}\right\rangle &&(n_{i_1} = 2,\quad n_{i_2} = 0),\\
&\left|\phi_{i_2}^{(1)}\right\rangle\left|\phi_{i_2}^{(2)}\right\rangle &&(n_{i_1} = 0,\quad n_{i_2} = 2),\\
\frac{1}{\sqrt{2}}&\left(\left|\phi_{i_1}^{(1)}\right\rangle\left|\phi_{i_2}^{(2)}\right\rangle + \left|\phi_{i_2}^{(1)}\right\rangle\left|\phi_{i_1}^{(2)}\right\rangle\right) &&(n_{i_1} = 1,\quad n_{i_2} = 1).
\end{aligned}$$

(b)   fermions

$$\frac{1}{\sqrt{2}}\left(\left|\phi_{i_1}^{(1)}\right\rangle\left|\phi_{i_2}^{(2)}\right\rangle - \left|\phi_{i_2}^{(1)}\right\rangle\left|\phi_{i_1}^{(2)}\right\rangle\right) \quad (n_{i_1} = 1,\quad n_{i_2} = 1).$$

Note that if the particles are considered to be distinguishable, the possible wave functions are

$$
\begin{aligned}
\left|\phi_{i_1}^{(1)}\right\rangle\left|\phi_{i_1}^{(2)}\right\rangle & \left(n_{i_1} = 2, \ n_{i_2} = 0\right), \\
\left|\phi_{i_2}^{(1)}\right\rangle\left|\phi_{i_2}^{(2)}\right\rangle & \left(n_{i_1} = 0, \ n_{i_2} = 2\right), \\
\left|\phi_{i_1}^{(1)}\right\rangle\left|\phi_{i_2}^{(2)}\right\rangle & \left(n_{i_1} = 1, \ n_{i_2} = 1\right), \\
\left|\phi_{i_2}^{(1)}\right\rangle\left|\phi_{i_1}^{(2)}\right\rangle & \left(n_{i_1} = 1, \ n_{i_2} = 1\right).
\end{aligned}
$$

Because a quantum system composed of identical particles is completely specified by the occupation numbers n_i of the one-particle quantum states, the symmetric or antisymmetric wave function may be represented in the form $|n_1, \ldots, n_i, \ldots\rangle$ (observe that, as has been shown in the previous example, in a system of distinguishable particles the mere specification of the occupation numbers n_i is not sufficient to determine the wave function of the system). The formalism of the particle number operator is based on the definition of creation and annihilation operators. For bosons, the annihilation operator $\hat{a}_i$ is defined as

$$
\hat{a}_i|n_1, \ldots, n_i, \ldots\rangle = \sqrt{n_i}|n_1, \ldots, n_i - 1, \ldots\rangle, \tag{1.88}
$$

since it destroys a particle in state i. The adjoint operator $a_i^\dagger$ verifies

$$
\hat{a}_i^\dagger|n_1, \ldots, n_i, \ldots\rangle = \sqrt{n_i + 1}|n_1, \ldots, n_i + 1, \ldots\rangle, \tag{1.89}
$$

and is called the creation operator, in view of the fact that it increases by one the number of particles in state i.

For fermions, where n_i can only take the values 0 and 1, the definitions of the annihilation and creation operators are

$$
\hat{a}_i|n_1, \ldots, n_i, \ldots\rangle = (-1)^{v_i} n_i|n_1, \ldots, 1 - n_i, \ldots\rangle, \tag{1.90}
$$

and

$$
\hat{a}_i^\dagger|n_1, \ldots, n_i, \ldots\rangle = (-1)^{v_i} (1 - n_i)|n_1, \ldots, 1 - n_i, \ldots\rangle, \tag{1.91}
$$

with

$$
v_i = \sum_{k=1}^{i-1} n_k.
$$

Note that, in this case, $\hat{a}_i^\dagger \hat{a}_i^\dagger = 0$, because no two fermions may be created in the same quantum state due to Pauli's exclusion principle.

From (1.88–1.91), the number operator is then defined as $\hat{n}_i = \hat{a}_i^\dagger \hat{a}_i$, so that

$$\hat{n}_i |n_1, \ldots, n_i, \ldots\rangle = n_i |n_1, \ldots, n_i, \ldots\rangle. \tag{1.92}$$

The total number of particles operator is

$$\hat{N} = \sum_i \hat{n}_i, \tag{1.93}$$

which commutes with the Hamiltonian operator

$$\left[\hat{N}, \hat{H}_N(\alpha) \right] = 0. \tag{1.94}$$

Equations (1.92–1.94) are applicable both to bosons and to fermions.

Further Reading

1. H. Goldstein, *Classical Mechanics*, 2nd edn. (Addison-Wesley, Cambridge, 1985). *A good introduction to all the theoretical aspects of classical mechanics.*
2. L.D. Landau, E.M. Lifshitz, *Mechanics* (Pergamon, Oxford, 1982). *A beautiful synthesis of classical mechanics.*
3. A. Messiah, *Quantum Mechanics* (Dover Publications, 1999). *A good introduction to all the theoretical aspects of quantum mechanics.*
4. L.D. Landau, E.M. Lifshitz, *Quantum Mechanics (Non-Relativistic Theory)* (Pergamon, Oxford, 1981). *An overview of quantum mechanics with many useful applications.*
5. P.A.M. Dirac, *The Principles of Quantum Mechanics* (Clarendon Press, Oxford, 1981). *A beautiful alternative presentation of quantum mechanics.*

Chapter 2
Thermodynamics

Abstract The systems which are the subject matter of statistical physics involve N particles, atoms or molecules, with $N \simeq 10^{23}$. Studying such systems using the laws of mechanics (either classical or quantum mechanical) is an impossible task since, for instance, to determine the mechanical state of a classical system at any instant of time it is necessary to solve Hamilton's equations and to know the initial mechanical state of the system. As will be seen later on, there exist even more profound reasons to abandon the mechanical study of systems composed of a very large number of particles or macroscopic systems. An alternative description of such systems is provided by thermodynamics, a theory which is briefly described in this chapter. For a more detailed study of this theory the readers may consult the texts included in the Further Reading.

2.1 Fundamental Equation

Thermodynamics is a theory that deals with the exchange of matter and energy between a system, which is described in terms of a reduced number of thermodynamic variables that characterize the system as a whole, and the external world, which is everything that does not belong to the system. A natural first thermodynamic variable is the volume of the system V which, as has been discussed in Chap. 1, is an external parameter. A second thermodynamic variable is the energy E, referred to as the internal energy in thermodynamics, a variable defined as the macroscopic manifestation of the law of conservation of energy. In the case of an equilibrium (the thermodynamic variables do not change with time), closed (the system does not exchange particles with the external world) and simple (i.e., a system whose only external parameter is the volume V) system, it is postulated in thermodynamics that the internal energy is a differentiable function of the volume and the entropy, namely

$$E = E(S, V), \tag{2.1}$$

where S is the entropy of the system, which is termed a thermal variable since, in contrast with the cases of V and E, it does not have a mechanical counterpart.

These three variables define the state of the system. Note that E and V are extensive variables, i.e., magnitudes that grow proportionally to the size of the system. It is further assumed that the entropy S is also an extensive and additive variable, namely that the entropy of a composite system is the sum of the entropies of the constituent macroscopic subsystems. The change in the energy of the system when the entropy and the volume are modified is, therefore, given by

$$dE = \frac{\partial E}{\partial S}dS + \frac{\partial E}{\partial V}dV, \tag{2.2}$$

which is the sum of two contributions. The first one is the change in energy due to a change in the entropy S, while the second is the mechanical work performed when the external parameter V is modified.

The partial derivatives in (2.2) are defined as

$$T = \frac{\partial E}{\partial S}, \quad p = -\frac{\partial E}{\partial V}, \tag{2.3}$$

where T is the absolute (or thermodynamic) temperature of the system and p its pressure, both of which are positive since it is postulated that $E(S, V)$ is an increasing function of S that also decreases with V. Equation (2.2) is valid when the system evolves reversibly (see below) between two close equilibrium states. This equation is a particular case of the first law of thermodynamics:

$$dE = d'W + d'Q, \tag{2.4}$$

which expresses the fact that the variation of the internal energy is the sum of the work $d'W$ performed against the pressure of the external world p_0, which in a reversible process is equal to the pressure of the system

$$d'W = -pdV, \tag{2.5}$$

(the minus sign is chosen so that $d'W < 0$ when $dV > 0$) and of the amount of heat $d'Q$ absorbed by the system from the external world at the temperature T_0, which in a reversible process is equal to the temperature of the system, i.e.,

$$d'Q = TdS \tag{2.6}$$

(the notation $d'W$ and $d'Q$ indicates that the differentials of work and heat are not exact differentials). Note that thermodynamics does not provide definitions of the magnitudes E, V, and S, but that it only establishes through (2.2) the relationship between the changes in these variables in a reversible process. The power of thermodynamics resides precisely in the fact that this relationship is independent of the specific nature of the system.

Sometimes it is necessary to generalize expression (2.5) in order to include other extensive external parameters $\{X_j\}$, namely

$$d'W = \sum_j \frac{\partial E}{\partial X_j} dX_j. \tag{2.7}$$

In the case of open systems, i.e., systems which may exchange particles with the external world, it is postulated that the state of the system is defined by the variables E, S, $\{X_j\}$, and $\{N_\alpha\}$, where N_α is the number of particles of species α. It is further admitted that the energy is a differentiable function of the entropy, of the extensive external parameters, and of the numbers of particles, namely

$$E = E(S, \{X_j\}, \{N_\alpha\}). \tag{2.8}$$

This equation is known as the fundamental equation of thermodynamics and it follows from it that in a differential reversible process connecting two equilibrium states, the variation in the energy of the system is

$$dE = TdS + \sum_j \frac{\partial E}{\partial X_j} dX_j + \sum_\alpha \mu_\alpha dN_\alpha, \tag{2.9}$$

where

$$\mu_\alpha = \frac{\partial E}{\partial N_\alpha}, \tag{2.10}$$

is the chemical potential of species α. In the case of a system that contains N particles of a single species, whose chemical potential is denoted by μ, and where the only external parameter is the volume V, one has

$$dE = TdS - pdV + \mu dN. \tag{2.11}$$

2.2 Intensive Variables

Note that the number of thermodynamic variables defining the state of a system in the fundamental Eq. (2.8) is always small. On the contrary, in the mechanical description the corresponding number of variables increases with N. This is due to the fact that the thermodynamic variables E, S, $\{X_j\}$, and $\{N_\alpha\}$ are magnitudes that characterize the system globally. Since all of them are extensive, if one denotes by $\{Y_k\} \equiv (S, X_1, X_2, ..., N_1, N_2, ...)$, from the fundamental Eq. (2.8) it follows that E is a homogeneous function of degree one in the variables Y_k, namely

$$E(\{\lambda Y_k\}) = \lambda E(\{Y_k\}), \tag{2.12}$$

where λ is an arbitrary positive real number. If the conjugate variable or conjugate force f_l of Y_l is defined through

$$f_l = \frac{\partial E}{\partial Y_l}, \tag{2.13}$$

it follows from (2.12) that

$$f_l(\{\lambda Y_k\}) = f_l(\{Y_k\}). \tag{2.14}$$

The above equation shows that the conjugate variables f_l are homogeneous functions of degree zero in the variables Y_k, i.e., they do not depend on the size of the system. Hence, they are called intensive variables. Note that the ratio of two extensive variables is an intensive variable like, for instance, the energy per particle E/N. It is common to write the extensive variables as capital letters while the intensive variables, with the only exception (for historical reasons) of the absolute temperature T, are usually written as small letters. An important relation may be obtained from the derivative of (2.12) with respect to λ by considering the result in the particular case where one sets $\lambda = 1$, namely

$$\sum_k f_k Y_k = E. \tag{2.15}$$

By differentiating this equation and comparing the result with (2.9), the Gibbs–Duhem relation is obtained:

$$\sum_k Y_k df_k = 0, \tag{2.16}$$

which shows that the changes in the intensive variables f_k are not independent.

Since, on the other hand, E is a differentiable function of the variables Y_k, one has that $\partial^2 E/\partial Y_l \partial Y_m$ is a symmetrical matrix. This property may also be expressed as

$$\frac{\partial f_l}{\partial Y_m} = \frac{\partial f_m}{\partial Y_l}, \tag{2.17}$$

which are the so-called Maxwell relations. Note that throughout the text and for the sake of simplifying the notation, it is understood that in the partial derivatives all the other variables remain constant.

The thermodynamic description may, therefore, be summarized as follows. To each macroscopic system one associates a variable E or internal energy. According to the fundamental equation, E is a function of a reduced number of extensive variables $\{Y_k\}$. For each Y_k there is, according to (2.13), an intensive variable f_k and so E

is referred to as a thermodynamic potential in view of the fact that all the intensive variables may be obtained by derivation from the fundamental equation $E = E(\{Y_k\})$. The structure of thermodynamics is, therefore, in a way reminiscent of that of classical mechanics.

2.3 Law of Entropy Increase

In order to introduce the law of entropy increase, note that (2.11) may be rewritten in the form:

$$dS = \frac{1}{T}dE + \frac{p}{T}dV - \frac{\mu}{T}dN, \qquad (2.18)$$

which expresses the fact that S may be considered as a function of the extensive variables E, V, and N. This formulation of thermodynamics in which $S = S(E, V, N)$ is the dependent variable is known as the "entropy representation," whereas the formulation (2.11), i.e., $E = E(S, V, N)$ is known as the "energy representation." From (2.18) it follows that the intensive variables in the entropy representation are

$$\frac{1}{T} = \frac{\partial S}{\partial E}, \quad \frac{p}{T} = \frac{\partial S}{\partial V}, \quad \frac{\mu}{T} = -\frac{\partial S}{\partial N}. \qquad (2.19)$$

From a mathematical point of view this change of representation consists in obtaining S as a function of E, V, and N from the fundamental equation $E = E(S, V, N)$.

Note that (2.18) describes the changes that take place in a differential reversible process that connects two equilibrium states. The second law of thermodynamics states that when the intensive variables of the external world (the temperature T_0, the pressure p_0, and the chemical potential μ_0) are constant, because the number of degrees of freedom of the external world is much greater than that of the system, then the following inequality always holds:

$$dS \geq \frac{1}{T_0}dE + \frac{p_0}{T_0}dV - \frac{\mu_0}{T_0}dN, \qquad (2.20)$$

which may be written in the form

$$T_0 d_i S \equiv T_0 dS - dE - p_0 dV + \mu_0 dN \geq 0, \qquad (2.21)$$

where $d_i S$ is the change of entropy due to the irreversible processes taking place in the system. Equation (2.21) indicates that the system evolves in such a way that the entropy production $d_i S$ is positive, until the system reaches a state of equilibrium with the external world for which $d_i S = 0$. In particular, if one considers an isolated system (which does exchange neither energy nor volume or particles with the external world) then it follows from (2.21) that

$$d_i S \equiv dS \geq 0, \tag{2.22}$$

which expresses that an isolated system evolves toward the state of maximum entropy.

From the second law of thermodynamics (2.20) and from (2.18), three important consequences can be derived:

1. In the natural evolution of a non-equilibrium system, the changes in the thermodynamic variables verify the following relation,

$$T_0 dS - dE - p_0 dV + \mu_0 dN \geq 0, \tag{2.23}$$

    until equilibrium is reached. In the equilibrium state, the intensive variables of the system and those of the external world become equal: $T = T_0$, $p = p_0$, and $\mu = \mu_0$.
2. If the differential changes of the thermodynamic variables take place between two equilibrium states, then

$$T dS - dE - p dV + \mu dN = 0, \tag{2.24}$$

    i.e., these changes are not independent so that a fundamental relation $S = S(E, V, N)$ or $E = E(S, V, N)$ exists.
3. If a state of equilibrium is transformed into a non-equilibrium state, then $d_i S > 0$ cannot hold, since this inequality corresponds to a natural evolution and the former is not. Neither can it happen that $d_i S = 0$, since this corresponds to differential changes between two equilibrium states. Therefore, $d_i S < 0$. Hence, to distinguish the natural or spontaneous evolution toward equilibrium and the forced or virtual transformation out of equilibrium, the natural changes are denoted by $dS, \ldots$ while the virtual ones are denoted by $\delta S, \ldots$ For any virtual change of the state of the system one, therefore, has

$$T_0 \delta S - \delta E - p_0 \delta V + \mu_0 \delta N < 0. \tag{2.25}$$

2.4 Thermodynamic Potentials

Although the basic aspects of thermodynamics are embodied in (2.21–2.25), there are equivalent formulations that involve the introduction of other thermodynamic potentials. As a first example, consider the fundamental equation in the energy representation, $E = E(S, V, N)$. If one wants that, instead of S, T be the independent variable, which is the conjugate variable of S with respect to E in (2.11), a Legendre transformation may be performed (see Appendix A) from E to a new potential, denoted by F and known as the Helmholtz free energy. To that end, the function $F = F(T, V, N)$ is defined through

$$F(T, V, N) = E(S, V, N) - TS, \qquad (2.26)$$

where $S = S(T, V, N)$ in (2.26) is obtained after solving the implicit equation $T = \partial E / \partial S$. Differentiating (2.26) and comparing with (2.11) one then has

$$dF = -SdT - pdV + \mu dN, \qquad (2.27)$$

so that in this representation

$$S = -\frac{\partial F}{\partial T}, \quad p = -\frac{\partial F}{\partial V}, \quad \mu = \frac{\partial F}{\partial N}. \qquad (2.28)$$

From (2.15) and (2.26) it follows that

$$F = -pV + \mu N, \quad f = -pv + \mu, \qquad (2.29)$$

where $f = F/N$ is the Helmholtz free energy per particle, $v = V/N = 1/\rho$ is the volume per particle or specific volume, and ρ is the density. Differentiating $F = Nf$ one finds $dF = Ndf + fdN = -SdT - pdV + \mu dN$, namely $Ndf = -SdT - pdV + (\mu - f)dN$, so that from (2.29) it follows that $Ndf = -SdT - p(dV - vdN) = -SdT - Npdv$, or

$$df = -sdT - pdv, \qquad (2.30)$$

where $s = S/N$ is the entropy per particle. This equation shows that the intensive variable f is a function of the intensive variables T and v.

In the thermodynamic analysis of a system, the choice of the thermodynamic potential, although arbitrary, is to a great extent determined by the physical conditions of the problem under study. Thus, since a phase transition takes place at constant temperature and pressure (see below), an adequate potential is the Gibbs free energy, defined through

$$G(T, p, N) = E(S, V, N) - TS + pV, \qquad (2.31)$$

which is the double Legendre transform of the energy with respect to S and V, where $S = S(T, p, N)$ and $V = V(T, p, N)$ are obtained from (2.11) after solving the implicit equations $T = \partial E / \partial S$ and $p = -\partial E / \partial V$. After differentiation of (2.31) and comparison with (2.11) one obtains:

$$dG = -SdT + Vdp + \mu dN, \qquad (2.32)$$

so that in this representation

$$S = -\frac{\partial G}{\partial T}, \quad V = \frac{\partial G}{\partial p}, \quad \mu = \frac{\partial G}{\partial N}. \qquad (2.33)$$

On the other hand, from (2.31) and (2.15) it follows that:

$$G = \mu N, \tag{2.34}$$

and hence the chemical potential is equal to the Gibbs free energy per particle. Upon differentiation of (2.34) one finds $dG = Nd\mu + \mu dN$ and after comparing with (2.32) it follows that

$$d\mu = -sdT + vdp, \tag{2.35}$$

which is equivalent to the Gibbs–Duhem relation (2.16).

To close this section, consider the grand potential or Landau free energy:

$$\Omega(T, V, \mu) = E(S, V, N) - TS - \mu N, \tag{2.36}$$

which is the double Legendre transform of the energy with respect to S and N, where $S = S(T, V, \mu)$ and $N = N(T, V, \mu)$ are obtained from (2.11) after solving the implicit equations $T = \partial E/\partial S$ and $\mu = \partial E/\partial N$. After differentiation of (2.36) and comparison with (2.11) one obtains

$$d\Omega = -SdT - pdV - Nd\mu, \tag{2.37}$$

so that in this representation

$$S = -\frac{\partial \Omega}{\partial T}, \quad p = -\frac{\partial \Omega}{\partial V}, \quad N = -\frac{\partial \Omega}{\partial \mu}. \tag{2.38}$$

On the other hand, from (2.36) and (2.15) it follows that

$$\Omega = -pV. \tag{2.39}$$

Upon differentiation of (2.39) and comparison of the result with (2.37) one finds

$$dp = \rho sdT + \rho d\mu, \tag{2.40}$$

which is an expression equivalent to (2.35).

2.5　Equilibrium Conditions

From (2.25) it follows that in a virtual process of an isolated system ($\delta E = 0$, $\delta V = 0$ and $\delta N = 0$), $\delta S < 0$, i.e., the entropy may only decrease and hence S must have a maximum value in the equilibrium state. This result may be transformed into an extremum principle. Thus, if $S(E, V, N; \{Z\})$ denotes the variational entropy of a system with constrains denoted by $\{Z\}$, the entropy of the equilibrium state is a maximum of $S(E, V, N; \{Z\})$ with respect to $\{Z\}$ for given values of E, V, and N:

$$S(E, V, N) = \left\{\max_{\{Z\}} S(E, V, N; \{Z\})\right\}_{E,V,N}. \tag{2.41}$$

Similarly, from (2.25) one can see that if $\delta S = 0$, $\delta V = 0$, and $\delta N = 0$, then $\delta E > 0$, i.e., the energy can only grow and hence E is a minimum in the equilibrium state. Therefore, if $E(S, V, N; \{Z\})$ is the variational energy of a system with constrains described by $\{Z\}$, one has

$$E(S, V, N) = \left\{\min_{\{Z\}} E(S, V, N; \{Z\})\right\}_{S,V,N}, \tag{2.42}$$

so that the energy of the equilibrium state is a minimum of $E(S, V, N; \{Z\})$ with respect to $\{Z\}$ for given values of S, V, and N.

These extremum conditions may also be expressed using the other thermodynamic potentials. Note that if $\delta V = 0$ and $\delta N = 0$, the extremum principle (2.25) can be rewritten as $T_0 \delta S - \delta E < 0$, so that if the temperature of the system remains constant and equal to the one of the external world, $T = T_0$, then $\delta F > 0$, and hence F is a minimum in the state of equilibrium. One then has

$$F(T, V, N) = \left\{\min_{\{Z\}} F(T, V, N; \{Z\})\right\}_{T=T_0,V,N}, \tag{2.43}$$

namely the Helmholtz free energy of the equilibrium state is a minimum of $F(T, V, N; \{Z\})$ with respect to $\{Z\}$.

If $\delta N = 0$, the extremum principle is written as $T_0 \delta S - \delta E - p_0 \delta V < 0$, so that if the temperature and the pressure of the system remain constant and equal to the temperature and pressure of the external world, $T = T_0$ and $p = p_0$, it follows that $\delta G > 0$, and hence G is a minimum in the equilibrium state. One then has

$$G(T, p, N) = \left\{\min_{\{Z\}} G(T, p, N; \{Z\})\right\}_{T=T_0,p=p_0,N}, \tag{2.44}$$

namely the Gibbs free energy of the equilibrium state is a minimum of $G(T, p, N; \{Z\})$ with respect to $\{Z\}$.

Finally, if $\delta V = 0$, the extremum principle reads $T_0 \delta S - \delta E + \mu_0 \delta N < 0$, so that if the temperature and chemical potential of the system remain constant and equal to the temperature and chemical potential of the external world, $T = T_0$ and $\mu = \mu_0$, it follows that $\delta \Omega > 0$, and hence Ω is a minimum in the equilibrium state. One then has

$$\Omega(T, V, \mu) = \left[\min_{\{Z\}} \Omega(T, V, \mu; \{Z\})\right]_{T=T_0,V,\mu=\mu_0}, \tag{2.45}$$

namely the Landau free energy of the equilibrium state is a minimum of $\Omega(T, V, \mu; \{Z\})$ with respect to $\{Z\}$.

2.6 Stability Conditions

In the previous section, the extremum principles involving the different thermodynamic potentials were examined. In order to provide a more detailed analysis of some of them, consider a simple closed system which exchanges energy and volume with the external world at the temperature T_0 and at the pressure p_0. By definition, the intensive parameters of the external world remain constant and are, in general, different from the temperature T and the pressure p of the system. The second law of thermodynamics implies that for a virtual process

$$\delta(E + p_0 V - T_0 S) = \delta E + p_0 \delta V - T_0 \delta S > 0. \tag{2.46}$$

In the energy representation $E = E(S, V)$, and so upon expanding E in a Taylor series up to second order one has

$$\delta E = \frac{\partial E}{\partial V}\delta V + \frac{\partial E}{\partial S}\delta S$$
$$+ \frac{1}{2}\left[\frac{\partial^2 E}{\partial V^2}(\delta V)^2 + 2\frac{\partial^2 E}{\partial V \partial S}\delta V \delta S + \frac{\partial^2 E}{\partial S^2}(\delta S)^2\right]. \tag{2.47}$$

Since $\partial E/\partial V = -p$, $\partial E/\partial S = T$ and the equilibrium state is an extremum, it follows from (2.46) and (2.47) that

$$p = p_0, \quad T = T_0, \tag{2.48}$$

which are the conditions for mechanical equilibrium and thermal equilibrium between the system and the external world. It should be clear that it is also possible to obtain in this way the condition for chemical equilibrium ($\mu = \mu_0$) in the case where, on top of the above, the system and the external world also exchange particles.

Since the extremum is a minimum, one then has

$$\frac{\partial^2 E}{\partial V^2}(\delta V)^2 + 2\frac{\partial^2 E}{\partial V \partial S}\delta V \delta S + \frac{\partial^2 E}{\partial S^2}(\delta S)^2 > 0, \tag{2.49}$$

which holds whenever the following inequalities are satisfied:

$$\frac{\partial^2 E}{\partial V^2} > 0, \quad \frac{\partial^2 E}{\partial V^2}\frac{\partial^2 E}{\partial S^2} - \left(\frac{\partial^2 E}{\partial V \partial S}\right)^2 > 0, \quad \frac{\partial^2 E}{\partial S^2} > 0. \tag{2.50}$$

These are known as the stability conditions (note that the first two inequalities imply the third one). From the first inequality it follows that

$$\frac{\partial^2 E}{\partial V^2} = -\frac{\partial p}{\partial V} = \frac{1}{V \chi_S}, \quad \chi_S \equiv -\frac{1}{V}\left(\frac{\partial V}{\partial p}\right)_S > 0, \tag{2.51}$$

namely that the isentropic compressibility coefficient χ_S is positive. In a similar way, from the third inequality one finds

$$\frac{\partial^2 E}{\partial S^2} = \frac{\partial T}{\partial S} = \frac{1}{T C_V}, \quad C_V \equiv T\left(\frac{\partial S}{\partial T}\right)_V > 0, \tag{2.52}$$

i.e., the constant volume heat capacity C_V is also positive.

The stability conditions may also be formulated using any other thermodynamic potential. Assume, for instance, that in (2.46) the independent variables are taken to be the volume V and the temperature T, which are the natural variables of the Helmholtz free energy. Then, since in this case $E = E(T, V)$ and $S = S(T, V)$, in the Taylor series expansion of (2.46) up to second order, the extremum conditions are

$$\frac{\partial E}{\partial T} - T_0 \frac{\partial S}{\partial T} = 0, \quad \frac{\partial E}{\partial V} + p_0 - T_0 \frac{\partial S}{\partial V} = 0. \tag{2.53}$$

The first one implies that

$$T_0 = \frac{\partial E}{\partial T}\frac{\partial T}{\partial S} = \frac{\partial E}{\partial S} = T, \tag{2.54}$$

which is the condition for thermal equilibrium, while the second one reads

$$p_0 = -\frac{\partial(E - TS)}{\partial V} = -\frac{\partial F}{\partial V} = p, \tag{2.55}$$

which is the condition for mechanical equilibrium. In deriving this condition, it has been taken into account that $T_0 = T$.

In order to determine the stability conditions, it is required that the quadratic form be positive definite. The coefficient that multiplies $(\delta V)^2$ is

$$\frac{\partial^2 E}{\partial V^2} - T \frac{\partial^2 S}{\partial V^2} = \frac{\partial^2 F}{\partial V^2} > 0, \tag{2.56}$$

which implies that the isothermal compressibility coefficient χ_T is positive:

$$\frac{\partial^2 F}{\partial V^2} = -\frac{\partial p}{\partial V} = \frac{1}{V \chi_T}, \quad \chi_T \equiv -\frac{1}{V}\left(\frac{\partial V}{\partial p}\right)_T > 0. \tag{2.57}$$

The coefficient that multiplies $(\delta T)^2$ is

$$\frac{\partial^2 E}{\partial T^2} - T\frac{\partial^2 S}{\partial T^2} > 0. \tag{2.58}$$

Since

$$\frac{\partial^2 E}{\partial T^2} = \frac{\partial}{\partial T}\left(T\frac{\partial S}{\partial T}\right) = \frac{\partial S}{\partial T} + T\frac{\partial^2 S}{\partial T^2}, \tag{2.59}$$

it follows that (2.58) may be rewritten as

$$\frac{\partial S}{\partial T} = -\frac{\partial^2 F}{\partial T^2} > 0, \tag{2.60}$$

which, like in (2.52), implies that C_V is positive.

From (2.50), (2.56) and (2.60) the following result may be inferred. When the stability conditions are expressed in terms of any arbitrary thermodynamic potential, the second derivative of the potential with respect to any of its natural variables is positive (the function is convex) if the variable is extensive and negative (the function is concave) if the variable is intensive. For instance,

$$\frac{\partial^2 G}{\partial T^2} < 0, \quad \frac{\partial^2 G}{\partial p^2} < 0, \tag{2.61}$$

and from (2.32) it follows that the first inequality in (2.61) implies that the constant pressure heat capacity,

$$C_p \equiv T\left(\frac{\partial S}{\partial T}\right)_p > 0, \tag{2.62}$$

is positive, while from the second one it follows that $\chi_T > 0$.

2.7 Coexistence Conditions

Thus far, the equilibrium state of the system is, as referred to in thermodynamics, a phase, i.e., the system is homogeneous (the intensive thermodynamic variables are uniform). The concept of a phase is an idealization of reality since the existence of external fields, such as the gravitational field, leads to inhomogeneities in the system. Note also that since the system is enclosed in a region R of volume V, the intensive variables cannot be uniform due to the influence of the walls of the container. In the thermodynamic study of a system it is assumed that these effects are small.

It is well known that there exist different phases of matter. A simple fluid of spherical molecules may, for instance, be found in a gas phase, in a liquid phase, or in a solid phase. Each of these phases is thermodynamically stable for given conditions of temperature and pressure. The paramagnetic and ferromagnetic phases of some materials or the isotropic or nematic phases of liquid crystals represent other kinds of phases of matter. In general, given two phases of a system, the phase stable at a lower temperature presents greater order. Thus, a solid is more ordered than a liquid because the former is not invariant under an arbitrary rotation or translation. Similarly, the ferromagnetic phase is more ordered than the paramagnetic one because the former is not invariant under an arbitrary rotation.

In some cases, phases may coexist. That is, under given conditions of temperature and pressure the system is not homogeneous. Examples of this are the liquid–vapor coexistence, the liquid–solid coexistence and the vapor–solid coexistence. In other cases, it is impossible to have phase coexistence. For instance, a material cannot be both ferromagnetic and paramagnetic at prescribed values of the temperature and pressure.

In Sect. 2.5 the equilibrium conditions between a system and the external world were obtained. The same conditions must hold between two macroscopic phases (in the following denoted 1 and 2) of a system in thermodynamic equilibrium. Indeed, consider an isolated equilibrium system of entropy $S = S_1 + S_2$, energy $E = E_1 + E_2$, volume $V = V_1 + V_2$ and $N = N_1 + N_2$ particles. In a virtual process, from (2.25) it follows that,

$$\delta S < 0, \tag{2.63}$$

since E, V, and N are constant. This condition implies that the entropies of the phases $S_1 = S_1(E_1, V_1, N_1)$ and $S_2 = S_2(E_2, V_2, N_2)$ satisfy

$$\delta S_1 + \delta S_2 < 0. \tag{2.64}$$

Since

$$\delta S_1 = \frac{1}{T_1}\delta E_1 - \frac{p_1}{T_1}\delta V_1 + \frac{\mu_1}{T_1}\delta N_1, \quad \delta S_2 = \frac{1}{T_2}\delta E_2 - \frac{p_2}{T_2}\delta V_2 + \frac{\mu_2}{T_2}\delta N_2,$$

and

$$\delta E_1 + \delta E_2 = 0, \quad \delta V_1 + \delta V_2 = 0, \quad \delta N_1 + \delta N_2 = 0,$$

(2.64) can then be written as

$$\left(\frac{1}{T_1} - \frac{1}{T_2}\right)\delta E_1 - \left(\frac{p_1}{T_1} - \frac{p_2}{T_2}\right)\delta V_1 + \left(\frac{\mu_1}{T_1} - \frac{\mu_2}{T_2}\right)\delta N_1 < 0. \tag{2.65}$$

Note that because δE_1, δV_1, and δN_1 are arbitrary and the equilibrium state is an extremum, from (2.65) the conditions for thermal equilibrium (equality of temperatures),

$$T_1 = T_2, \tag{2.66}$$

for mechanical equilibrium (equality of pressures),

$$p_1 = p_2, \tag{2.67}$$

and for chemical equilibrium (equality of chemical potentials),

$$\mu_1 = \mu_2, \tag{2.68}$$

follow.

2.8 Phase Diagrams

The thermodynamic equilibrium states in which phases coexist are usually represented in the so-called phase diagrams that may be obtained from any thermodynamic potential. The method to be followed to obtain a phase diagram depends, however, on the choice of thermodynamic potential. In this section, the phase diagrams are derived from the Helmholtz free energy per particle $f = F(T, V, N)/N = f(T, v)$ and from the Gibbs free energy per particle $g = G(T, p, N)/N = g(T, p)$, which in a one-component system coincides with the chemical potential $\mu = \mu(T, p)$ (see Sect. 2.4). The main difference between $\mu(T, p)$ and $f(T, v)$ is that in the former the two independent variables, p and T, appear explicitly in the equilibrium conditions (2.66–2.68), while in the latter only one of its independent variables, T, appears in these conditions.

2.8.1 Gibbs Free Energy

To construct the phase diagram from the Gibbs free energy, note that, since $\mu = \mu(T, p)$, from the equilibrium conditions $T_1 = T_2 = T$ and $p_1 = p_2 = p$ one has

$$\mu_1(T, p) = \mu_2(T, p), \tag{2.69}$$

which defines the implicit relation

$$p = p(T), \tag{2.70}$$

that the independent variables p and T must satisfy in order to have coexistence between phases 1 and 2.

According to (2.35) a phase may be represented by a surface $\mu = \mu(T, p)$ in the space (μ, T, p) or, as is usually done, by the projection of this surface onto isobaric or isothermal planes, as shown in Figs. 2.1 and 2.2.

From the thermodynamic relations

$$\frac{\partial \mu}{\partial T} = -s, \quad \frac{\partial \mu}{\partial p} = v, \tag{2.71}$$

and

$$\frac{\partial^2 \mu}{\partial T^2} = -\frac{c_p}{T}, \quad \frac{\partial^2 \mu}{\partial p^2} = -v\chi_T, \tag{2.72}$$

Fig. 2.1 Projection onto the isobaric plane, $p = p_0$, of the Gibbs free energy per particle $\mu = \mu(T, p)$, with $\mu_0 = \mu(T_0, p_0)$. The curve $\mu(T, p_0)$ is concave ($c_p > 0$) and decreases with T

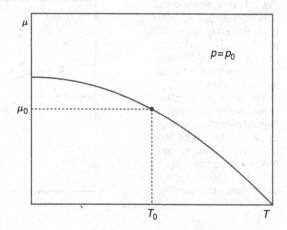

Fig. 2.2 Projection onto the isothermal plane, $T = T_0$, of the Gibbs free energy per particle $\mu = \mu(T, p)$, with $\mu_0 = \mu(T_0, p_0)$. The curve $\mu(T_0, p)$ is concave ($\chi_T > 0$) and increases with p

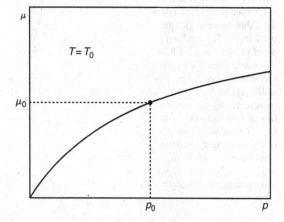

where χ_T is the isothermal compressibility coefficient and $c_p = C_p/N$ the constant pressure specific heat, it follows that the projection of the surface $\mu = \mu(T, p)$ onto an isobaric plane is a concave curve, because $c_p > 0$, and a decreasing function of T, while the projection onto an isothermal plane is a concave function, since $\chi_T > 0$, and an increasing function of p.

Phase coexistence is obtained from (2.69), i.e., from the intersection of the surfaces that represent each phase or, as usual, from the intersection of the curves resulting from the projection of these surfaces onto isobaric and isothermal planes (Figs. 2.3 and 2.4).

In the point of intersection both phases coexist, but the slopes of the curves are in general different since the two phases differ both in entropy ($s_1 \neq s_2$) and in density ($v_1 \neq v_2$). On each side of the point of intersection one of the phases has a lower Gibbs free energy and is thus the stable phase, which has been represented by a

Fig. 2.3 Projection onto the isobaric plane, $p = p_0$, of the Gibbs free energies per particle, $\mu_1(T, p)$ and $\mu_2(T, p)$, of phases 1 and 2. Phase 1 is stable when $T > T_0$ and metastable when $T < T_0$. The stable regions have been represented by a *continuous line* and the metastable ones by a *broken line*. At the point of coexistence (T_0, p_0) the two phases are stable, $\mu_0 \equiv \mu_1(T_0, p_0) = \mu_2(T_0, p_0)$ and exchange their stability

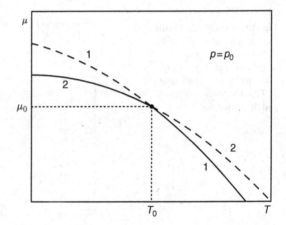

Fig. 2.4 Projection onto the isothermal plane, $T = T_0$, of the Gibbs free energies per particle, $\mu_1(T, p)$ and $\mu_2(T, p)$, of phases 1 and 2. Phase 1 is stable when $p < p_0$ and metastable when $p > p_0$. The stable regions have been represented by a *continuous line* and the metastable ones by a *broken line*. At the point of coexistence (T_0, p_0) the two phases are stable, $\mu_0 \equiv \mu_1(T_0, p_0) = \mu_2(T_0, p_0)$, and exchange their stability

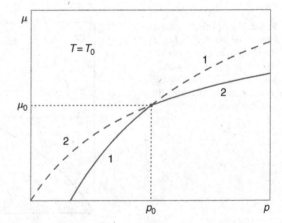

Fig. 2.5 Phase diagram (T, p) corresponding to Figs. 2.3 and 2.4. The regions of stability of the phases (indicated by 1 and 2) are separated by the coexistence line on which both phases are stable and coexist

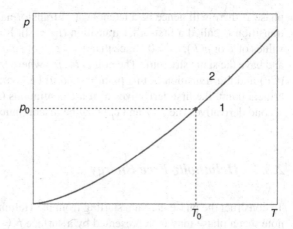

continuous line. A stable phase ($\chi_T > 0$, $c_p > 0$) which does not have the lower Gibbs free energy is said to be metastable and has been represented by a broken line.

Since the concave curves of Figs. 2.3 and 2.4 can only intersect in one point, there is a one-to-one correspondence between the points of intersection and the corresponding point in the (T, p)-plane. In this way, all the information about the intersection of the surfaces $\mu_1(T, p)$ and $\mu_2(T, p)$, that describes the equilibrium between two phases, is transferred from the space (μ, T, p) to the plane (T, p).

In the (T, p) phase diagram of Fig. 2.5, the coexistence line (2.70) separates the two regions of this plane where either one of the phases is stable. If the coexistence line is crossed, there will be a phase transition during which the stable phase becomes metastable and vice versa, and the two phases exchange their stability. On the coexistence line both phases have the same Gibbs free energy per particle and, therefore, are simultaneously stable although the derivative of μ (with respect to T or p) is, in general, different at the point of coexistence.

Upon substitution of (2.70) in (2.69) it follows that on the coexistence line:

$$\mu_1(T, p(T)) = \mu_2(T, p(T)). \tag{2.73}$$

Taking the derivative of (2.73) with respect to T yields

$$\frac{\partial \mu_1}{\partial p}\frac{dp(T)}{dT} + \frac{\partial \mu_1}{\partial T} = \frac{\partial \mu_2}{\partial p}\frac{dp(T)}{dT} + \frac{\partial \mu_2}{\partial T}, \tag{2.74}$$

while from (2.71), one then obtains the Clausius–Clapeyron equation,

$$\frac{dp(T)}{dT} = \frac{s_1 - s_2}{v_1 - v_2}, \tag{2.75}$$

which relates the slope of the coexistence curve (2.70) to the discontinuities in the entropy per particle and the volume per particle. At the transition from phase 1 to phase 2, there will hence be a latent heat of transformation $l_Q = T(s_2 - s_1)$ and the transition is called a first-order transition ($l_Q \neq 0$). It may happen that, for certain values of T or $p(T)$, $l_Q = 0$. Since then $s_1 = s_2$, both phases are entropically identical and have the same structure. The point (T_c, p_c) where $l_Q = 0$ is called a critical point (CP) and the transition at this point is said to be continuous (see Chap. 9). At the critical point, the first derivatives of μ are continuous ($s_1 = s_2$ and $v_1 = v_2$), but the second derivatives (i.e., χ_T and c_p) may be discontinuous.

2.8.2 Helmholtz Free Energy

To construct the phase diagram starting from the Helmholtz free energy per particle, note that a phase may be represented by a surface $f = f(T, v)$ in the (f, T, v) space or, as usually done, by the projection of this surface onto isochoric and isothermal planes.

From the thermodynamic relations

$$\frac{\partial f}{\partial T} = -s, \quad \frac{\partial f}{\partial v} = -p, \tag{2.76}$$

and

$$\frac{\partial^2 f}{\partial T^2} = -\frac{c_V}{T}, \quad \frac{\partial^2 f}{\partial v^2} = \frac{1}{v\chi_T}, \tag{2.77}$$

where $c_V = C_V/N$ is the constant volume specific heat. It follows that the projection of the surface $f = f(T, v)$ onto an isochoric plane is a concave curve, since $c_V > 0$, and a decreasing function of T, whereas the projection onto an isothermal plane is a convex curve, since $\chi_T > 0$, which decreases with v. These curves are shown schematically in Figs. 2.6 and 2.7.

Consider the intersection of the surfaces $f_1(T, v)$ and $f_2(T, v)$ that represent each of the phases or, alternatively, the intersection of the curves that result from projecting these surfaces onto isochoric and isothermal planes. Note that since at such a point of intersection onto an isothermal plane, $T = T_0$, the two phases have the same specific volume v_0 (Fig. 2.8), this is not a point of coexistence because, in general, when phases coexist they have different densities ($v_1 \neq v_2$). For the same reason, the projection onto an isochoric plane does not provide any information about coexistence.

Note, however, that from the projection onto an isothermal plane one may derive the values of p and μ at this temperature. In fact, from (2.76) it follows that the slope of the curve $f(T, v)$ at constant temperature is minus the pressure. On the other hand, if $f(T, 0)$ is the intercept of the tangent to the curve at the point (v, f), the equation of this line is $f(T, v) - f(T, 0) = -pv$, namely $f(T, 0) = f(T, v) + pv = \mu$, where (2.29) has been taken into account.

Fig. 2.6 Projection onto the isochoric plane, $v = v_0$, of the Helmholtz free energy per particle $f = f(T, v)$, with $f_0 = f(T_0, v_0)$. The curve $f(T, v_0)$ is concave ($c_V > 0$) and a decreasing function of T

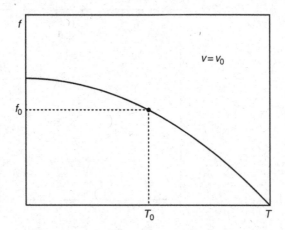

Fig. 2.7 Projection onto the isothermal plane, $T = T_0$, of the Helmholtz free energy per particle $f = f(T, v)$, with $f_0 \equiv f(T_0, v_0)$. The curve $f(T_0, v)$ is convex ($\chi_T > 0$) and decreases with v

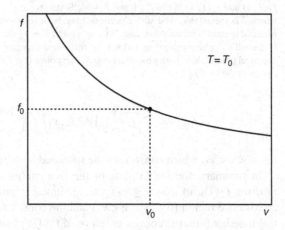

The previous results led Maxwell to devise a geometrical method to locate the coexistence between two phases. In Fig. 2.8 the projection of the curves $f_1(T_0, v)$ and $f_2(T_0, v)$ (which intersect at $v = v_0$) onto an isothermal plane ($T = T_0$) are represented. It is observed from this figure that phase 2 is metastable with respect to phase 1 when $v > v_0$ and that phase 1 is metastable with respect to phase 2 when $v < v_0$. Let v_1 and v_2 be the ordinates of the points of the curves $f_1(T_0, v)$ and $f_2(T_0, v)$ with the same tangent, which is unique since the curves are convex, and called Maxwell's double tangent. Note that along the line that joins the points $(v_1, f_1(T_0, v_1))$ and $(v_2, f_2(T_0, v_2))$, the pressure and the chemical potential remain constant, so that these points correspond to phase coexistence. Since, on the other hand, the free energy along the rectilinear segment is lower than the free energy of each phase, the thermodynamic state in which both phases coexist is the stable state. Along this segment, $f(T_0, v)$ is a linear combination of $f_1(T_0, v_1)$ and $f_2(T_0, v_2)$, namely.

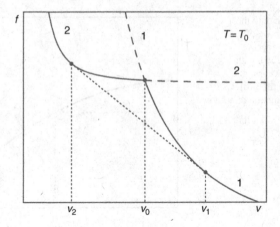

Fig. 2.8 Projection onto the isothermal plane, $T = T_0$, of the Helmholtz free energies per particle $f_1(T, v)$ and $f_2(T, v)$ of phases 1 and 2, which intersect at $v = v_0$. Phase 1 is stable when $v > v_0$, where 2 is metastable. The stable portions have been represented with a *continuous line* and the metastable ones with a *broken line*. At $v = v_1$ and $v = v_2$ the two curves have the same tangent (Maxwell's double tangent) and along the rectilinear segment (*dotted line*) the pressures and the chemical potentials of both phases are equal. The points (v_1, T_0) and (v_2, T_0) belong to the binodal curves in the (v, T)-plane

$$f(T_0, v) = \left(\frac{v - v_2}{v_1 - v_2}\right) f_1(T_0, v_1) + \left(\frac{v_1 - v}{v_1 - v_2}\right) f_2(T_0, v_2), \qquad (2.78)$$

for $v_2 < v < v_1$, which constitutes the so-called lever rule.

In summary, the stable parts of the free energy of the system are the convex function $f_2(T_0, v)$ (for $v \leq v_2$), the rectilinear segment of the double tangent or coexistence region (for $v_2 < v < v_1$) and the convex function $f_1(T_0, v)$ (for $v_1 \leq v$) that together form the convex envelope of $f_1(T_0, v)$ and $f_2(T_0, v)$.

The phase diagrams that result in the planes (v, T) and (v, p) are shown in Figs. 2.9 and 2.10. Observe that coexistence (a line in the (T, p)-plane of Fig. 2.5) in these planes occurs in regions delimited by two curves called binodals.

2.8.3 van der Waals Loop

It may happen that at a temperature $T = T_0$ the free energy $f(T_0, v)$ is formed by two convex branches ($v < v_2'$ and $v > v_1'$) separated by a concave branch $\left(v_2' < v < v_1'\right)$, such as illustrated in Fig. 2.11.

In this last branch the stability condition $\chi_T > 0$ is not satisfied and so the system is unstable. Note that if v_2 and v_1 are the abscissa of the points of the double tangent of the convex branches then $v_2 < v_2' < v_1' < v_1$. For that reason, the region of instability in the phase diagrams (v, T) and (v, p), which is delimited by two curves called

Fig. 2.9 Phase diagram in the (v, T)-plane in which we have indicated the regions where phases 1 and 2 are stable, the binodal curves (*continuous lines*) and the binodal region where the two phases coexist $(2 + 1)$

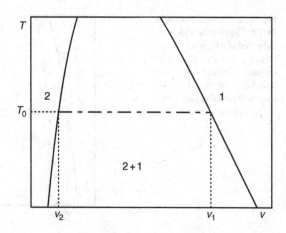

Fig. 2.10 Phase diagram in the (v, p)-plane in which are indicated the regions where phases 1 and 2 are stable, the binodal curves (*continuous lines*) and the binodal region where the two phases coexist $(2 + 1)$. In the diagram an isotherm $T = T_0$ (*dash-dotted line*) has also been represented

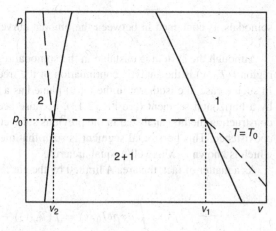

Fig. 2.11 van der Waals loop of the Helmholtz free energy per particle $f(T, v)$ at a temperature $T = T_0$. In the concave branch $v_2' < v < v_1'$ (*broken line*) the system is unstable $(\chi_T < 0)$. At $v = v_1$ and $v = v_2$ the curve has the same tangent (*dotted line*)

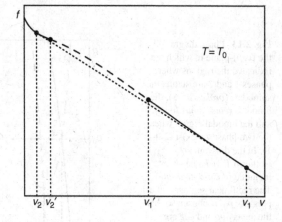

Fig. 2.12 Phase diagram in the (v, T)-plane in which are indicated the regions where phases 1 and 2 are stable, the spinodals (*broken lines*), the binodals (*continuous lines*), and the binodal region where the two phases coexist (2 + 1)

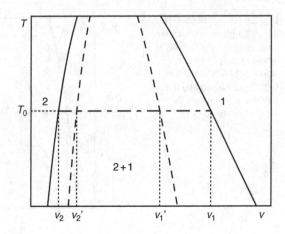

spinodals, is contained in between the binodal curves, as shown in Figs. 2.12 and 2.13.

Although the system is unstable in the spinodal region, it is admitted that in this region $f(T_0, v)$ is the analytic continuation of the free energy of the stable regions. In such a case, the isotherm in the (v, p)-plane has a loop, which must be replaced by a horizontal segment (see Fig. 2.13), because according to the double tangent construction method analyzed in the previous subsection, the pressures are equal at coexistence. This horizontal segment is such that the areas abc and cde are equal, which is known as Maxwell's equal area rule.

As a matter of fact, the area A limited by the curve $p = p(T_0, v)$ is

$$A = \int_{v_2}^{v_1} dv\, p(T_0, v) = f(T_0, v_2) - f(T_0, v_1), \tag{2.79}$$

Fig. 2.13 Phase diagram in the (v, p)-plane in which are indicated the regions where phases 1 and 2 are stable, the spinodals (*broken lines*), the binodals (*continuous lines*), and the binodal region where the two phases coexist (2 + 1). In the diagram an isotherm $T = T_0$ has been included (*dash-dotted line*). The rectilinear segment, at a pressure p_0, is chosen so that the areas abc and cde are equal

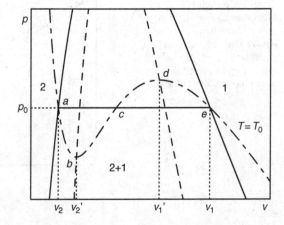

where the fact that $p = -\partial f/\partial v$, even inside the instability region, has been accounted for. If one subtracts from A the area of the rectangle $p_0(v_1 - v_2)$, where $p_0 = p_0(T_0)$ is the pressure at coexistence, and it is required that this difference vanishes, then.

$$f(T_0, v_2) - f(T_0, v_1) - p_0(v_1 - v_2) = 0, \qquad (2.80)$$

i.e., $\mu_1(T_0, p_0) = \mu_2(T_0, p_0)$, which is the condition for chemical equilibrium of the phases. This then proofs Maxwell's rule.

2.8.4 An Example of a Phase Diagram

The phase diagram of a system contains, in general, more than two phases. To obtain the phase diagram one has to analyze separately the coexistence of each pair of phases, following the methods described in the previous subsections, and their relative stability.

As a typical example, Figs. 2.14 and 2.15 show the phase diagrams of a simple system containing three phases: solid (S), liquid (L), and vapor (V).

Note that these three phases may coexist simultaneously at a point (T, p) of the phase diagram called the triple point (T_t, p_t), which is the solution of the following system of equations,

$$\mu_1(T_t, p_t) = \mu_2(T_t, p_t) = \mu_3(T_t, p_t), \qquad (2.81)$$

in which the three phases coexist. Observe that in these diagrams the vapor–liquid coexistence line ends at a critical point (T_c, p_c) where the two phases have the same structure and merge into the so-called fluid phase (F). Note that the liquid only occupies a small region of the diagrams, between the triple point and the critical

Fig. 2.14 Phase diagram of a simple system in the (T, p)-plane. The phases are S (solid), L (liquid), V (vapor), and F (fluid). T_c and p_c are the temperature and pressure of the liquid–vapor critical point. T_t and p_t are temperature and pressure of the triple point in which the three phases S, L, and V coexist

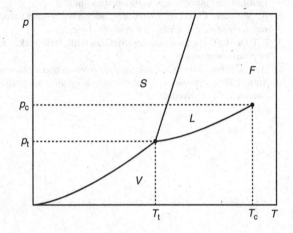

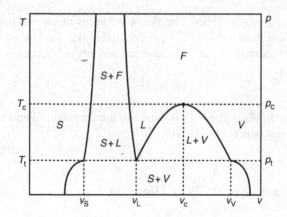

Fig. 2.15 Phase diagram of a simple system in the (v, T) and (v, p) planes. The phases are S (solid), L (liquid), V (vapor), and F (fluid). T_c, p_c, and v_c are the temperature, pressure, and volume per particle of the liquid–vapor critical point. T_t and p_t are the temperature and pressure of the triple point in which the three phases S, L, and V coexist. v_S, v_L, and v_V are the volumes per particle at coexistence. Note the discontinuity in the slopes of the binodals for $T = T_t$

point, and that there is no solid–fluid critical point. All transitions are first-order transitions. $(l_Q \neq 0)$ except at the critical point where the transition is continuous $(l_Q = 0)$. Included in the (v, T) and (v, p) diagrams are the binodals and the coexistence regions. vc is the volume per particle at the critical point and the volumes per particle of the phases S, L, and V at the triple point are denoted by v_S, v_L, and v_V.

Further Reading

1. M.W. Zemansky, R.H. Dittman, *Heat and Thermodynamics*, 7th edn. (McGraw-Hill, New York, 1996). *A first introduction to thermodynamics.*
2. A. Munster, *Classical Thermodynamics* (Wiley-Interscience, New York, 1970). *A classic summary of the principles of thermodynamics.*
3. R.T. De Hoff, *Thermodynamics* (McGraw-Hill, New York, 1993). *An introduction with a special emphasis on mixtures.*
4. H. B. Callen, *Thermodynamics and an Introduction to Thermostatistics*, 2nd edn. (Wiley, New York, 1985). *Provides a good discussion of the foundations of both phenomenological and statistical thermodynamics.*

Chapter 3
Statistical Physics

Abstract From the analysis given in the first two chapters the reader may have noticed the fundamental difference between the mechanical and the thermodynamic description of a system. In the first case, the mechanical state of N material points at time t is determined by $3N$ generalized coordinates and their $3N$ conjugate momenta (classical mechanics) or by the wave function (quantum mechanics). In the second instance, the equilibrium state of a simple closed system is specified by only two independent variables, such as the volume and the entropy. It is well known that the laws of mechanics provide an adequate description of the motion in systems with a small number of degrees of freedom. This is indeed the case of the two-body problem whose most representative examples are the planetary motion in classical mechanics and the hydrogen atom in quantum mechanics. The integrability of mechanical systems, namely the possibility of predicting their time evolution given some prescribed initial conditions, is, however, the exception rather than the rule. The most evident manifestation of the complexity of mechanical motions is the chaotic behavior of their evolution even for systems with only a few degrees of freedom. The origin of chaos lies in the nonlinear character of the equations of motion which, as a consequence, implies that a small change of the initial condition completely alters the time evolution. Assume that an experiment is performed in which one measures the final pressure of a gas initially occupying one half of a cubic box of volume V, at the temperature T which is held constant throughout the experiment. Once the wall separating the two halves of the box is removed, the gas, which starts from a non-equilibrium initial condition, expands all over the box until, after a certain time, the pressure reaches a stationary equilibrium value. When the system is initially prepared, its mechanical state is defined, within classical mechanics, by some generalized coordinates and their conjugate momenta. If the experiment is repeated, the result of the new measurement, except for small fluctuations, will be the same as that of the first experiment. On the other hand, it is clear that the initial mechanical state of the second experiment will, in general, be different from the one that the gas had in the first experiment. It should be noted that even in the case where both initial mechanical states are rather close to each other, their time evolution will, in general, be totally different due to the instability of the solutions to Hamilton's equations. The question, thus, arises as to why the same macroscopic result is reproduced by the different experiments.

© The Author(s), under exclusive license to Springer Nature Switzerland AG 2021   47
M. Baus and C. F. Tejero, *Equilibrium Statistical Physics*,
https://doi.org/10.1007/978-3-030-75432-7_3

3.1   Dynamical Functions and Fields

Consider a classical fluid of N particles of mass m contained in a closed region R. As was discussed in Chap. 1, the mechanical state of the fluid is determined by $3N$ generalized coordinates and their $3N$ conjugate momenta. If the description of the mechanical state of the fluid is made using the Cartesian coordinates of the particles, $\mathbf{r}^N$, and their conjugate momenta, $\mathbf{p}^N$, such a state being indicated by $(q, p) = (\mathbf{r}^N, \mathbf{p}^N)$, the corresponding Hamilton's equations are

$$\dot{\mathbf{r}}_i = \frac{1}{m}\mathbf{p}_i, \quad \dot{\mathbf{p}}_i = -\frac{\partial U_N(\mathbf{r}^N)}{\partial \mathbf{r}_i}, \quad (i = 1, \ldots, N), \tag{3.1}$$

where it has been admitted that the Hamiltonian is given by (1.9).

In order to analyze, within the framework of classical mechanics, the law of conservation of the number of particles, consider a volume element $d\mathbf{r}$ in the Euclidean space centered at a point $\mathbf{r}$. Let $\rho_1(q, p; \mathbf{r})$ be the dynamical function that represents the local density of particles at $\mathbf{r}$. One may then write

$$\rho_1(q, p; \mathbf{r}) \equiv \sum_{i=1}^{N} \delta(\mathbf{r} - \mathbf{r}_i), \tag{3.2}$$

where $\delta(\mathbf{r} - \mathbf{r}_i)$ is the Dirac delta function. Equation (3.2) expresses that particle i does not contribute to the local density at $\mathbf{r}$ if $\mathbf{r}_i \neq \mathbf{r}$ and that it contributes an infinite amount (since it has been assumed that the particles are point particles) if $\mathbf{r}_i = \mathbf{r}$. Upon integration of (3.2) over the region R in which the fluid is contained, one obtains

$$\int_R d\mathbf{r}\,\rho_1(q, p; \mathbf{r}) = N. \tag{3.3}$$

The time evolution of the dynamical function $\rho_1(q, p; \mathbf{r})$ is, according to (1.23),

$$\dot{\rho}_1(q, p; \mathbf{r}) = \{\rho_1, H_N(\alpha)\} = \sum_{i=1}^{N} \frac{\mathbf{p}_i}{m} \cdot \nabla_i \delta(\mathbf{r} - \mathbf{r}_i)$$

$$= -\nabla \cdot \sum_{i=1}^{N} \mathbf{v}_i \delta(\mathbf{r} - \mathbf{r}_i), \tag{3.4}$$

where $\nabla_i = \partial/\partial \mathbf{r}_i$, $\nabla = \partial/\partial \mathbf{r}$, $\mathbf{v}_i = \mathbf{p}_i/m$ and use has been made of the fact that $\nabla_i \delta(\mathbf{r} - \mathbf{r}_i) = -\nabla \delta(\mathbf{r} - \mathbf{r}_i)$. The above equation may be written in the form

$$\dot{\rho}_1(q, p; \mathbf{r}) = -\nabla \cdot \mathbf{j}(q, p; \mathbf{r}), \tag{3.5}$$

where the dynamical function current density $\mathbf{j}(q, p; \mathbf{r})$ has been defined through

$$\mathbf{j}(q, p; \mathbf{r}) \equiv \sum_{i=1}^{N} \mathbf{v}_i \delta(\mathbf{r} - \mathbf{r}_i). \tag{3.6}$$

At this stage, two observations should be made. The first one is that the dynamical functions $\rho_1(q, p; \mathbf{r})$ and $\mathbf{j}(q, p; \mathbf{r})$ are irregularly varying functions of space and time, since their evolution depends on the values of the coordinates r, and of the velocities v, of all the particles in the fluid. The second one is that since the physical trajectories of the mechanical systems are, in general, unstable, a small modification in the initial mechanical state (q^0, p^0) notably changes the evolution of $\rho_1(q, p; \mathbf{r})$ and $\mathbf{j}(q, p; \mathbf{r})$, although, for each initial state of the fluid, these dynamical functions must comply with the conservation law (3.5).

These observations contrast appreciably with the description of the fluid using the mechanics of continuous media. In the latter, the fluid is characterized by variables that, in the simplest cases, are continuous functions of space and time. Examples of such variables are the density field $\rho_1(\mathbf{r}, t)$ and the velocity field $\mathbf{u}(\mathbf{r}, t)$. The first one is the density of particles, namely $\rho_1(\mathbf{r}, t)d\mathbf{r}$ is the number of particles that at time t are contained in the volume element $d\mathbf{r}$ centered at $\mathbf{r}$, such that upon integration over the region R that contains the fluid one obtains

$$\int_R d\mathbf{r}\rho_1(\mathbf{r}, t) = N. \tag{3.7}$$

The time evolution of the density field may be derived from the following reasoning. Let R_1 be an open region contained in R and Ω_1 its surface. The variation of the number of particles in R_1 is due to the flow of particles through Ω_1, namely

$$\frac{d}{dt} \int_{R_1} d\mathbf{r}\rho_1(\mathbf{r}, t) = \int_{R_1} d\mathbf{r} \frac{\partial \rho_1(\mathbf{r}, t)}{\partial t} = -\int_{\Omega_1} d\Omega \cdot \mathbf{u}(\mathbf{r}, t)\rho_1(\mathbf{r}, t)$$

$$= -\int_{R_1} d\mathbf{r}\nabla \cdot [\rho_1(\mathbf{r}, t)\mathbf{u}(\mathbf{r}, t)], \tag{3.8}$$

where the velocity field or mean fluid velocity at the point $\mathbf{r}$ and at time t, $\mathbf{u}(\mathbf{r}, t)$, has been introduced, and the surface integral has been transformed into a volume integral. Since (3.8) must be satisfied for any R_1, it follows that

$$\frac{\partial \rho_1(\mathbf{r}, t)}{\partial t} = -\nabla \cdot \mathbf{j}(\mathbf{r}, t), \tag{3.9}$$

where $\mathbf{j}(\mathbf{r}, t) = \rho_1(\mathbf{r}, t)\mathbf{u}(\mathbf{r}, t)$ is the current density field. In the mechanics of continuous media, (3.9) is known as the continuity equation. Equations (3.5) and (3.9) are, of course, two equivalent forms of the law of conservation of the number of particles.

In contrast with the first form, the second one does not make any reference to the mechanical state of the fluid. If both equations represent the same physical process, the question arises as to how to relate one to the other.

It seems clear that to the dynamical function $\rho_1\left(q^t, p^t; \mathbf{r}\right)$, where q^t, p^t indicate the mechanical state of the system at time t, one has to associate the density field $\rho_1(\mathbf{r}, t)$ and in general to a dynamical function $a\left(q^t, p^t; \mathbf{r}\right)$ a field $a(\mathbf{r}, t)$ will be associated. If the mechanical state q^t, p^t is expressed in terms of the initial mechanical state q^0, p^0, which in what follows will be indicated by q, p, and of time t, one may write

$$a\left(q^t, p^t; \mathbf{r}\right) = a(q, p; \mathbf{r}, t) \to a(\mathbf{r}, t) \equiv \langle a(q, p; \mathbf{r}, t)\rangle. \tag{3.10}$$

This correspondence has to meet two requirements. The first one is that the irregular spatiotemporal evolution of the dynamical function must be smoothed out when the corresponding field is obtained. The second condition is that the field cannot depend on the initial mechanical state of the system. If one defines

$$a(\mathbf{r}, t) \equiv \int dq \int dp a(q, p; \mathbf{r}, t)\rho(q, p), \tag{3.11}$$

the correspondence is established upon multiplication of $a(q, p; \mathbf{r}, t)$ by a function $\rho(q, p)$ and subsequent integration over all the initial mechanical states (q, p). If the function $\rho(q, p)$ is positive definite in phase space,

$$\rho(q, p) \geq 0, \tag{3.12}$$

and normalized

$$1 = \int dq \int dp \rho(q, p), \tag{3.13}$$

then (3.11–3.13) associate a dynamical function with a field, which is the average value of the dynamical function with a probability density $\rho(q, p)$ in the phase space of the system (see Appendix B). This is the first postulate of classical statistical physics. It fulfills the two required conditions. On the one hand, multiplication of the dynamical function by the probability density and integration over the complete phase space allow the field variable so defined to be continuous in space and time. On the other hand, the field is independent of the initial mechanical state of the system because in (3.11) one averages over all possible initial states.

3.2 Liouville's Equation

It is important to point out that, according to (3.11), the time evolution of the field $a(\mathbf{r}, t)$ is induced by that of the dynamical function $a(q, p; \mathbf{r}, t)$. From (1.32), it follows that

$$a(q, p; \mathbf{r}, t) = e^{-\mathcal{L}_N t} a(q, p; \mathbf{r}, 0), \tag{3.14}$$

(note that the Liouville operator $\mathcal{L}_N$ acts upon the variables (q, p)), and hence (3.11) may be written as

$$a(\mathbf{r}, t) = \int dq \int dp \left[e^{-\mathcal{L}_N t} a(q, p; \mathbf{r}, 0) \right] \rho(q, p)$$
$$= \int dq \int dp a(q, p; \mathbf{r}, 0) \left[e^{\mathcal{L}_N t} \rho(q, p) \right], \tag{3.15}$$

where to go from the first to the second expression, an integration by parts has been performed (since $\mathcal{L}_N$ is a linear differential operator), and it has been admitted that $\rho(q, p)$ and all of its derivatives vanish at the limits of integration. Equation (3.15) may also be written as

$$a(\mathbf{r}, t) = \int dq \int dp a(q, p; \mathbf{r}, 0) \rho(q, p; t), \tag{3.16}$$

where the following definition has been introduced

$$\rho(q, p; t) \equiv e^{\mathcal{L}_N t} \rho(q, p). \tag{3.17}$$

From (3.17), it follows that the evolution of the probability density is different from that of the dynamical function (3.14). Indeed, while $\rho(q, p; t)$ is obtained by allowing the operator $e^{\mathcal{L}_N t}$ act upon the probability density at the initial instant, the operator that produces a displacement of the dynamical function in a time t is $e^{-\mathcal{L}_N t}$. After derivation of (3.17) with respect to t, one obtains

$$\frac{\partial \rho(q, p; t)}{\partial t} = \mathcal{L}_N \rho(q, p; t) = \{ H_N(\alpha), \rho(t) \}, \tag{3.18}$$

which is known as the Liouville equation. Note that by transforming (3.11) into (3.16) the time evolution of the field $a(\mathbf{r}, t)$ is induced in the latter expression by that of the probability density $\rho(q, p; t)$. In this way, the Liouville equation plays a key role in classical statistical physics.

3.3   Systems in Equilibrium

Note that in thermodynamics the intensive variables (the particle density, the pressure, etc.) are uniform and time-independent magnitudes, which represent a particular case of (3.16) in which $a(\mathbf{r}, t)$ does depend explicitly neither on $\mathbf{r}$ nor on t. This implies that, in order to derive the intensive variables of a system in thermodynamic equilibrium, the probability density has to be stationary (it cannot depend explicitly on time) and, therefore, from (3.18) it follows that

$$\{H_N(\alpha), \rho\} = 0. \tag{3.19}$$

This equation is, of course, much simpler than (3.18). As a matter of fact, according to the first postulate of classical statistical physics, $\rho(q, p; t)$ can be any non-negative normalizable function. In the case of systems in thermodynamic equilibrium, (3.19) indicates that $\rho(q, p)$ is a function of the integrals of motion of the system. This fact, although in itself a great simplification, does not specify the probability density of a system in equilibrium, since this probability density may be any function of the integrals of motion (provided the conditions given in (3.12) and (3.13) are satisfied). As indicated in Chap. 1, in general every mechanical system in the absence of external forces has only seven additive integrals of motion: energy, linear momentum, and angular momentum. Note that, in view of the fact that linear momentum and angular momentum are associated with a global translation and a global rotation of the system, respectively (which should not affect the thermodynamics of the system), it may be admitted that the only relevant integral of motion is the energy. Therefore, the second postulate of classical statistical physics is that in a system in equilibrium, the probability density is a function of the Hamiltonian of the system, namely

$$\rho(q, p) = \rho((H_N(q, p; \alpha)), \tag{3.20}$$

and hence (3.16) is written as

$$\langle a \rangle = \int dq \int dp \, a(q, p) \rho(H_N(q, p; \alpha)). \tag{3.21}$$

Observe that in order for (3.21) to be useful, one still has to know explicitly the function $\rho(H_N(q, p; \alpha)$. This question is addressed in the following chapters.

Once the connection between mechanics and thermodynamics has been established, the energy E of a system of N particles in equilibrium whose only external parameter is the volume V is, according to (3.21), the average value of the Hamiltonian, namely

$$E = \int dq \int dp H_N(q, p; V) \rho(H_N(q, p; V)). \qquad (3.22)$$

Although this definition seems natural, it must still comply with the requirement that the energy of the system is a function of the extensive variables entropy, S, volume, V, and number of particles, N, as it is postulated in the fundamental equation of thermodynamics (2.8). It is clear that E is a function of the number of particles N. The dependence upon the volume V of the system is also evident since, in order to confine the particles within a region R, the Hamiltonian of the system must include an external potential (see (1.20)), which makes it explicitly dependent on V. In order to determine how the energy depends on the entropy one must recall that, as indicated in Chap. 2, S is an extensive variable with no mechanical counterpart. According to Gibbs, it is postulated that the entropy of a system of N particles and energy E contained in a region R of volume V is given by the expression:

$$S \equiv -k_B \int dq \int dp \ln\left[h^{3N}\rho(H_N(q, p; V))\right]\rho(H_N(q, p; V))$$
$$= \left\langle -k_B \ln\left[h^{3N}\rho(H_N(q, p; V))\right]\right\rangle, \qquad (3.23)$$

where k_B is Boltzmann's constant and h Planck's constant. Note that S is the average value of a non-mechanical property, so that the entropy is, in this sense, a thermal variable (it is not the average value of a dynamical function). Since, according to (3.22), E depends on the choice of $\rho(H_N(q, p; V))$, E is related, albeit indirectly, with S. One refers to each particular choice of $\rho(H_N(q, p; V))$ as a "Gibbs ensemble," and some particular cases will be studied in the following chapters. In one of these ensembles, the microcanonical ensemble, in which the energy of the system is constant, it will be shown in Chap. 4 that (3.23) is the fundamental equation of thermodynamics $S = S(E, V, N)$.

Up till now the postulates of classical equilibrium statistical physics have been introduced, and they may be summarized as follows. In the first, one associates a field variable to every dynamical function, and this field variable is defined as the average value of the dynamical function over a Gibbs ensemble. The thermodynamics of the system is obtained from the second postulate when the probability density is stationary and a function of the Hamiltonian. Although these postulates have been derived using intuitive ideas, their rigorous justification is neither simple nor well known. In spite of that, it seems gradually more evident that the statistical physics of equilibrium systems is a well-established theory. In the last three sections of this chapter, some basic questions related to these postulates are analyzed whose rigorous study would require a level of mathematics exceeding the one of this text. But before doing so, in the next section the postulates of quantum statistical physics are introduced.

3.4   Density Operator

The mechanical state of a quantum system at a given time $t = 0$ is completely specified if the values of a complete set of commuting observables are known. Such a state is described by a wave function (or vector in the Hilbert space $\mathcal{H}|\Psi\rangle \equiv \Psi(q)$, which is an eigenvector of the observables and is assumed to be normalizable, namely $\langle \Psi|\Psi \rangle = 1$. This is the maximum information that may be obtained in quantum mechanics, and hence, in this instance it is said that the system is in a pure state. The expectation value of an operator $\hat{a}$ in this state is given by $\langle \Psi|\hat{a}|\Psi \rangle$.

Note the analogy between the evolution of a pure state, represented in quantum mechanics by a vector $|\Psi\rangle$, and the physical trajectory in classical mechanics. In the first case, according to Schrodinger's equation (1.56), a pure state only changes its phase, so that its time evolution is determined by the complete set of commuting observables. In the second case, the trajectory in phase space is completely determined by the intersection of the $6N - 1$ "surfaces" $C_n(q, p) = C_n^*$, where $C_n(q, p)$ is the nth integral of motion whose value is C_n^*. The complete set of commuting observables in quantum mechanics plays, therefore, a role equivalent to that of the integrals of motion in classical mechanics.

As it also occurs in some systems in classical mechanics, where only a few integrals of motion are known, it is likely that in a quantum system all the required information to determine the eigenvector $|\Psi\rangle$ may not be known, i.e., it may happen that not all the values of a complete set of commuting observables are available. In such a case, assume that the system may be found in pure states $|\Psi_i\rangle = \Psi_i(q)$ (which although normalizable are not necessarily orthogonal) with probabilities pi. It is then postulated that the expectation value of an operator $\hat{a}$ is given by the expression

$$\langle \hat{a} \rangle = \sum_i p_i \langle \Psi_i|\hat{a}|\Psi_i \rangle, \tag{3.24}$$

where the probabilities $\{p_i\}$ verify the following conditions:

$$p_i \geq 0, \tag{3.25}$$

and

$$\sum_i p_i = 1. \tag{3.26}$$

Equations (3.24–3.26) are the quantum mechanical equivalent to (3.11–3.13) and constitute the first postulate of quantum statistical physics. Observe that $\langle \hat{a} \rangle$ in (3.24) contains two types of averages. The first one, $\langle \Psi_i|\hat{a}|\Psi_i \rangle$, is of quantum origin and represents the expectation value of the operator $\hat{a}$ in the state $|\Psi_i\rangle$ while the second average, which is of a statistical nature, associates a probability p_i to the state $|\Psi_i\rangle$. When in (3.24) all the p_i vanish except the one of a particular state, one obtains the expectation value of the operator in a pure case so that, for the sake of distinguishing it from the latter, in (3.24) it is said that the state of the system is a mixed state.

The analogy between classical and quantum statistics may be more clearly seen if one defines the density operator or density matrix $\hat{\rho}$ as

$$\hat{\rho} = \sum_i p_i |\Psi_i\rangle\langle\Psi_i|. \tag{3.27}$$

Given an orthonormal basis set $|\mathbf{b}_n\rangle$ and an operator $\hat{a}$, one has

$$\begin{aligned}
\mathrm{Tr}(\hat{a}\hat{\rho}) &= \sum_n \langle\mathbf{b}_n|\hat{a}\hat{\rho}|\mathbf{b}_n\rangle = \sum_n \sum_i p_i\langle\mathbf{b}_n|\hat{a}|\Psi_i\rangle\langle\Psi_i|\mathbf{b}_n\rangle \\
&= \sum_n \sum_i p_i\langle\Psi_i|\mathbf{b}_n\rangle\langle\mathbf{b}_n|\hat{a}|\Psi_i\rangle \\
&= \sum_i p_i\langle\Psi_i|\hat{a}|\Psi_i\rangle = \langle\hat{a}\rangle,
\end{aligned} \tag{3.28}$$

where the spectral resolution of the identity (1.52) has been used. It may also be shown further that $\hat{\rho}$ is self-adjoint $\hat{\rho}^\dagger = \hat{\rho}$ (note that since $\hat{\rho}$ contains an infinite number of terms, one may not affirm that the adjoint of the sum is equal to the sum of the adjoints, nevertheless the proof is simple), positive (see (1.55)),

$$\langle\mathbf{b}_n|\hat{\rho}|\mathbf{b}_n\rangle = \sum_i p_i\langle\mathbf{b}_n|\Psi_i\rangle\langle\Psi_i|\mathbf{b}_n\rangle = \sum_i p_i|\langle\mathbf{b}_n|\Psi_i\rangle|^2 \geq 0, \tag{3.29}$$

and of unit trace,

$$\begin{aligned}
\mathrm{Tr}\hat{\rho} &= \sum_n \langle\mathbf{b}_n|\hat{\rho}|\mathbf{b}_n\rangle = \sum_n \sum_i p_i\langle\mathbf{b}_n|\Psi_i\rangle\langle\Psi_i|\mathbf{b}_n\rangle \\
&= \sum_n \sum_i p_i\langle\Psi_i|\mathbf{b}_n\rangle\langle\mathbf{b}_n|\Psi_i\rangle \\
&= \sum_i p_i\langle\Psi_i|\Psi_i\rangle = \sum_i p_i = 1,
\end{aligned} \tag{3.30}$$

where the spectral resolution of the identity and the fact that the states $|\Psi_i\rangle$ are normalized have been accounted for. Equations (3.28–3.30) are the quantum equivalent to (3.11–3.13).

Observe that since in the Schrodinger picture the operators $\hat{a}$ do not change with time and the states evolve according to Schrodinger's Eq. (1.56), and the time dependence of $\langle\hat{a}\rangle$ is induced by the evolution of the density operator. This evolution equation is written as

$$i\hbar\frac{\partial\hat{\rho}(t)}{\partial t} = i\hbar\sum_i p_i\left[\frac{\partial|\Psi_i(t)\rangle}{\partial t}\langle\Psi_i(t)| + |\Psi_i(t)\rangle\frac{\partial\langle\Psi_i(t)|}{\partial t}\right]$$

$$= \sum_i p_i \left[\hat{H}_N(\alpha) |\Psi_i(t)\rangle \langle \Psi_i(t)| - |\Psi_i(t)\rangle \langle \Psi_i(t)| \hat{H}_N(\alpha) \right]$$

$$= \left[\hat{H}_N(\alpha), \hat{\rho}(t) \right], \tag{3.31}$$

which is known as von Neumann's equation (which is the quantum equivalent to Liouville's equation). In von Neumann's equation, the commutator $\left[\hat{H}_N(\alpha), \hat{\rho}(t) \right]$ plays the role of the Poisson bracket $\{ H_N(\alpha), \rho(t) \}$ in (3.18).

In a system in equilibrium, the average values $\langle \hat{a} \rangle$ do not depend on time, and hence, the density operator is stationary (it cannot depend explicitly on time). This implies that, according to (3.31), $\hat{\rho}$ commutes with $\hat{H}_N(\alpha)$. In analogy with the classical case, it is postulated that the density operator of a system in equilibrium is only a function of the Hamiltonian operator, i.e.,

$$\hat{\rho} = \hat{\rho} \left(\hat{H}_N(\alpha) \right). \tag{3.32}$$

This second postulate of quantum statistical physics, by which it is admitted that $\hat{\rho}$ is not a function of operators that represent conserved variables but only of $\hat{H}_N(\alpha)$, is justified in a certain sense by the fact that all mechanical systems conserve energy. The different choices of $\hat{\rho} \left(\hat{H}_N(\alpha) \right)$ give rise to the Gibbs ensembles of quantum statistical physics.

3.5   Ergodicity

Consider a classical system of N particles in equilibrium whose Hamiltonian is $H_N(q, p; \alpha)$. According to (3.21), the thermodynamic variable $\langle a \rangle$ is the average value of a dynamical function $a(q, p)$ taken over a Gibbs ensemble, namely

$$\langle a \rangle = \int dq \int dp a(q, p) \rho(H_N(q, p; \alpha)), \tag{3.33}$$

where $\rho(H_N(q, p; \alpha))$ is a stationary probability density. If the postulates are correct, the result of (3.33) should reproduce the experimental value of $\langle a \rangle$.

Note that since the majority of experimental measurements are performed during a small but finite interval of time, as the measurement proceeds a system of N particles evolves through a series of mechanical states (q^t, p^t) with $t_i < t < t_f$, where t_i is the time at which the measurement is initiated and tf the time at which the measurement is completed. Let (q, p) be the mechanical state of the system at time $t = 0$. According to Boltzmann, since every experimental measurement requires a finite time (compared, say, with the collision time between particles), one may reasonably wonder whether the average statistical value (3.33) is equal to the average temporal value of the dynamical function $a(q, p; t)$ where the average is taken over a period of time $\tau = t_f - t_i$, i.e.,

$$\overline{a}_\tau(q, p) = \frac{1}{\tau} \int_0^\tau dt\, a(q, p; t). \tag{3.34}$$

There is an important branch of statistical physics, referred to as "Ergodic Theory" which studies the equivalence between (3.33) and (3.34). The most important conclusion of this theory is that if the phase space trajectory densely covers the energy surface $H_N(q, p; \alpha) = E$, then the system is ergodic, namely

$$\overline{a} \equiv \lim_{\tau \to \infty} \overline{a}_\tau(q, p) = \langle a \rangle, \tag{3.35}$$

which establishes the equivalence of the average over the microcanonical Gibbs ensemble (i.e., the ensemble whose probability density is uniform on the energy surface and which will be analyzed in Chap. 4) and the time average over an infinite time interval, which is independent of the initial condition. The latter may be obtained approximately from "Molecular Dynamics" simulations (see Chap. 7) in which $\overline{a}_\tau(q, p)$ is determined over a time interval τ much greater than the collision time between particles. The interpretation of (3.35) is the following. Assume that one starts from an initial mechanical state of a conservative system. This state is on the energy surface $H_N(q, p; \alpha) = E$ and during the time evolution remains on this surface. When $\tau \to \infty$, one should expect that the trajectory has visited almost every point of the surface (except for a set of measure zero). If at any time the mechanical state of the system is represented by a point, the picture that emerges, after a time $\tau \to \infty$, is that of a cloud of points covering densely the surface. The time average of a dynamical function is equivalent, in this case, to the average value of the same dynamical function over the Gibbs ensemble whose probability density is uniform on the energy surface (microcanonical ensemble).

Another fundamental aspect of (3.35) is that when the statistical and time averages are equal, everything seems to indicate that the statistical average (3.33) is nothing else but an alternative form to evaluate (3.34) with the advantage that in the former it is not required to solve the equations of motion of the system. There are, however, two important ideas to be emphasized. By definition, the time average of a dynamical function (3.34) when $\tau \to \infty$ is independent of time. The average over a Gibbs ensemble (3.34) may either be independent of time if the probability density is stationary, or a function of time if the probability density depends explicitly on time. Since it has been postulated that the connection between dynamical functions and fields is made through (3.16), the mechanical justification of (3.35) is only possible for systems in equilibrium. Recall further that the ergodic theorem is based on the fact that the physical trajectory is regular enough on the energy surface $H_N(q, p; \alpha) = E$. It may be shown that this is indeed true not only in some systems of $N \gg 1$ particles, but also in some systems with a reduced number of degrees of freedom. Ergodicity seems, thus, to be related with the instability of the physical trajectories of mechanical systems and not with the number of particles that constitute the system. In the following section it is shown, however, that in order to obtain the intensive variables of thermodynamics it is necessary to study the system in the so-called thermodynamic limit in which the number of particles of the system $N \to \infty$. Therefore, the ergodic problem turns out to be unrelated to the study of the thermodynamics of the system.

3.6 Thermodynamic Limit

According to (3.33), the intensive thermodynamic variables are the average values of certain dynamical functions over a Gibbs ensemble. Since there exist different ensembles, as will be seen in the following chapters, the average values in each ensemble will, in general, be different. Under certain conditions, in what is known as the "Thermodynamic Limit" (TL), it is possible to show that the intensive thermodynamic variables are independent of the probability density with which one performs the average in (3.33).

In the next chapters the microcanonical, canonical, and grand canonical ensembles will be introduced. These ensembles are defined by a probability density in the phase space of the system and describe an isolated, a closed, and an open system, respectively. To each ensemble, a thermodynamic potential will be associated. The corresponding thermodynamic potentials are the entropy $S(E, V, N)$ (microcanonical), the Helmholtz free energy $F(T, V, N)$ (canonical), and the grand potential $\Omega(T, V, \mu)$ (grand canonical). Before introducing the thermodynamic limit, let us first consider some properties of these thermodynamic potentials. As indicated in Chap. 2, all these potentials are extensive variables. That means, for example, the entropy $S(E, V, N)$ cannot be an arbitrary function of the extensive variables E, V, and N. Indeed, the extensivity condition reads

$$S(\lambda E, \lambda V, \lambda N) = \lambda S(E, V, N), \tag{3.36}$$

where λ is an arbitrary positive real number. Setting $\lambda = 1/V$ in (3.36) leads to

$$S(E, V, N) = V S(\bar{e}, 1, \rho) \equiv V \bar{s}(\bar{e}, \rho), \tag{3.37}$$

where $\bar{s}(\bar{e}, \rho)$ is the entropy per unit volume, which is a function of the intensive variables $\bar{e} = E/V$ and $\rho = N/V$.

Since the Helmholtz free energy is a function of the extensive variables V and N, the extensivity condition reads in this case

$$F(T, \lambda V, \lambda N) = \lambda F(T, V, N), \tag{3.38}$$

and setting $\lambda = 1/V$ in (3.38) yields

$$F(T, V, N) = V F(T, 1, \rho) \equiv V \bar{f}(T, \rho), \tag{3.39}$$

where the Helmholtz free energy per unit volume $\overline{f}(T, \rho)$ depends on the intensive variables T and ρ.

Finally, the extensivity condition for the grand potential is

$$\Omega(T, \lambda V, \mu) = \lambda \Omega(T, V, \mu), \tag{3.40}$$

since it only depends on the extensive variable V. Again, setting $\lambda = 1/V$ in (3.40) leads to

$$\Omega(T, V, \mu) = V \Omega(T, 1, \mu) \equiv -V p(T, \mu), \tag{3.41}$$

where, after (2.39), $p(T, \mu)$ is the pressure, which is a function of the intensive variables T and μ. In summary, the extensivity of the three thermodynamic potentials implies that they can be written as the product of the volume of the system by a function of two intensive thermodynamic variables.

In order to introduce the thermodynamic limit, note that the average (3.33) is a function of the volume V of the region R in which the system is contained. Moreover, since the limits of integration of the coordinates of the particles go up to the boundary of the region R, the result is also dependent on the particular geometry of the region under consideration. For the sake of avoiding this dependence, one may consider the following limiting process. Assume that the average (3.33) is determined for a sequence of regions of the same geometry but with increasing volumes $V = V_1 < V_2 < V_3 < \ldots$ Let $\langle a \rangle_k$ ($k = 1, 2, \ldots$) denote the values of the intensive variables in each of the systems of the sequence. It seems evident that as the volume of the region R grows, the influence of the geometry of R should become smaller, and so one could expect that the intensive variables $\langle a \rangle_k$ may be written in the form

$$\langle a \rangle_k = \tilde{a} + \tilde{a}_k(V_k), \tag{3.42}$$

where the first term is independent of the geometry of the systems in the sequence, and the second one verifies the condition

$$\lim_{V_k \to \infty} \tilde{a}_k(V_k) = 0. \tag{3.43}$$

If these conditions are met, the variable $\tilde{a}$ in (3.42) may be interpreted as the "intrinsic value" of the intensive variable, which may, therefore, be obtained by computing the limit

$$\lim_{V_k \to \infty} \langle a \rangle_k = \tilde{a}. \tag{3.44}$$

Observe that if the systems in the sequence only differ in their volume V_k, the systems "scale" with the original system if V is the unique independent extensive variable of the thermodynamic potential associated with the ensemble (grand ensemble). In general, in order for the systems in the sequence to be equivalent, it is necessary to enforce that all the extensive variables grow proportionally to the volumes V_k ($k = 1, 2, \ldots$), thereby leading to a sequence of systems that scale with the original system. For instance, if $\rho = N/V$ is the density of particles of the original system, the numbers of particles of the systems in the sequence $N = N_1 < N_2 < N_3 < \ldots$ have to grow in such a way that the sequence $\rho_k = N_k/V_k$ ($k = 1, 2, \ldots$) fulfills the condition

$$\lim_{V_k \to \infty} \rho_k = \rho. \tag{3.45}$$

Therefore, when V and N are the independent extensive variables of the thermodynamic potential associated with the ensemble (canonical ensemble), the thermodynamic limit is defined by

$$\text{TL}: \quad V_k \to \infty, \quad N_k \to \infty, \quad \rho_k \to \rho < \infty, \tag{3.46}$$

while when the independent extensive variables of the thermodynamic potential associated to the ensemble are V, N, and E (microcanonical ensemble), the thermodynamic limit is defined by

$$\text{TL}: \quad V_k \to \infty, \quad N_k \to \infty, \quad E_k \to \infty, \quad \rho_k \to \rho < \infty, \quad \bar{e}_k \to \bar{e} < \infty, \tag{3.47}$$

where $\bar{e}$ is the energy per unit volume of the original system and $\bar{e}_k = E_k/V_k$, with $E = E_1 < E_2 < E_3 \ldots$ the energy of the systems in the sequence.

It is clear that since in the thermodynamic limit, one only imposes the requirement that in the system of infinite volume a reduced number of variables, ρ_k in (3.46) and ρ_k and $\bar{e}_k$ in (3.47), be the same as that of the original system, this does not guarantee that all the intensive variables (3.33) approach to finite limits $\tilde{a}$ in the thermodynamic limit and neither that these limits are equal to the values of the intensive variables of the original system. One would have to prove that these limiting values do not depend neither on the geometry of the chosen regions which form the sequence, nor on the chosen sequence, nor on the probability density with which one performs the averages of the dynamical functions. If one could prove all this, the idea of applying the thermodynamic limit would, therefore, be to obtain out of the original system a "scaled" system in which the thermodynamic variables (3.33) are independent of the ensemble and of the particular geometry of the region R. Although the mathematical formulation of the thermodynamic limit is not simple, in what follows some known results are summarized.

Since the average values over a Gibbs ensemble are a function of the Hamiltonian of the system, the existence of the thermodynamic limit depends on some properties of $H_N^{\text{int}}(q, p; \alpha)$. The first one, known as the "stability condition" is that the interaction energy must be bounded from below by a limit which is proportional to N, i.e.,

$$H_N^{\text{int}}(q, p; \alpha) \geq -AN, \tag{3.48}$$

where A is a positive constant independent of N. Physically, (3.48) expresses the fact that the system cannot collapse, namely that the energy per particle will not become infinitely negative when N increases. The second property is that the interaction pair potential $V(r)$ (assuming that the interactions are of that kind) has to meet the "weak decay condition"

$$V(r) \leq \frac{B}{r^{d+\varepsilon}}, (r \geq R > 0), \tag{3.49}$$

where d is the dimensionality of space, and B and ε are positive constants. This equation expresses the fact that $V(r)$ may not be too repulsive at large distances, since otherwise the system would explode. Thus, $V(r)$ has to decrease when $r \geq R$ in such a way that

$$\int_{r \geq R} d\mathbf{r} V(r) < \infty. \tag{3.50}$$

An important exception to (3.50) is the Coulomb potential $V(r) = e^2/r$, where e is the electric charge of the particles, since in this case (3.50) yields for $d = 3$:

$$\int_{r \geq R} d\mathbf{r} V(r) = \int_R^\infty d\mathbf{r} V(r) = 4\pi e^2 \int_R^\infty dr r = \infty. \tag{3.51}$$

Observe, however, that for systems of electrically charged particles, there must be at least two kinds of particles, e.g., electrons and positive ions, in order to maintain the overall electroneutrality. For mobile charges (plasmas and colloidal dispersions), the electroneutrality condition also holds locally and the particles adjust themselves in such a way that there is a screening of the Coulomb interactions at large distances (see Sects. 7.3 and 8.10.2). The resulting effective interaction, which decays as $e^{-\kappa r}/r$, with κ the Debye screening parameter, verifies both the stability condition and the weak decay condition.

In general, however, the exact pair potential is not known, or is too complicated, and one then uses instead some simple model for $V(r)$. The explicit expression of such a model pair potential must, however, be compatible with the existence of a thermodynamic limit for the model system. An exact result which is helpful in the construction of acceptable $V(r)$ functions is provided by the following statement. If $V(r)$ does admit a finite Fourier transform

$$\tilde{V}(k) = \int d\mathbf{r} e^{-i\mathbf{k}\cdot\mathbf{r}} V(r) < \infty, \tag{3.52}$$

then a necessary and sufficient condition for the existence of the thermodynamic limit is

$$\tilde{V}(k) \geq 0 \quad (\forall k), \tag{3.53}$$

including $k = 0$. For instance, when as above $V(r) = e^2/r$, then $\tilde{V}(k) = 4\pi e^2/k^2$ and hence $\tilde{V}(0)$ does not exist, in agreement with (3.51). But, on the contrary, when $V(r) = e^2 e^{-\kappa r}/r$, then $\tilde{V}(k) = 4\pi e^2/(k^2 + \kappa^2)$, and $0 < \tilde{V}(k) < \infty(\forall k)$ including $k = 0$, when $\kappa \neq 0$. Consider next a potential function of the form

$$V(r) = \varepsilon \left(\frac{e^{-ar}}{ar} - \frac{e^{-br}}{br} \right), \tag{3.54}$$

which in contradistinction with the previous examples, which are repulsive, contains a repulsion of range $1/a$ and an attraction of range $1/b$, with ε setting the scale of the interaction energy. In this case

$$\tilde{V}(k) = \frac{4\pi\varepsilon}{a} \left(\frac{1}{k^2 + a^2} - \frac{a}{b} \frac{1}{k^2 + b^2} \right), \tag{3.55}$$

and hence $\tilde{V}(k) \geq 0$ implies here $b > a$. In other words, a system with such a pair potential will possess a thermodynamic limit only when the attractions decay more rapidly than the repulsions. Of course, not all the plausible $V(r)$ functions will have a finite $\tilde{V}(k)$. For instance, for the Lennard–Jones (LJ) model:

$$V(r) = 4\varepsilon \left[\left(\frac{\sigma}{r} \right)^{12} - \left(\frac{\sigma}{r} \right)^6 \right], \tag{3.56}$$

often used to model atomic systems (see Sect. 8.2.4), $\tilde{V}(k)$ does not exist. Nevertheless, the Lennard–Jones potential does satisfy (3.48) and (3.50) guaranteeing thereby the existence of the thermodynamic limit for this system. The reason why $\tilde{V}(k)$ does not exist is easily seen to be due to the strongly repulsive character, $\sim(\sigma/r)^{12}$, of the Lennard–Jones potential $V(r)$ at short distances, i.e., for $r \to 0$. This is often the case for potentials modeling atomic matter. On the contrary, in the case of soft matter (see Chap. 8) the constitutive particles often experience only a much softer repulsion at short distances leading, e.g., to a finite value for $V(0)$. For instance, the potential defined as

$$V(r) = 4\varepsilon \left[\left(\frac{\sigma}{r + \gamma\sigma} \right)^{12} - \left(\frac{\sigma}{r + \gamma\sigma} \right)^6 \right], \tag{3.57}$$

verifies $0 < V(0) < \infty$ for $0 < y < 1$, whereas for $\gamma = 0$ the Lennard–Jones model is recovered. In this case $\tilde{V}(k)$ does exist when $\gamma \neq 0$, but it is easily verified that $\tilde{V}(0) < 0$ for $\gamma > (2/33)^{1/6}$, and therefore, the thermodynamic limit will only exist when the potential does not become too soft at $r = 0$, i.e., for $0 < y < (2/33)^{1/6}$, otherwise the system will collapse in the thermodynamic limit. In summary, the thermodynamic limit will exist only when the pair potential $V(r)$ is sufficiently repulsive at short distances and decays to zero sufficiently rapidly at large distances.

A third condition is that the thermodynamic limit has to be taken in such a way that the system becomes infinite in all directions. This excludes, for instance, systems which are confined between walls. Finally, the thermodynamic limit cannot be taken when the density of the system $\rho = N/V$ is very small (very dilute system), since in this case the particle–wall interactions dominate over the particle–particle interactions.

Under these conditions it can be shown (from a mathematical point of view) that for a large number of dynamical functions $a(q, p)$, the average value (3.33) has a thermodynamic limit $\tilde{a}$:

$$\mathrm{TL}[\langle a \rangle] = \tilde{a}, \qquad (3.58)$$

which is independent of the Gibbs ensemble, namely of the probability density with which the average of the dynamical function is taken (it is understood here that in (3.58) the variable $\langle a \rangle$ is intensive). In other words, the thermodynamics can be obtained from any Gibbs ensemble provided that, at the end of the calculations, (3.58) is applied.

In particular, in the canonical ensemble it can be rigorously shown that by defining

$$F(T, V, N) = -k_B T \ln \left[\frac{1}{N! h^{3N}} \int d\mathbf{r}^N \int d\mathbf{p}^N e^{-\beta H_N(\mathbf{r}^N, \mathbf{p}^N; V)} \right], \qquad (3.59)$$

where $\beta = 1/k_B T$, then

$$\mathrm{TL}\left[\frac{1}{V} F(T, V, N) \right] = \overline{f}(T, \rho) < \infty, \qquad (3.60)$$

i.e., the limit of $F(T, V, N)/V$ is finite in the thermodynamic limit (3.46), when the Hamiltonian $H_N(\mathbf{r}^N, \mathbf{p}^N; V)$ in (3.59) satisfies the stability condition (3.48) and the weak decay condition (3.50). The function $\overline{f}(T, \rho)$ in (3.60) can then be identified with the Helmholtz free energy per unit volume in (3.39). This function is moreover concave in T and convex in ρ. On the other hand, the Helmholtz free energy per particle $f(T, v) = v\overline{f}(T, \rho)$, where $v = 1/\rho$, is a concave decreasing function of T and a convex decreasing function of v (see Figs. 2.6 and 2.7). Observe that (3.59) and (3.60) are, therefore, the mathematically rigorous justification of (3.39), which has been derived from the extensivity condition of the thermodynamic potential. Proceeding analogously with the microcanonical and macrocanonical ensembles, a mathematically rigorous justification of (3.37) and (3.41) can also be found, leading to the final demonstration of the equivalence of the Gibbs ensembles. Observe that from the thermodynamic point of view the equivalence of the thermodynamic potentials is based, on the contrary, on the properties of Legendre's transformation.

It should be noted that the thermodynamic limit, which seems to have only a simple operative interest in statistical physics, has also other much deeper implications. In fact, from a mathematical point of view, only in the thermodynamic limit it is possible to show the existence of a phase transition. Moreover, in this limit a thermodynamic system cannot be unstable, and hence, for instance, the existence of van der Waals loops is not possible (see Fig. 2.11).

In reality, however, all systems studied, in the laboratory or on a computer, are always finite. In this case, what is really measured or computed is $\langle a \rangle_k$ of (3.42). The quantity which is needed to, say, verify the statistical theory is $\tilde{a}$ of (3.44). Therefore, specific precautions have to be taken to ensure that $\tilde{a}_k(V_k)$ of (3.42) can be neglected so that $\langle a \rangle_k \simeq \tilde{a}$. In the laboratory, one usually realizes (3.43) approximately by considering only large, macroscopic, systems. When studying a system on a computer, one usually imposes periodic boundary conditions, so as to avoid a too strong dependence of the results on the geometry of the surface of V, and one estimates then $\langle a \rangle_k$ for different system-sizes and extrapolates the results to infinite size by postulating a simple law for (3.43).

3.7   Symmetry Breaking

Two phases of a system may or may not differ in their symmetries as the liquid and solid phases or the liquid and vapor phases, respectively. Normally, one of the phases has the same symmetries as the Hamiltonian and the other phase has a smaller number of symmetries, in which case it is said that in the latter phase some of the symmetries have been broken, i.e., they no longer exist. For instance, the symmetries of a Hamiltonian which is invariant under translation and rotation are the same as those of the vapor and the liquid, which are uniform and isotropic phases. This fact is evident since the intensive variables of these two phases are, according to the postulates of statistical physics, average values of dynamical functions with a probability density in phase space $\rho(H_N(q, p; \alpha))$. A solid, however, does not have the symmetries of translation and rotation of the Hamiltonian (it only has partial symmetries). Therefore, in a crystal there are three crystallographic axes along which the invariance under translation is broken and is reduced to a set of discrete translations. At first sight, the existence of this symmetry breaking is not compatible with the postulates, since the mere integration over the phase space cannot change the symmetries of the Hamiltonian. Because in nature every phase has some given symmetries, the question arises as to how to deduce the existence, say of a solid, from the postulates of statistical physics.

One way to solve this problem is due to Bogoliubov. Assume that to the Hamiltonian of the system $H_N(q, p; \alpha)$ one adds the contribution of an external field, $H_N^{\text{ext}}(q, p)$, with less symmetry than $H_N(q, p; \alpha)$, and thus referred to as the symmetry breaking field. Consider the Hamiltonian $H_N(q, p; \alpha) + \lambda H_N^{\text{ext}}(q, p)$, where λ $(0 \leq \lambda \leq 1)$ is a real parameter which measures the strength of the external field. It is clear that the average,

$$\langle a \rangle_\lambda = \int dq \int dp a(q, p) \rho \big(H_N(q, p; \alpha) + \lambda H_N^{\text{ext}}(q, p) \big), \qquad (3.61)$$

will have the same symmetries as $H_N(q, p; \alpha) + \lambda H_N^{\text{ext}}(q, p)$. When $\lambda \to 0$ the system, provided it is finite, cannot have any broken symmetry, but if this limit and the thermodynamic limit are taken in the order

$$\text{TL}\left[\lim_{\lambda \to 0} \langle a \rangle_\lambda \right] = \text{TL}[\langle a \rangle] = \tilde{a}, \qquad (3.62)$$

or in the inverse order

$$\lim_{\lambda \to 0} \text{TL}[\langle a \rangle_\lambda] = \lim_{\lambda \to 0} \tilde{a}_\lambda = \tilde{a}_0, \qquad (3.63)$$

there are two possibilities. If $\tilde{a}_0 = \tilde{a}$, it is said that the system does not admit any symmetry breaking due to the field $H_N^{\text{ext}}(q, p)$. It may happen, on the other hand, that $\tilde{a}_0 \neq \tilde{a}$ in which case it is said that there exists in the system a phase with a broken symmetry.

As a first example, consider a system of N magnetic moments or spins in a region R of volume V in the absence of a magnetic field. If the Hamiltonian $H_N(q, p; V)$ contains an isotropic interaction term between the magnetic moments (for instance, a pair interaction proportional to the scalar product of two magnetic moments, as in the so-called Heisenberg model which is analyzed in Chap. 9), the total magnetic moment of the system (which is the average value of the total magnetic moment dynamical function over a Gibbs ensemble with a probability density $\rho(H_N(q, p; V))$) will be zero, hence the system is paramagnetic. The reason is that, since the Hamiltonian is invariant under rotations, there can exist no privileged direction in space. Since in nature some materials, at temperatures below the so-called Curie temperature, are ferromagnetic (they have a total magnetic moment different from zero in the absence of a magnetic field), the question arises as to how one can obtain this phase in statistical physics. As indicated in the previous section, in order to show the existence of a phase transition (in this case the paramagnetic–ferromagnetic transition), it is necessary to take the thermodynamic limit (3.46). Observe, however, that since in all the systems in the sequence of increasing volume the invariance of the Hamiltonian under rotations is maintained, the total magnetic moment in all of them is always zero, even in the system of infinite volume. Therefore, the thermodynamic limit is a necessary but not sufficient condition to show the existence of the transition. Bogoliubov's method is based on "helping" the system to find a privileged direction in space (the ordered phase). Such "help" has to be weak in order to ensure that the transition is a spontaneous phenomenon and not a forced one. To that end, assume that one includes in the Hamiltonian of the system a term accounting for the interaction between a magnetic field and the spins on the surface and such that these surface spins become oriented in the same direction. Note that normally surface effects are small in macroscopic systems and tend to zero in the thermodynamic limit. This interaction term denoted by $H_N^{\text{ext}}(q, p)$ in Bogoliubov's method includes an external parameter (the magnetic field) that does not act on all the spins of the system but only on those on the surface. It is worth noting that although this magnetic field is weak, in all of the systems of the sequence of increasing volume (including the system of infinite volume) there is now a privileged direction. Once the thermodynamic limit has been taken, in order to guarantee that the phenomenon is spontaneous, one must also take the limit of zero field. According to Bogoliubov, these two limits do not always commute and hence the results of (3.62) and (3.63) may turn out to be different. In the first order of taking the limits (3.62) the system is paramagnetic while in the second one the system may be paramagnetic, $\tilde{a}_0 = \tilde{a}$ (if the temperature is greater than the Curie temperature), or ferromagnetic, $\tilde{a}_0 \neq \tilde{a}$ (if the temperature is less than the Curie temperature). The reason is that the thermal fluctuation may or may not destroy the order created by the magnetic field on the surface.

As a second example, consider a liquid at a temperature T. Assume that the Hamiltonian $H_N(q, p; V)$ contains, as in the system of spins, an isotropic pair interaction and let $H_N^{\text{ext}}(q, p)$ be a periodic external field that acts only on the atoms or molecules contained in an enclosure R' of volume $V'(V' \ll V)$. This external field is, therefore, a crystalline "seed" which is introduced into the system to help it crystallize in much the same way as it is done in experiments of crystal growth. In the absence of the seed, the system in the thermodynamic limit is a fluid. With the seed present, the systems in the sequence have a crystalline structure, including the system of infinite volume (note that the thermodynamic limit has to be taken again before the weak field limit $V'/V \to 0$). In this case, the symmetry of the phase may be that of a liquid or that of a crystal, depending on whether the temperature is greater or less than the crystallization temperature.

In summary, Bogoliubov's method of weak external fields shows, according to the experiments which have served as a guide to develop it, that the postulates of statistical physics are not incompatible with the existence of phases having less symmetry than the Hamiltonian of the system. Since this method is rather complex, in the study of such phase transitions one does not, in general, introduce an external field, but instead usually imposes the symmetry breaking directly on the relevant physical variables (see, for instance, Chap. 10).

Further Reading

1. R. Balescu, *Equilibrium and Nonequilibrium Statistical Mechanics* (Wiley, New York, 1975). *A general treatment of the statistical physics of both equilibrium and non-equilibrium systems within the same framework.*
2. A. I. Khinchin, *Mathematical Foundations of Statistical Mechanics* (Dover Publications, New York, 1949). *Provides a mathematical discussion of the ergodic theory.*
3. D. Ruelle, *Statistical Mechanics* (W. A. Benjamin, New York, 1969). *Contains a rigorous treatment of the thermodynamic limit.*

4. N.N. Bogoliubov, Quasi-averages in problems of statistical mechanics, in *Lectures on Quantum Statistics*, vol. 2. (Gordon and Breach, New York, 1970). *Introduces the notion of symmetry breaking into statistical physics.*
5. A.S. Wightman, *Statistical Mechanics at the Turn of the Decade*, Dekker (New York, 1971). *A general discussion of the foundations of statistical physics.*

Part II
Ideal Systems

Chapter 4
Microcanonical Ensemble

Abstract In this chapter the microcanonical ensemble, which describes an isolated system, is introduced. In this ensemble the entropy and the temperature, i.e. the fundamental thermal variables of thermodynamics, are defined.

4.1 Classical Microcanonical Ensemble

The second postulate of classical statistical physics states that the probability density in phase space $\rho(q, p)$ of an equilibrium system is a function of the Hamiltonian $H_N(q, p; \alpha)$. Note that the particular form of $\rho(q, p)$ not only depends on $H_N(q, p; \alpha)$, but also on the interaction between the system and the external world. Indeed, according to the second law of thermodynamics, a non-equilibrium system which exchanges energy, volume, and particles with the external world evolves toward an equilibrium state. In this state the values of the intensive variables of the system, namely temperature T, pressure p, and chemical potential μ, become equal to those of the corresponding variables of the external world; thus, strictly speaking, the probability density at equilibrium should be indicated by $\rho((H_N(q, p; \alpha); T, p, \mu))$. This interaction with the external world is what may favor some regions of phase space with respect to others. The simplest case is, therefore, the one in which the system does exchange neither energy nor matter with the external world (isolated system), and so the probability density in phase space $\rho(q, p)$ only depends on $H_N(q, p; \alpha)$.

Consider then an isolated system of finite energy E. In what follows it will be assumed that $E \geq 0$, which may always be taken for granted due to the arbitrary choice of the zero of the potential energy. The average over the Gibbs ensemble has to account for the fact that all those mechanical states (q, p) for which $H_N(q, p; \alpha) \neq E$ cannot contribute to the average value over the ensemble, since they do not conserve energy. Assume that one draws in the phase space of the system the "surface" of constant energy $H_N(q, p; \alpha) = E$ which, as will be shown in this chapter, is a regular surface of finite volume. It has been already pointed out that $\rho(H_N(q, p; \alpha)) = 0$ when $H_N(q, p; \alpha) \neq E$. The question is now how to determine the probability density on the energy surface. Intuitively, there is no physical reason for which some

© The Author(s), under exclusive license to Springer Nature Switzerland AG 2021   71
M. Baus and C. F. Tejero, *Equilibrium Statistical Physics*,
https://doi.org/10.1007/978-3-030-75432-7_4

regions on the energy surface should be more probable than others, and thus it can be argued that the probability density must be uniform over this surface. It is then postulated that the probability density in phase space of an equilibrium system of energy E is given by

$$\rho(q, p) = \frac{1}{h^{3N}} \frac{1}{\omega(E, \alpha, N)} \delta(E - H_N(q, p; \alpha)), \tag{4.1}$$

where h is Planck's constant and $\omega(E, \alpha, N)$ is a function which can be obtained from the normalization condition (3.13), namely

$$\omega(E, \alpha, N) = \frac{1}{h^{3N}} \int dq \int dp\, \delta(E - H_N(q, p; \alpha)). \tag{4.2}$$

The Gibbs ensemble described by (4.1) and (4.2) is called the microcanonical ensemble which, by definition, is the one that describes an isolated system. (Note that the introduction of Planck's constant in (4.1) and (4.2) is arbitrary. This point will be examined in the following chapters.) The average value of a dynamical function $a(q, p)$ in the microcanonical ensemble (4.1) and (4.2) is in turn given by

$$\langle a \rangle = \frac{1}{h^{3N}} \frac{1}{\omega(E, \alpha, N)} \int dq \int dp\, a(q, p)\delta(E - H_N(q, p; \alpha)). \tag{4.3}$$

From (4.3) it follows that if the external parameter is the volume of the system ($\alpha = V$), all the variables $\langle a \rangle$ are functions of the energy E, the volume V, and the number of particles N. As stated in Chap. 2, these are the independent variables in the entropy representation of a one-component thermodynamic system whose only external parameter is the volume. According to (4.3) all the macroscopic variables depend on $\omega(E, \alpha, N)$, which is a function of the same extensive variables as the entropy depends upon in thermodynamics. It is, therefore, logical to think that both functions are related. Before establishing this relationship, consider the volume of phase space $\phi(E, \alpha, N)$ limited by the constant energy surface $H_N(q, p; \alpha) = E$:

$$\phi(E, \alpha, N) = \frac{1}{h^{3N}} \int dq \int dp\, \Theta(E - H_N(q, p; \alpha)), \tag{4.4}$$

where $\Theta(x)$ is the Heaviside step function, and the volume $\Omega(E, \alpha, N; \Delta E)$ of phase space contained between the constant energy surfaces $H_N(q, p; \alpha) = E + \Delta E$ and $H_N(q, p; \alpha) = E$:

$$\Omega(E, \alpha, N; \Delta E) = \phi(E + \Delta E, \alpha, N) - \phi(E, \alpha, N). \tag{4.5}$$

From (4.2), (4.4), and (4.5) it follows that

$$\omega(E, \alpha, N) = \frac{\partial \phi(E, \alpha, N)}{\partial E}, \quad \lim_{\Delta E \to 0} \frac{\Omega(E, \alpha, N; \Delta E)}{\Delta E} = \omega(E, \alpha, N), \tag{4.6}$$

since $\Theta'(x) = \delta(x)$. Note that $\omega(E, \alpha, N)$, $\phi(E, \alpha, N)$, and $\Omega(E, \alpha, N, \Delta E)$ are functions associated to the Gibbs ensemble that cannot be expressed as average values of dynamical functions, and this is why they are referred to as thermal variables. In classical statistical physics the entropy $S(E, \alpha, N)$ is defined as

$$S(E, \alpha, N) = k_B \ln \phi(E, \alpha, N), \qquad (4.7)$$

where k_B is the Boltzmann constant. Note that this statistical definition of entropy allows one to obtain absolute values of this quantity, provided one can determine the volume of phase space $\phi(E, \alpha, N)$, which is only possible in some simple systems. In the following sections it is shown that the statistical entropy (4.7) has all the specific properties of the thermodynamic entropy (see Chap. 2).

It is important to point out that when the external parameter α is an intensive variable, the statistical entropy (4.7) is not a function of the extensive variables of the system as it is postulated in the fundamental equation of thermodynamics. As shown in Sect. 4.7 the thermodynamic entropy is, in such cases, the Legendre transform of (4.7), whereby both functions contain the same physical information (see Appendix A).

4.2 Classical Ideal Gas

Consider a classical ideal gas of N particles of mass m contained in a closed region R of volume V. The Hamiltonian of the gas may be written as

$$H_N\left(\mathbf{r}^N, \mathbf{p}^N; V\right) = \sum_{j=1}^{N} \left(\frac{\mathbf{p}_j^2}{2m} + \phi_R(\mathbf{r}_j)\right), \qquad (4.8)$$

where $\mathbf{p}_j^2/2m$ is the kinetic energy of particle j, and $\phi_R(\mathbf{r}_j)$ is the external potential that keeps it inside the closed region, namely

$$\phi_R(\mathbf{r}_j) = \begin{cases} 0, \mathbf{r}_j \in R \\ \infty, \mathbf{r}_j \notin R \end{cases}. \qquad (4.9)$$

The volume of phase space limited by the surface $H_N\left(\mathbf{r}^N, \mathbf{p}^N; V\right) = E$ is, according to (4.4),

$$\phi(E, V, N) = \frac{1}{h^{3N}} \int d\mathbf{r}^N \int d\mathbf{p}^N \Theta\left(E - \sum_{j=1}^{N}\left\{\frac{\mathbf{p}_j^2}{2m} + \phi_R(\mathbf{r}_j)\right\}\right). \qquad (4.10)$$

Observe that although the domain of integration of each of the variables $\mathbf{r}_j$ in (4.10) is all of the Euclidean space, the inclusion of the potential $\phi_R(\mathbf{r}_j)$ in the Hamiltonian restricts the domain of integration to the region R, and so one obtains a factor V for each integral, i.e.,

$$\phi(E, V, N) = \left(\frac{V}{h^3}\right)^N \int d\mathbf{p}^N \Theta\left(2mE - \sum_{j=1}^{N} \mathbf{p}_j^2\right)$$

$$= \left(\frac{V}{h^3}\right)^N \mathcal{V}_{3N}(\sqrt{2mE}), \tag{4.11}$$

where $\mathcal{V}_{3N}(R)$ is the volume of the $3N$-dimensional sphere of radius R defined by

$$\mathcal{V}_n(R) = \int dx^n \Theta\left(R^2 - \sum_{j=1}^{n} x_j^2\right). \tag{4.12}$$

Note that, with the change of variable $x_i = Ry_i (i = 1, 2 \ldots n)$, (4.12) may be expressed as

$$\mathcal{V}_n(R) = R^n \int dy^n \Theta\left(1 - \sum_{j=1}^{n} y_j^2\right) = \mathcal{V}_n(1) R^n, \tag{4.13}$$

where $\mathcal{V}_n(1)$ is a constant that only depends on the dimensionality of space. To evaluate this constant, consider the integral

$$I \equiv \int_0^\infty du e^{-u} \mathcal{V}_n(\sqrt{u}), \tag{4.14}$$

which, according to (4.13), may be written in the form

$$I = \mathcal{V}_n(1) \int_0^\infty du e^{-u} u^{n/2} = \mathcal{V}_n(1) \Gamma\left(\frac{n}{2} + 1\right), \tag{4.15}$$

where $\Gamma(n)$ is the Euler gamma function:

$$\Gamma(n) = \int_0^\infty du e^{-u} u^{n-1}, \tag{4.16}$$

which, when n is an integer, is given by

$$\Gamma(n+1) = n!, \quad \Gamma\left(n + \frac{1}{2}\right) = \frac{1 \cdot 3 \cdot 5 \ldots (2n-1)}{2^n}\sqrt{\pi}. \tag{4.17}$$

On the other hand, according to (4.12),

$$I = \int_0^\infty du\, e^{-u} \int dx^n \Theta\left(u - \sum_{j=1}^n x_j^2\right)$$

$$= \int dx^n \int_0^\infty du\, e^{-u} \Theta\left(u - \sum_{j=1}^n x_j^2\right)$$

$$= \int dx^n e^{-(x_1^2 + \cdots + x_n^2)} = \pi^{n/2}. \qquad (4.18)$$

Therefore, from (4.15) and (4.18) it follows that

$$\mathcal{V}_n(1) = \frac{\pi^{n/2}}{\Gamma\left(\frac{n}{2} + 1\right)}. \qquad (4.19)$$

Simple examples are $\mathcal{V}_2(R) = \pi R^2$, which is the area of a circle, and $\mathcal{V}_2(R) = 4\pi R^3/3$, which is the volume of a sphere.

4.3 Entropy and the Gibbs Paradox

The volume of phase space $\phi(E, V, N)$ of the classical ideal gas is then given by

$$\phi(E, V, N) = \left(\frac{V}{h^3}\right)^N \frac{1}{\Gamma\left(\frac{3N}{2} + 1\right)} (2\pi m E)^{3N/2}. \qquad (4.20)$$

In order to derive the expression for the entropy from (4.7), the logarithm of the Euler gamma function in (4.20) has to be evaluated. To that end, use may be made of the Stirling's approximation

$$\Gamma(n) \simeq e^{-n} n^{n-1/2} \sqrt{2\pi} \left(1 + \frac{1}{12n} + \ldots\right), \qquad (4.21)$$

and so from (4.7), (4.20), and (4.21) it follows that

$$S(E, V, N) = Ns(e, \rho) + Nk_B(\ln N - 1) + O(\ln N), \qquad (4.22)$$

where $e = E/N$ is the energy per particle, $\rho = N/V$ is the density and

$$s(e, \rho) = k_B\left(\frac{3}{2} \ln\left(\frac{4\pi m e}{3h^2 \rho^{2/3}}\right) + \frac{5}{2}\right). \qquad (4.23)$$

It is seen from (4.22) that in the thermodynamic limit $N \to \infty$, $V \to \infty$, $E \to \infty$ with e $< \infty$ and $\rho < \infty$, the entropy per particle, $S(E,V,N)/N$, is

$$\mathrm{TL}\left[\frac{1}{N}S(E, V, N)\right] = \mathrm{TL}[s(\mathrm{e}, \rho) + k_B(\ln N - 1)] = \infty, \qquad (4.24)$$

where it has been taken into account that

$$\mathrm{TL}\left[\frac{O(\ln N)}{N}\right] = 0.$$

This anomalous result (the entropy is not an extensive variable in the thermodynamic limit) is known as the Gibbs paradox. There is an ad hoc criterion that solves the problem. Assume that when $\alpha = V$, (4.1) is written as

$$\rho(q, p) = \frac{1}{N!h^{3N}}\frac{1}{\omega^*(E, V, N)}\delta(E - H_N(q, p; V)), \qquad (4.25)$$

where $\omega^*(E, V, N)$ is obtained from the normalization condition of the probability density, i.e.,

$$\omega^*(E, V, N) = \frac{1}{N!h^{3N}}\int dq \int dp\, \delta(E - H_N(q, p; V))$$

$$= \frac{1}{N!}\omega(E, V, N). \qquad (4.26)$$

Observe that the average values of the dynamical functions (4.3) are the same irrespective of whether the averages are performed with (4.1) and (4.2) or with (4.25) and (4.26), the fundamental difference being that $\omega(E, V, N)$ has been replaced by $\omega^*(E, V, N)$. From (4.6) it follows that, for consistency, the functions (4.4) and (4.5) must be replaced by

$$\phi^*(E, V, N) = \frac{1}{N!}\phi(E, V, N), \quad \Omega^*(E, V, N; \Delta E) = \frac{1}{N!}\Omega(E, V, N; \Delta E).$$

Then, if the entropy of the gas is defined by the expression

$$S(E, V, N) = k_B \ln \phi^*(E, V, N) = k_B \ln\left(\frac{\phi(E, V, N)}{N!}\right), \qquad (4.27)$$

from (4.20) and (4.27) one has

$$S(E, V, N) = Ns(\mathrm{e}, \rho) + O(\ln N), \qquad (4.28)$$

where $N! = \Gamma(N+1)$ has been determined using the Stirling's approximation (4.21). Note that in this instance the entropy per particle, $S(E, V, N)/N$, is a finite quantity in the thermodynamic limit, namely

$$\text{TL}\left[\frac{1}{N}S(E, V, N)\right] = s(e, \rho). \tag{4.29}$$

The Gibbs' paradox may be illustrated through the following reasoning. Take a closed region of volume V split into two equal parts by a wall. On each side one has the same ideal gas of $N/2$ particles and energy $E/2$ in equilibrium. If the wall is removed, the gas attains an equilibrium state in which the N particles, of energy E, occupy the volume V. If the variation of entropy taking place in the process is determined from (4.22), one has

$$\text{TL}\left[\frac{1}{N}S(E, V, N) - \frac{2}{N}S\left(\frac{E}{2}, \frac{V}{2}, \frac{N}{2}\right)\right] = k_B \ln 2, \tag{4.30}$$

i.e. there is an increase of entropy upon removal of the wall. From a thermodynamic point of view, in this process the entropy cannot increase since, when the wall is put back into place, one recovers the initial state. On the other hand, note that from (4.28) it follows that there is no entropy increase in the process. In order to understand why the factor $1/N!$ solves the Gibbs paradox, notice that the volume in phase space (4.10) is a $3N$-dimensional integral extended to all the mechanical states of the system. Since in classical mechanics the particles are considered to be distinguishable, the mechanical states that correspond to permutations of two or more particles are different states and they have been considered so in (4.10). From a classical point of view, this definition of entropy is, therefore, correct. Nevertheless, since classical mechanics is only an approximation of quantum mechanics, where the particles are indistinguishable, the Gibbs factor in (4.27) has the purpose of reducing by $N!$, i.e., in the number of permutations that one may perform with the N particles of the gas, the mechanical states considered in (4.10). Any such permutation does not give rise to a new state in quantum mechanics, and in this way the Gibbs factor becomes a correction of quantum origin in a classical system.

Note that in classical physics it is not always the case that particles may be permuted as it occurs in a fluid (either gas or liquid). In such instances it is not required to introduce the Gibbs factor. Take, e.g., a system of N particles whose Hamiltonian is

$$H_N(\mathbf{r}^N, \mathbf{p}^N; \omega) = \sum_{j=1}^{N}\left(\frac{\mathbf{p}_j^2}{2m} + \frac{1}{2}m\omega^2(\mathbf{r}_j - \mathbf{R}_j)^2\right), \tag{4.31}$$

where $\mathbf{p}_j$ is the momentum of particle j, $\mathbf{r}_j$ its position vector, and $\mathbf{R}_j$ a constant. This Hamiltonian represents the harmonic oscillations of the atoms in a solid around the equilibrium positions $\mathbf{R}_j$ of a crystalline lattice. Note that in (4.31) it has been assumed that all the atoms vibrate with the same frequency. This simple model for the Hamiltonian of a solid is referred to as the classical Einstein model. The volume of phase space delimited by the surface of energy E is, in this case,

$$\phi(E, \omega, N) = \frac{1}{h^{3N}} \int d\mathbf{r}^N \int d\mathbf{p}^N \Theta\left(E - \sum_{j=1}^{N} H_N(\mathbf{r}^N, \mathbf{p}^N; \omega)\right), \quad (4.32)$$

which, with the change of variables,

$$\mathbf{x}_j = \frac{\mathbf{p}_j}{\sqrt{2m}}, \quad \mathbf{x}_{j+N} = \sqrt{\frac{m\omega^2}{2}}(\mathbf{r}_j - \mathbf{R}_j), \quad (j = 1, 2, \ldots, N), \quad (4.33)$$

may be written in the form

$$\phi(E, \omega, N) = \left(\frac{2}{h\omega}\right)^{3N} \int d\mathbf{x}^{2N} \Theta\left(E - \sum_{j=1}^{2N} \mathbf{x}_j^2\right)$$

$$= \left(\frac{2}{h\omega}\right)^{3N} \mathcal{V}_{6N}(\sqrt{E})$$

$$= \frac{1}{\Gamma(3N+1)} \left(\frac{2\pi E}{h\omega}\right)^{3N}. \quad (4.34)$$

The entropy per particle $S(E, \omega, N)/N$ in the thermodynamic limit $N \to \infty$, $E \to \infty$ with $e = E/N < \infty$ (note that in this system the external parameter is not the extensive variable V but the frequency ω of the oscillators, which is an intensive variable) is

$$\text{TL}\left[\frac{1}{N} S(E, \omega, N)\right] = 3k_B\left(1 + \ln\left(\frac{2\pi e}{3h\omega}\right)\right). \quad (4.35)$$

In the classical Einstein model it is, therefore, not necessary to divide the phase space volume $\phi(E, \omega, N)$ by $N!$ when determining the entropy. This is due to the fact that since the Hamiltonian (4.31) describes the vibrations of the atoms around their equilibrium positions, these atoms are localized, and it is not possible to permute them.

In view of the results obtained for these ideal systems, it will be assumed that, in general, the entropy (4.7) of a system of N particles and energy E whose Hamiltonian is $H_N(q, p; \alpha)$ may be written in the form

$$S(E, \alpha, N) = Ns(e, \alpha) + 0(\ln N), \quad (N \gg 1), \quad (4.36)$$

where $s(e, \alpha)$ is the entropy per particle. Note that if $\alpha = V$, (4.36) must be written as

$$S(E, V, N) = Ns(e, \rho) + 0(\ln N), \quad (N \gg 1), \quad (4.37)$$

i.e., the arguments of the entropy per particle are, in this case, the energy per particle and the density. With the previous proviso and in order to have a uniform notation, the expression (4.36) will henceforth be used.

From (4.7) and (4.36) it follows that

$$\phi(E, \alpha, N) = e^{Ns(e,\alpha)/k_B + O(\ln N)}, \tag{4.38}$$

and so, according to (4.6), one has

$$\omega(E, \alpha, N) = \frac{1}{k_B} s'(e, \alpha) e^{Ns(e,\alpha)/k_B + O(\ln N)}, \tag{4.39}$$

where $s'(e, \alpha) = \partial s(e, \alpha)/\partial e$. If, in analogy with (4.7), one defines

$$\hat{S}(E, \alpha, N) = k_B \ln \Omega(E, \alpha, N; \Delta E), \tag{4.40}$$

from (4.6) and (4.39) it follows that

$$\hat{S}(E, \alpha, N) = Ns(e, \alpha) + k_B \ln\left[\frac{\Delta E}{k_B} s'(e, \alpha)\right] + O(\ln N) + O((\Delta E)^2), \tag{4.41}$$

or, alternatively,

$$\mathrm{TL}\left[\frac{1}{N}\hat{S}(E, \alpha, N)\right] = s(e, \alpha), \tag{4.42}$$

where it has been assumed that

$$\mathrm{TL}\left[\frac{O((\Delta E)^2)}{N}\right] = 0.$$

Therefore, the definitions of the entropy given in (4.40) and (4.7) are equivalent in the thermodynamic limit.

4.4 Temperature and Thermal Equilibrium

Consider a system of N particles and energy E formed by two subsystems of N_1 and N_2 particles ($N = N_1 + N_2$) whose Hamiltonian is additive, namely

$$H_N(q, p; \alpha_1, \alpha_2) = H_{N_1}(q_1, p_1; \alpha_1) + H_{N_2}(q_2, p_2; \alpha_2), \tag{4.43}$$

where (q_i, p_i) ($i = 1, 2$) are the generalized coordinates and conjugate momenta of the subsystems of external parameters α_1 and α_2, and $(q, p) \equiv (q_1, q_2, p_1, p_2)$

being the mechanical state of the system. For the sake of simplifying the notation, the Hamiltonians that appear in (4.43) will be denoted in this section by H, H_1, and H_2.

Since the energy E is constant, the probability density in the phase space of the system is given by the microcanonical ensemble, namely

$$\rho(q, p) = \frac{1}{h^{3N}} \frac{1}{\omega(E, \alpha, N)} \delta(E - H_1 - H_2), \tag{4.44}$$

$$\omega(E, \alpha, N) = \frac{1}{h^{3N}} \int dq \int dp\, \delta(E - H_1 - H_2), \tag{4.45}$$

with $\alpha \equiv (\alpha_1, \alpha_2)$.

The probability density $\rho_1(q_1, p_1)$ of subsystem 1, which exchanges energy with subsystem 2, is the marginal probability density (see Appendix B):

$$
\begin{aligned}
\rho_1(q_1, p_1) &= \frac{1}{h^{3N}} \frac{1}{\omega(E, \alpha, N)} \int dq_2 \int dp_2\, \delta(E - H_1 - H_2) \\
&= \frac{1}{h^{3N_1}} \frac{\omega_2(E - H_1, \alpha_2, N_2)}{\omega(E, \alpha, N)},
\end{aligned} \tag{4.46}
$$

where (4.2) has been taken into account.

If one considers the equation $E_1 = H_1$, E_1 is a random variable (since it is a function of the mechanical state of subsystem 1) whose probability density may be determined from (4.46) as (see Appendix B)

$$
\begin{aligned}
\rho_1(E_1) &= \int dq_1 \int dp_1 \rho_1(q_1, p_1) \delta(E_1 - H_1) \\
&= \frac{1}{h^{3N_1}} \int dq_1 \int dp_1 \frac{\omega_2(E - H_1, \alpha_2, N_2)}{\omega(E, \alpha, N)} \delta(E_1 - H_1) \\
&= \frac{\omega_1(E_1, \alpha_1, N_1) \omega_2(E - E_1, \alpha_2, N_2)}{\omega(E, \alpha, N)}.
\end{aligned} \tag{4.47}
$$

The normalization condition of $\rho_1(E_1)$ $\left(\int_0^E dE_1 \rho_1(E_1) = 1 \right)$ implies that

$$\omega(E, \alpha, N) = \int_0^E dE_1 \omega_1(E_1, \alpha_1, N_1) \omega_2(E - E_1, \alpha_2, N_2), \tag{4.48}$$

which is the composition (convolution) law for $\omega(E, \alpha, N)$. Note that in order to derive (4.48) the only requirement is the additivity of the energy (4.43).

Upon integration of (4.48) with respect to E one has

$$\phi(E, \alpha, N) = \int_0^E dE_1 \omega_1(E_1, \alpha_1, N_1) \phi_2(E - E_1, \alpha_2, N_2), \qquad (4.49)$$

since $\phi_2(E - E_1, \alpha_2, N_2) = 0$ when $E - E_1 < 0$.

If the subsystems are macroscopic, i.e., if $N_1 = \gamma_1 N$ and $N_2 = \gamma_2 N$, with $\gamma_1 = O(1)$ and $\gamma_2 = O(1)$ ($\gamma_1 + \gamma_2 = 1$) and $N \gg 1$, the entropies of the subsystems are of $O(N)$, and from (4.6) and (4.7) it follows that (4.49) may be written as

$$\phi(E, \alpha, N) = \int_0^E dE_1 g(E_1) e^{Nf(E_1)}, \qquad (4.50)$$

where

$$k_B f N(E_1) = S_1(E_1, \alpha_1, N_1) + S_2(E - E_1, \alpha_2, N_2)$$
$$= k_B \ln[\phi_1(E_1, \alpha_1, N_1)\phi_2(E - E_1, \alpha_2, N_2)], \qquad (4.51)$$

and $k_B g(E_1) = \partial S_1(E_1, \alpha_1, N_1)/\partial E_1$, which may be approximated by (see Appendix B)

$$\phi(E, \alpha, N) \simeq \sqrt{\frac{2\pi}{N|f''(\tilde{E}_1)|}} g(\tilde{E}_1) e^{Nf(\tilde{E}_1)}, \qquad (4.52)$$

where $\tilde{E}_1$ is the global maximum of $f(E_1)$ in the interval $(0, E)$. This maximum is unique since $Nf(E_1)$ is the logarithm of the product of a monotonously increasing function of E_1, $\phi_1(E_1, \alpha_1, N_1)$ and a monotonously decreasing function of E_1, $\phi_2(E - E_1, \alpha_2, N_2)$. From (4.7) it follows that

$$S(E, \alpha, N) = S_1\left(\tilde{E}_1, \alpha_1, N_1\right) + S_2\left(E - \tilde{E}_1, \alpha_2, N_2\right) + O(\ln N), \qquad (4.53)$$

i.e., the entropy is additive. The condition of being an extremum of $Nf(E_1)$ is written as

$$\left(\frac{\partial S_1(E_1, \alpha_1, N_1)}{\partial E_1}\right)_{E_1=\tilde{E}_1} = \left(\frac{\partial S_2(E_2, \alpha_2, N_2)}{\partial E_2}\right)_{E_2=\tilde{E}_2}, \qquad (4.54)$$

with $\tilde{E}_2 = E - \tilde{E}_1$. If the absolute temperature T is defined through

$$\frac{1}{T} = \frac{\partial S(E, \alpha, N)}{\partial E}, \qquad (4.55)$$

from which the energy equation $E = E(T, \alpha, N)$ follows, the extremum condition (4.54) expresses the thermal equilibrium (equality of temperatures) of the subsystems. In statistical physics, thermal equilibrium takes place in the state of maximum probability, where the entropy is additive. Note that (4.55) and (2.19) are formally analogous, although in (4.55) the entropy has been defined through the microcanonical ensemble by (4.7). On the other hand, thermal equilibrium corresponds to the one obtained in (2.66), although in the latter derivation it was obtained from the extremum condition of the thermodynamic entropy.

It is convenient at this point to make the following observation. As was analyzed in Chap. 2, the exchange of energy between two subsystems is a necessary condition for them to reach thermal equilibrium. In order for this to happen, there must exist an interaction energy between the subsystems which, apparently, has not been accounted for in (4.43). This is true for each mechanical state of the Gibbs ensemble but, since the only restriction imposed on H_1 and H_2 is that $H_1 + H_2 = E$, the different mechanical states of the subsystems have different energy in the ensemble. Therefore, contrary to what may be suggested by (4.43), the subsystems exchange energy, and thus they reach the thermal equilibrium of (4.54).

4.5   Ideal Systems

The convolution law (4.48) is based, exclusively, on the additivity of the energy of the subsystems. In the case of an ideal system, this law may be used iteratively to obtain $\omega(E, \alpha, N)$ from the volume of the phase space of a single particle $\omega(E, \alpha, 1)$. As an example, consider a free particle of mass m in a closed region R of volume V. If its energy is E, the volume of the phase space of the particle delimited by the surface of energy $\mathbf{p}^2/2m = E$ is given by

$$\phi(E, V, 1) = \frac{1}{h^3} \int d\mathbf{r} \int d\mathbf{p}\, \Theta\left(E - \frac{\mathbf{p}^2}{2m} - \phi_R(\mathbf{r})\right)$$
$$= \frac{V}{h^3} V_3(\sqrt{2mE}) = \frac{V}{h^3} \frac{1}{\Gamma(5/2)}(2\pi mE)^{3/2}, \qquad (4.56)$$

yielding

$$\omega(E, V, 1) = \frac{3V}{2h^3} \frac{1}{\Gamma(5/2)}(2\pi m)^{3/2}\sqrt{E}. \qquad (4.57)$$

In this way, from (4.48) one obtains

$$\omega(E, V, N) = \int_0^E dE_1 \omega_1(E_1, V, N - 1)\omega_2(E - E_1, V, 1), \qquad (4.58)$$

and from the expression

$$\int_0^1 dx\, x^{n-1}(1-x)^{m-1} = \frac{\Gamma(n)\Gamma(m)}{\Gamma(n+m)}, \tag{4.59}$$

it may be easily demonstrated that after two or three iterations in (4.58), a recurrence law may be derived which allows one to obtain $\omega(E, V, N)$ for arbitrary N. This function is the derivative of $\phi(E, V, N)$ (see (4.20)) with respect to the energy. The interest of this result lies in the fact that the thermodynamic potential of the gas (the entropy) may be obtained in classical statistical physics from $\omega(E, V, 1)$. In Sect. 4.9 it will be shown that this result does not hold in quantum statistical physics.

4.6 Equipartition Theorem

From a classical point of view, the temperature is a magnitude associated with the kinetic energy of the particles of a system. According to (4.55) the inverse of the absolute temperature in classical statistical physics is the derivative of the entropy of the system with respect to the energy. These two definitions of the same variable may be related through the equipartition theorem.

Consider a system of N particles of mass m contained in a closed region R of volume V. The Hamiltonian of the system is

$$H_N(\mathbf{r}^N, \mathbf{p}^N; V) = \sum_{j=1}^N \frac{\mathbf{p}_j^2}{2m} + H_N^{\text{int}}(\mathbf{r}^N) + \sum_{j=1}^N \phi_R(\mathbf{r}_j), \tag{4.60}$$

where the first two terms correspond to the kinetic energy and the interaction potential energy, while the last one is the potential energy (4.9) that keeps the particles inside the region. In order to simplify the notation, in this section these terms will be denoted by H_N^{id}, H_N^{int}, and H_N^R. The average value of the kinetic energy of the system in the microcanonical ensemble (4.1 and 4.2) is given by

$$\langle H_N^{\text{id}} \rangle = \frac{1}{h^{3N}} \frac{1}{\omega(E, V, N)} \int d\mathbf{r}^N \int d\mathbf{p}^N\, H_N^{\text{id}} \delta(E - H_N^{\text{id}} - H_N^{\text{int}} - H_N^R). \tag{4.61}$$

From the identity

$$\frac{\partial}{\partial \lambda} \Theta(E - \lambda H_N^{\text{id}} - H_N^{\text{int}} - H_N^R) = -H_N^{\text{id}} \delta(E - \lambda H_N^{\text{id}} - H_N^{\text{int}} - H_N^R), \tag{4.62}$$

the integral I in (4.61) may be written as

$$I = -\left(\frac{\partial}{\partial\lambda}\int d\mathbf{r}^N \int d\mathbf{p}^N \Theta\left(E - \lambda H_N^{\text{id}} - H_N^{\text{int}} - H_N^R\right)\right)_{\lambda=1}, \tag{4.63}$$

or, alternatively, as

$$I = -\left(\frac{\partial}{\partial\lambda}\int d\mathbf{r}^N \mathcal{V}_{3N}\left(\sqrt{\frac{2m}{\lambda}[E - H_N^{\text{int}} - H_N^R]}\right)\right)_{\lambda=1}, \tag{4.64}$$

where $\mathcal{V}_{3N}(R)$ is the volume of the $3N$-dimensional sphere of radius R. The derivative with respect to λ in (4.64) may be determined from (4.13), leading to

$$\begin{aligned}
\langle H_N^{\text{id}}\rangle &= \frac{3N}{2}\frac{1}{h^{3N}}\frac{1}{\omega(E,V,N)}\int d\mathbf{r}^N \mathcal{V}_{3N}\left(\sqrt{2m[E - H_N^{\text{int}} - H_N^R]}\right) \\
&= \frac{3N}{2}\frac{1}{h^{3N}}\frac{1}{\omega(E,V,N)}\int d\mathbf{r}^N \int d\mathbf{p}^N \Theta\left(E - H_N^{\text{id}} - H_N^{\text{int}} - H_N^R\right) \\
&= \frac{3N}{2}\frac{\phi(E,V,N)}{\omega(E,V,N)}. \tag{4.65}
\end{aligned}$$

Since

$$\frac{\phi(E,V,N)}{\omega(E,V,N)} = \left(\frac{\partial\ln\phi(E,V,N)}{\partial E}\right)^{-1} = k_B\left(\frac{\partial S(E,V,N)}{\partial E}\right)^{-1} = k_B T,$$

one finally obtains

$$\langle H_N^{\text{id}}\rangle = \frac{3}{2}N k_B T, \tag{4.66}$$

which expresses that the average kinetic energy ($k_B T/2$ per degree of freedom) is proportional to the temperature, which is the usual notion in classical statistical physics.

4.7 Equation of State

Throughout this chapter it has been assumed that the Hamiltonian of the system is a function of some external parameters. If one varies the external parameter α by $d\alpha$, the result is

$$\langle dH_N(q,p;\alpha)\rangle = \left\langle\frac{\partial H_N(q,p;\alpha)}{\partial\alpha}d\alpha\right\rangle = -A_\alpha d\alpha, \tag{4.67}$$

where A_α is the conjugate variable of α, namely

$$A_\alpha \equiv \left\langle -\frac{\partial H_N(q, p; \alpha)}{\partial \alpha} \right\rangle. \tag{4.68}$$

As was shown in Chap. 1, if α is a uniform and constant magnetic field $\mathbf{B}$, $A_{\mathbf{B}}$ is the total magnetic moment of the system $\mathbf{M}$, while if α is the volume V, A_V is the pressure p.

The conjugate variable of the external parameter α in the microcanonical ensemble may be determined as follows:

$$
\begin{aligned}
A_\alpha &= -\frac{1}{h^{3N}} \frac{1}{\omega(E, \alpha, N)} \int dq \int dp \frac{\partial H_N(q, p; \alpha)}{\partial \alpha} \delta(E - H_N(q, p; \alpha)) \\
&= \frac{1}{h^{3N}} \frac{1}{\omega(E, \alpha, N)} \int dq \int dp \frac{\partial}{\partial \alpha} \Theta(E - H_N(q, p; \alpha)) \\
&= \frac{1}{\omega(E, \alpha, N)} \frac{\partial \phi(E, \alpha, N)}{\partial \alpha},
\end{aligned} \tag{4.69}
$$

which can be expressed as

$$
\begin{aligned}
A_\alpha &= \frac{\phi(E, \alpha, N)}{\omega(E, \alpha, N)} \frac{\partial \ln \phi(E, \alpha, N)}{\partial \alpha} = \left(\frac{\partial \ln \phi(E, \alpha, N)}{\partial E} \right)^{-1} \frac{\partial \ln \phi(E, \alpha, N)}{\partial \alpha} \\
&= \left(\frac{\partial S(E, \alpha, N)}{\partial E} \right)^{-1} \frac{\partial S(E, \alpha, N)}{\partial \alpha} = T \frac{\partial S(E, \alpha, N)}{\partial \alpha},
\end{aligned} \tag{4.70}
$$

which is the equation of state corresponding to the parameter α.

As it was already pointed out in Sect. 4.1 when the external parameter is intensive, the entropy (4.7) is not a function of the extensive variables of the system as it is postulated in the fundamental equation of thermodynamics. Nevertheless, the entropy (4.7) contains all the thermodynamic information about the system (the energy Eq. (4.55) and the equation of state (4.70)). Consider, for instance, that α is the magnetic field $\mathbf{B}$, in which case (4.70) reads

$$\mathbf{M} = T \frac{\partial S(E, \mathbf{B}, N)}{\partial \mathbf{B}}. \tag{4.71}$$

From (4.71) and (4.55) it follows that the differential of the statistical entropy when E and $\mathbf{B}$ change (at constant N) is given by

$$T dS(E, \mathbf{B}, N) = dE + \mathbf{M} \cdot d\mathbf{B}. \tag{4.72}$$

If the thermodynamic entropy $S(E, \mathbf{M}, N)$ is defined as the Legendre transform (see Appendix A) of $S(E, \mathbf{B}, N)$ with respect to $\mathbf{B}$ (note that in order to simplify the notation both functions S are denoted by the same letter), one has

$$T S(E, \mathbf{M}, N) = T S(E, \mathbf{B}, N) - \mathbf{M} \cdot \mathbf{B}, \tag{4.73}$$

where in (4.73) $\mathbf{B} = \mathbf{B}(E, \mathbf{M}, N)$ is obtained from the equation of state (4.71). Therefore,

$$TdS(E, \mathbf{M}, N) = dE - \mathbf{B} \cdot d\mathbf{M}, \tag{4.74}$$

which is the usual expression in thermodynamics, and so the equation of state,

$$\mathbf{B} = -T \frac{\partial S(E, \mathbf{M}, N)}{\partial \mathbf{M}}, \tag{4.75}$$

contains the same information as (4.71).

4.8 Entropy and Irreversibility

The second law of thermodynamics establishes that, in an irreversible process, an isolated system evolves toward the state with maximum entropy compatible with the external conditions. Consider, for instance, the free expansion of an ideal gas. As initial condition, one has a container of volume V divided into two regions of volume $V/2$, one of which is occupied by a classical ideal gas of N particles and energy E. If the wall separating the two regions is removed, experience shows that the gas expands freely until it reaches a final state in which the gas occupies the whole container. According to (4.20) and (4.27) the variation of the entropy in the free expansion turns out to be

$$\Delta S = Nk_B \left(\ln V - \ln\left(\frac{V}{2}\right) \right) > 0, \tag{4.76}$$

in agreement with the second law of thermodynamics. Note, however, that (4.76) is not a proof of the law of entropy increase. As a matter of fact, the statistical definition of entropy (4.27) is only applicable to an equilibrium system. In order to derive (4.76) it has been assumed that, starting from an initial equilibrium state, which becomes a non-equilibrium state after the wall is removed, the gas expands freely and ultimately reaches another final equilibrium state. Upon comparison of the entropies of the initial and final states, it is found that the entropy of the final state is greater than the one of the initial state. But this argument involves an essential point which has not been derived, namely that the experiments show that the gas expands when the wall is removed. If, on the other hand, the everyday experience indicated that, starting from an initial state in which the gas occupies the volume V, the gas would contract freely until it occupied a volume $V/2$, from the definition in (4.27) it would follow that the entropy would diminish in the process. The second law of thermodynamics establishes that an isolated system evolves toward a state of maximum entropy, which in turn determines the arrow of time, as is the case of the free expansion. This arrow of time is what has not been derived to determine the result contained in (4.76).

Consider again the Gibbs definition of entropy (3.23), particularized to the microcanonical ensemble. Note that the probability density in phase space may be written as

$$\rho(q, p) = \frac{1}{h^{3N}} \frac{\Theta(E + \Delta E - H_N(q, p; \alpha)) - \Theta(E - H_N(q, p; \alpha))}{\Omega(E, \alpha, N; \Delta E)}, \qquad (4.77)$$

which leads to (4.1) in the limit $\Delta E \to 0$ because

$$\lim_{\Delta x \to 0} \frac{\Theta(x + \Delta x) - \Theta(x)}{\Delta x} = \delta(x). \qquad (4.78)$$

According to (4.77) the probability density is constant in the region of phase space in which $E < H_N(q, p; \alpha) < E + \Delta E$, and so the Gibbs entropy is given by

$$S = -k_B \langle \ln[h^{3N} \rho(H_N(q, p; \alpha))] \rangle = k_B \ln \Omega(E, \alpha, N; \Delta E), \qquad (4.79)$$

which coincides with the definition (4.40). The interest of this result is that since the Gibbs entropy is defined as a function of the probability density in phase space, it may be readily generalized to a non-equilibrium system by replacing the stationary probability density $\rho(H_N(q, p; \alpha))$ with the time-dependent probability density $\rho(q, p; t)$, i.e.,

$$S(t) = -k_B \int dq \int dp \ln[h^{3N} \rho(q, p; t)] \rho(q, p; t). \qquad (4.80)$$

The evolution equation of $S(t)$ is thus given by

$$\dot{S}(t) = -k_B \int dq \int dp \{1 + \ln[h^{3N} \rho(q, p; t)]\} \frac{\partial \rho(q, p; t)}{\partial t}. \qquad (4.81)$$

Note that

$$\int dq \int dp \frac{\partial \rho(q, p; t)}{\partial t} = \frac{\partial}{\partial t} \int dq \int dp \, \rho(q, p; t) = 0, \qquad (4.82)$$

because the probability density is normalized at all times. On the other hand, from the Liouville Eq. (3.18) one finds

$$I = \int dq \int dp \ln \rho(q, p; t) \frac{\partial \rho(q, p; t)}{\partial t}$$

$$= \int dq \int dp \ln \rho(q, p; t) \{H_N(\alpha), \rho(t)\}$$

$$= \int dq \int dp \, H_N(q, p; \alpha)\{\rho(t), \ln \rho(t)\} = 0, \qquad (4.83)$$

where it has been taken into account that $\rho(q, p; t)$ vanishes on the surface of the phase space and that $\{p(t), \ln \rho(t)\} = 0$. Therefore, when generalizing the Gibbs entropy to a system out of equilibrium one has

$$\dot{S}(t) = 0, \qquad (4.84)$$

i.e., the Gibbs entropy is constant in the time evolution, and so it is not possible to derive from it the second law of thermodynamics. The reason for this is that the first postulate of classical statistical physics associates a field (3.16) to a dynamical function. This postulate is statistical with respect to the initial conditions and does not contradict the laws of mechanics. As a matter of fact, the time evolution of the field is induced by that of the dynamical function, and this is in turn induced by the Hamilton equations. Since the equations of mechanics are invariant under the change $t \to -t$, it is not possible to derive from them a non-mechanical evolution, such as the one predicted by the second law of thermodynamics. On the other hand, it must be pointed out that, while the thermodynamic limit allows one to derive equilibrium thermodynamics (see Sects. 3.6 and 3.7), this limit is necessary but not sufficient to justify irreversibility.

4.9   Quantum Microcanonical Ensemble

As was analyzed in Chap. 3, the density operator $\hat{\rho}$ of an equilibrium system commutes with the Hamiltonian operator $\hat{H}_N(\alpha)$. The additional postulate that $\hat{\rho}$ is a function of $\hat{H}_N(\alpha)$ is in a certain way justified because in all mechanical systems energy is conserved (note that, as in the classical case, the density operator of an equilibrium system depends also on the intensive parameters T, p, and μ if the system exchanges energy, volume, and particles with the external world). When $\left[\hat{\rho}, \hat{H}_N(\alpha)\right] = 0$, it is always possible to find a basis of the Hilbert space in which both operators are diagonal. This basis is not necessarily the one formed with the eigenvectors of $\hat{H}_N(\alpha)$, whose eigenvalue equation reads

$$\hat{H}_N(\alpha)|\Psi_n\rangle = E_n^{(N)}(\alpha)|\Psi_n\rangle, \qquad (4.85)$$

where it has been assumed that the spectrum is discrete. Note however that, contrary to what occurs in classical physics, one cannot define an isolated system as the one whose energy is E, since it is conceivable that there may not exist a quantum state n of (4.85) whose energy is such that $E_n^{(N)}(\alpha) = E$. In order to avoid this, in quantum mechanics an isolated system will be defined as one whose energy lies between E and $E + \Delta E$, with $\Delta E \ll E$, although the interval is assumed to be large enough so that there exist a great number of quantum states of energies $E_n^{(N)}(\alpha)$ such that $E < E_n^{(N)}(\alpha) < E + \Delta E$. The number of these quantum states, which will be denoted by $\Omega(E, \alpha, N; \Delta E)$, may be written as

$$\Omega(E, \alpha, N; \Delta E) = \sum_n \Theta\big(E + \Delta E - E_n^{(N)}(\alpha)\big)$$
$$- \sum_n \Theta\big(E - E_n^{(N)}(\alpha)\big). \tag{4.86}$$

The density of quantum states $\omega(E, \alpha, N)$ is defined by the equation

$$\omega(E, \alpha, N) = \sum_n \delta\big(E - E_n^{(N)}(\alpha)\big), \tag{4.87}$$

and the number of quantum states whose energy is less than E, $\phi(E, \alpha, N)$, is defined as

$$\phi(E, \alpha, N) = \sum_n \Theta\big(E - E_n^{(N)}(\alpha)\big). \tag{4.88}$$

Since the system has an energy between E and $E + \Delta E$, those quantum states of (4.85) such that $E_n^{(N)}(\alpha) < E$ or $E_n^{(N)}(\alpha) > E + \Delta E$ should not contribute to the average value of the operator (3.28) in the statistical mixing. On the other hand, from the different quantum states n of (4.85) with energies $E < E_n^{(N)}(\alpha) < E + \Delta E$ in which the system may be found, there is no physical reason forcing one to be more probable than the others. In the microcanonical ensemble of a quantum system, a probability of zero is assigned to those states that are not compatible with the energy of the system and the same probability to those states that are compatible. For that reason it is postulated that the density operator of an isolated system is given by the following expression:

$$\langle \Psi_m | \hat{\rho} | \Psi_n \rangle = \delta_{mn} \frac{\Theta\big(E + \Delta E - E_n^{(N)}(\alpha)\big) - \Theta\big(E - E_n^{(N)}(\alpha)\big)}{\Omega(E, \alpha, N; \Delta E)}. \tag{4.89}$$

That is, it is a diagonal matrix in the energy representation whose only nonzero elements are those corresponding to quantum states whose energy $E_n^{(N)}(\alpha)$ verifies $E < E_n^{(N)}(\alpha) < E + \Delta E$. The probability of each of these quantum states is $\Omega^{-1}(E, \alpha, N; \Delta E)$.

The average value of an operator a in the quantum microcanonical ensemble is then given by

$$\langle \hat{a} \rangle = \sum_n \langle \Psi_n | \hat{a} \hat{\rho} | \Psi_n \rangle = \sum_n \sum_m \langle \Psi_n | \hat{a} | \Psi_m \rangle \langle \Psi_m | \hat{\rho} | \Psi_n \rangle$$

$$= \sum_n \langle \Psi_n | \hat{a} | \Psi_n \rangle \frac{\Theta\left(E + \Delta E - E_n^{(N)}(\alpha)\right) - \Theta\left(E - E_n^{(N)}(\alpha)\right)}{\Omega(E, \alpha, N; \Delta E)}, \qquad (4.90)$$

whose interpretation is simple: The average value in the statistical mixing is the sum of the expectation values of the operator, $\langle \Psi_n | \hat{a} | \Psi_n \rangle$, for all the eigenstates $|\Psi_n\rangle$ of $\hat{H}_N(\alpha)$ whose energy lies in the interval $(E, E + \Delta E)$, divided by the number of quantum states in the interval, $\Omega(E, \alpha, N; \Delta E)$. This average gives, therefore, the same statistical weight to all these quantum states.

Once more, if the external parameter is the volume, the intensive variables (4.90) are functions of the energy E, the volume V, and the number of particles N (the dependence on ΔE tends to zero in the thermodynamic limit). By the same line of reasoning followed in the classical case, the entropy of a quantum system may be defined through the following expressions:

$$S(E, V, N) = k_B \ln \phi(E, V, N), \qquad (4.91)$$

and

$$\hat{S}(E, V, N) = k_B \ln \Omega(E, V, N; \Delta E), \qquad (4.92)$$

which are equivalent in the thermodynamic limit. In the case of a fluid, the functions $\phi(E, V, N)$ and $\Omega(E, V, N; \Delta E)$ are increasing functions of the energy, and thus the absolute temperature, defined by the thermodynamic relation (4.55), is nonnegative. Note further that, since in a quantum fluid the particles are indistinguishable, one only has to count the number of quantum states whose wave function has a given symmetry (bosons or fermions). This is the reason why it is not easy to compute $\phi(E, V, N)$ and $\Omega(E, V, N; \Delta E)$ even when the system is ideal. In order to get a better insight in the complexity of ideal quantum systems, one may consider the following example. The convolution in the case of a quantum system formed by two subsystems that exchange energy in an additive way is again (4.48) where $\omega(E, \alpha, N)$ corresponds now to the density of quantum states (4.87). When the subsystems are macroscopic, this law leads to the additivity of the entropy and to the thermal equilibrium between the subsystems (see Sect. 4.4). Assume now that one wants to apply this convolution law to determine the thermodynamic properties of an ideal quantum gas of N particles contained in a closed region R of volume V, and whose Hamiltonian operator is the sum of one-particle operators, namely

$$\omega(E, V, N) = \int_0^E dE_1 \sum_{n_1} \delta\left(E_1 - E_{n_1}^{(N-1)}(V)\right) \sum_{n_2} \delta\left(E - E_1 - E_{n_2}^{(1)}(V)\right)$$

$$= \sum_{n_1} \sum_{n_2} \delta\left(E - E_{n_1}^{(N-1)}(V) - E_{n_2}^{(1)}(V)\right). \qquad (4.93)$$

It is clear that when combining a quantum state n_1 of the system of $N - 1$ particles with a quantum state n_2 of the N-th particle, the result is not a quantum state of the system of N particles. The reason is that the product of the wave function of the system of $N - 1$ particles times the wave function of the N-th particle is not a wave function of the system of N particles (see Sect. 1.9). Therefore, the law (4.93) cannot be used to derive the thermodynamics of an ideal quantum gas from $\omega(E, V, 1)$, as one can do in the case of a classical ideal gas (see Sect. 4.5). The thermodynamic properties of a quantum ideal gas, which are not as simple as those of its classical counterpart, will be analyzed in Chap. 6.

4.10 Absolute Negative Temperatures

In much the same way as it happens for the classical microcanonical ensemble, the determination of the entropy of a system in the quantum microcanonical ensemble may only be achieved in some simple cases. The reason is that one has to select from all the quantum states of the system only those whose energy lies in the interval between E and $E + \Delta E$. The existence of such a restriction is what makes the evaluation of the entropy difficult.

As a simple example, consider an ideal gas of N particles of total angular momentum $\mathbf{j}$ under the action of a constant and uniform magnetic field $\mathbf{B}$ in the direction of the z-axis. The energy of a particle is quantized according to (1.82). Therefore, the energy of the system is given by

$$E^{(N)}_{\{m_j\}}(B) = -g\mu_B B \sum_{j=1}^{N} m_j, \tag{4.94}$$

which in the case of $j = 1/2$ reads

$$E^{(N)}_{\{m_j\}}(B) = -\overline{\mu} B \sum_{j=1}^{N} \overline{m}_j, \tag{4.95}$$

with $\overline{\mu} = g\mu_B/2$ and $\overline{m}_j = \pm 1$, states that are denoted by $+$ and $-$. If the value of the energy of the system is fixed, $E^{(N)}_{\{m_j\}}(B) = E$, it is univocally determined by the number of particles in each state, N_+ and N_-, which verify the following relations:

$$E = -\overline{\mu} B (N_+ - N_-), \quad N = N_+ + N_-. \tag{4.96}$$

Note that there exists a degeneration in the energy levels if the energy is specified through N_+ and N_-. Such degeneration turns out to be

$$\frac{N!}{N_+! N_-!},\tag{4.97}$$

and when the numbers N_+ and N_- change by one unit, the variation in the absolute value of the energy is $2\overline{\mu}B$. The number of quantum states $\Omega(E, B, N; \Delta E)$ is, therefore,

$$\Omega(E, B, N; \Delta E) = \frac{\Delta E}{2\overline{\mu}B} \frac{N!}{N_+! N_-!}$$

$$= \frac{\Delta E}{2\overline{\mu}B} \frac{N!}{\left(\frac{N}{2} - \frac{E}{2\overline{\mu}B}\right)! \left(\frac{N}{2} + \frac{E}{2\overline{\mu}B}\right)!}.\tag{4.98}$$

Note that in (4.98) it has been assumed that the degeneration (4.97) is the same in the whole interval $(E, E + \Delta E)$. The justification of this approximation is based on the fact that the variation of the number of particles with orientation $-$, ΔN_- $= -\Delta N_+$, when the energy changes from E to $E + \Delta E$, is $\Delta N_- = \Delta E/2\overline{\mu}B$. If one admits that $\Delta E \ll E = O(N)$, then $\Delta N_- \ll N_-$ in the whole interval.

From (4.92) it follows that the entropy per particle in the thermodynamic limit,

$$\hat{s}(e, B) = \mathrm{TL}\left[\frac{1}{N} k_B \ln \Omega(E, B, N; \Delta E)\right],$$

is given by

$$\hat{s}(e, B) = -k_B \left\{ \left(\frac{1-x}{2}\right) \ln\left(\frac{1-x}{2}\right) + \left(\frac{1+x}{2}\right) \ln\left(\frac{1+x}{2}\right) \right\},\tag{4.99}$$

where $e = E/N < \infty$ is the energy per particle, $x = e/\overline{\mu}B$, and Stirling's approximation (4.21) has been used (Fig. 4.1).

Fig. 4.1 Entropy per particle, in units of the Boltzmann constant, of an ideal system of magnetic moments ($j = 1/2$) placed in a magnetic field B as a function of the variable $x = e/\overline{\mu}B$. When $x > 0$, the functions $\hat{s}(x)$ and $s(x)$ are represented, respectively, by a *broken line* and by a *dotted line*. When $x < 0$, the two functions are equal (*continuous line*)

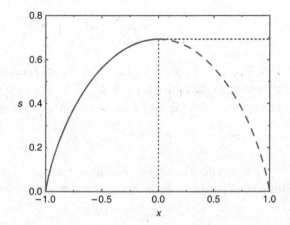

From (4.55) it follows that the absolute temperature is given by

$$\frac{1}{T} = \frac{\partial \hat{s}(e, B)}{\partial e} = \frac{k_B}{2\overline{\mu} B} \ln\left(\frac{1-x}{1+x}\right). \tag{4.100}$$

Note that if $1 - x > 1 + x$, the energy per particle is negative, and the absolute temperature is positive. On the other hand, if $1 - x < 1 + x$, the energy per particle is positive, and the absolute temperature is negative. When $x \to 0_-$, $T \to \infty$, and when $x \to 0_+$, $T \to -\infty$ Hence, the absolute negative temperatures are greater than $+\infty$. This peculiarity of a system of N particles of total angular momentum $\mathbf{j}$ is due to the fact that the entropy $\hat{S}(E, B, N)$ is not an increasing function of the energy for all E, since it has a maximum at $E = 0$.

Consider now the definition of entropy (4.91). From (4.87) and (4.88) one has

$$\phi(E, B, N) = \frac{2\overline{\mu} B}{\Delta E} \sum_{E' \leq E} \Omega(E', B, N; \Delta E), \tag{4.101}$$

which when $N > 1$ may be written as

$$\phi(E, B, N) = \frac{2\overline{\mu} B}{\Delta E} \sum_{e' \leq e} e^{N\hat{s}(e', B) + O(\ln N)}, \tag{4.102}$$

where $\hat{s}(e, B)$ is given by (4.99). Note that in the thermodynamic limit the only term that contributes to the sum in (4.102) is the term that makes the entropy per particle to be maximum, namely

$$s(e, B) = \text{TL}\left[\frac{1}{N} k_B \ln \phi(E, B, N)\right] = \max_{e' \leq e} \hat{s}(e', B). \tag{4.103}$$

When $\hat{s}(e', B)$ increases with e' (negative energies) the maximum corresponds to $e = e'$, and both definitions of entropy coincide $s(e, B) = \hat{s}(e, B)(e < 0)$. When $e' \geq 0$, $\hat{s}(e', B)$ is a decreasing function of the energy, and the global maximum is located at $e' = 0$. Therefore, $s(e, B) = \hat{s}(0, B) = k_B \ln 2(e \geq 0)$. When the energy is nonnegative, the absolute temperature defined through $s(e, B)$ is positive and infinite. This particular feature of the system of magnetic moments by which the two definitions of entropy differ is a consequence of the fact that, when $e \geq 0$, the number of quantum states does not grow upon an increase in energy (see Fig. 4.1).

Further Reading

1. J.W. Gibbs, *The Scientific Papers* (Dover Publications, New York, 1961). *Difficult to read but of overwhelming historic influence.*

2. R. Becker, *Theory of Heat,* 2nd edn. (Springer, Berlin, 1967). *A good introduction to statistical physics.*
3. A. Munster, *Statistical Thermodynamics,* vol. 1 (Springer, Berlin, 1969). *An advanced treatise of equilibrium statistical mechanics.*
4. S.G. Brush, *Statistical Physics and the Atomic Theory of Matter* (Princeton University Press, Princeton, 1983). *Contains a short history of statistical physics.*
5. C. Cercignani, *Ludwig Boltzmann: The Man Who Trusted Atoms* (Oxford University Press, London, 1998). *A biography of one of the founding fathers of statistical physics.*

Chapter 5
Canonical Ensemble

Abstract In the previous chapter the basic concepts of the microcanonical ensemble, which describes a system whose energy is constant, have been analyzed. The entropy is the thermodynamic potential associated to this ensemble. The microcanonical ensemble presents two types of difficulties. The first one is mathematical since, even in the simplest systems, determining the entropy is not a trivial issue. The second difficulty is that the systems studied in thermodynamics are, in general, not isolated. Note that, for instance, thermal equilibrium is attained when two systems exchange energy. In this chapter the statistical physics of systems whose energy is variable is analyzed.

5.1 Classical Canonical Ensemble

In the previous chapter it was shown that when a system of N_1 particles and Hamiltonian $H_{N_1}(q_1, p_1; \alpha_1)$ exchanges energy with the external world, of N_2 particles and Hamiltonian $H_{N_2}(q_2, p_2; \alpha_2)$, the probability density of the energy of the system is given by (4.47), i.e.,

$$\rho_1(E_1) = \frac{\omega_1(E_1, \alpha_1, N_1)\omega_2(E - E_1, \alpha_2, N_2)}{\omega(E, \alpha, N)}$$
$$= \omega_1(E_1, \alpha_1, N_1)\frac{\Omega_2(E - E_1, \alpha_2, N_2; \Delta E)}{\Omega(E, \alpha, N; \Delta E)} \tag{5.1}$$

where $\alpha \equiv (\alpha_1, \alpha_2)$, and in the second equality of (5.1) it has been assumed that $\Delta E \ll E - E_1$. When $N_1 \gg 1$ and $N_2 \gg 1$, $\rho_1(E_1)$ is a function with a very pronounced maximum for a certain value $\tilde{E}_1$. In this state of maximum probability, the temperatures are equal (see (4.54)). Assume now that $N_2 \gg N_1$, which is the normal situation in an experiment in physics in which the system has much less degrees of freedom than the external world. In this case it is to be expected that whenever $\rho_1(E_1)$ differs appreciably from zero, the energy E_2 is such that $E_2 \gg E_1$. Although in these circumstances it would seem logical to expand the function $\Omega_2(E - E_1, a_2, N_2; \Delta E)$ of (5.1) in a Taylor series around $E_2 = E$, such an expansion converges very slowly since this function grows very rapidly with the energy when N_2 $\gg 1$ (for instance, for an ideal gas $\Omega_2(E - E_1, \alpha_2, N_2; \Delta E) \sim (E - E_1)^{3N_2/2-1}$). In contrast, the Taylor series expansion of the logarithm of this function converges more rapidly. Therefore, let us consider the following expansion of (5.1):

$$\ln\left(\frac{\Omega_2(E - E_1, \alpha_2, N_2; \Delta E)}{\Omega_2(E, \alpha_2, N_2; \Delta E)}\right) = -\beta_2 E_1 + \cdots, \tag{5.2}$$

with

$$\beta_2 \equiv \left(\frac{\partial \ln \Omega_2(E_2, \alpha_2, N_2; \Delta E)}{\partial E_2}\right)_{E_2=E}, \tag{5.3}$$

where $T_2 = 1/k_B \beta_2$ is the absolute temperature of the external world (see (4.55)).

The probability density of the energy of the system may then be approximated by the expression:

$$\rho_1(E_1) = \frac{\Omega_2(E, \alpha_2, N_2; \Delta E)}{\Omega(E, \alpha, N; \Delta E)} \omega_1(E_1, \alpha_1, N_1) e^{-\beta_2 E_1}. \tag{5.4}$$

Taking into account the normalization condition of $\rho_1(E_1)$, (5.4) may be rewritten in the form

$$\rho_1(E_1) = \frac{1}{Z_1(\beta_2, \alpha_1, N_1)} \omega_1(E_1, \alpha_1, N_1) e^{-\beta_2 E_1}, \tag{5.5}$$

where

$$Z_1(\beta_2, \alpha_1, N_1) = \int_0^E dE_1 \omega_1(E_1, \alpha_1, N_1) e^{-\beta_2 E_1} \tag{5.6}$$

is the classical partition function of the system. Note that, except for β_2, all the variables appearing in (5.5) and (5.6) are system variables. In what follows and for the sake of simplifying the notation, all subindexes will be omitted, i.e.,

$$\rho(E) = \frac{1}{Z(\beta, \alpha, N)} \omega(E, \alpha, N) e^{-\beta E}, \tag{5.7}$$

with

$$Z(\beta, \alpha, N) = \int_0^\infty dE \omega(E, \alpha, N) e^{-\beta E}, \tag{5.8}$$

where the variables E, α, and N correspond to the system and the parameter β to the external world. Note that, in view of the pronounced maximum of $\rho(E)$, in the partition function (5.8) the upper limit in the integral (the total energy of the system) has been replaced by infinity. The ensemble described by (5.7) and (5.8) is known as the canonical ensemble and represents a system in thermal contact (i.e., which exchanges energy) with another system (the external world) which has a much greater number of degrees of freedom.

By a similar reasoning, according to (4.46),

$$\rho_1(q_1, p_1) = \frac{1}{h^{3N_1}} \frac{\Omega_2(E - H_1, \alpha_2, N_2; \Delta E)}{\Omega(E, \alpha, N; \Delta E)}, \tag{5.9}$$

so that upon performing the Taylor expansion

$$\ln\left(\frac{\Omega_2(E - H_1, \alpha_2, N_2; \Delta E)}{\Omega_2(E, \alpha_2, N_2; \Delta E)}\right) = -\beta_2 H_1 + \cdots, \tag{5.10}$$

imposing the normalization condition to $\rho_1(q_1, p_1)$, and, finally, dropping all the subindexes, which correspond in all cases to variables of the system except for the parameter β, which refers to the external world, one has

$$\rho(q, p) = \frac{1}{h^{3N}} \frac{1}{Z(\beta, \alpha, N)} e^{-\beta H_N(q, p; \alpha)}, \tag{5.11}$$

with

$$Z(\beta, \alpha, N) = \frac{1}{h^{3N}} \int dq \int dp\, e^{-\beta H_N(q, p; \alpha)}, \tag{5.12}$$

where the integration of the partition function (5.12) has been extended to the whole phase space of the system without any restriction. Note that, in order to make the partition function (5.12) dimensionless, in both $\rho(q, p)$ and $Z(\beta, \alpha, N)$ a factor $1/h^{3N}$ has been introduced. Equations (5.11) and (5.12) are the expressions of the canonical ensemble in the phase space of the system.

An immediate conclusion that shows that the canonical ensemble is mathematically simpler than the microcanonical ensemble follows from the convolution law (4.48). Since the partition function (5.8) is the Laplace transform of $\omega(E, \alpha, N)$, one has

$$Z(\beta, \alpha, N) = Z_1(\beta, \alpha_1, N_1) Z_2(\beta, \alpha_2, N_2), \tag{5.13}$$

i.e., the partition function of a system formed by two subsystems which exchange energy in an additive way is the product of the partition functions of the subsystems. Clearly, this composition law is much simpler than (4.48).

5.2 Mean Values and Fluctuations

In a system described by the canonical ensemble the energy is a random variable whose probability density is (5.7). The Hamiltonian of the system, $H_N = H_N(q, p; \alpha)$, is a function of the mechanical state of the system whose probability density in phase space is given by (5.11). One thus has

$$
\begin{aligned}
\langle H_N \rangle &= \frac{1}{h^{3N}} \frac{1}{Z(\beta, \alpha, N)} \int dq \int dp\, H_N e^{-\beta H_N} \\
&= \frac{1}{h^{3N}} \frac{1}{Z(\beta, \alpha, N)} \left(-\frac{\partial}{\partial \beta} \right) \int dq \int dp\, e^{-\beta H_N},
\end{aligned}
\tag{5.14}
$$

i.e.,

$$
\langle H_N \rangle = -\frac{1}{Z(\beta, \alpha, N)} \left(\frac{\partial}{\partial \beta} \right) Z(\beta, \alpha, N),
\tag{5.15}
$$

which is known as the energy equation.

In a similar fashion, the mean square value of H_N is determined as follows:

$$
\langle H_N^2 \rangle = \frac{1}{Z(\beta, \alpha, N)} \left(\frac{\partial^2}{\partial \beta^2} \right) Z(\beta, \alpha, N),
\tag{5.16}
$$

so that the fluctuation in the energy of the system is

$$
\langle H_N^2 \rangle - \langle H_N \rangle^2 = \frac{\partial}{\partial \beta} \left(\frac{1}{Z(\beta, \alpha, N)} \frac{\partial Z(\beta, \alpha, N)}{\partial \beta} \right),
\tag{5.17}
$$

or, alternatively,

$$
\langle H_N^2 \rangle - \langle H_N \rangle^2 = -\frac{\partial}{\partial \beta} \langle H_N \rangle.
\tag{5.18}
$$

Note that the conjugate variable A_α of the external parameter α is

$$
A_\alpha = -\left\langle \frac{\partial H_N}{\partial \alpha} \right\rangle,
\tag{5.19}
$$

i.e.,

$$
\begin{aligned}
A_\alpha &= -\frac{1}{h^{3N}} \frac{1}{Z(\beta, \alpha, N)} \int dq \int dp\, \frac{\partial H_N}{\partial \alpha} e^{-\beta H_N} \\
&= \frac{1}{\beta h^{3N}} \frac{1}{Z(\beta, \alpha, N)} \left(\frac{\partial}{\partial \alpha} \right) \int dq \int dp\, e^{-\beta H_N},
\end{aligned}
\tag{5.20}
$$

yielding

$$A_\alpha = \frac{1}{\beta Z(\beta, \alpha, N)} \frac{\partial Z(\beta, \alpha, N)}{\partial \alpha}, \tag{5.21}$$

which is known as the equation of state corresponding to the parameter α. Two particular cases of (5.21) are

$$p = k_B T \frac{\partial \ln Z(\beta, V, N)}{\partial V}, \quad \mathbf{M} = k_B T \frac{\partial \ln Z(\beta, \mathbf{B}, N)}{\partial \mathbf{B}}, \tag{5.22}$$

when the external parameter is the volume V of the system or the magnetic field $\mathbf{B}$.

5.3 Helmholtz Free Energy

As has been shown in the previous section, the energy equation and the equations of state are determined in the canonical ensemble if one knows the partition function of the system which, in this way, is a thermodynamic potential. In particular, when the external parameter is the volume V, the independent variables β, V, and N are the same as those of the Helmholtz free energy (2.26) (although in thermodynamics one takes the temperature T as independent variable, in statistical physics it is common to consider the parameter $\beta = 1/k_B T$ as the independent variable, but this should not lead to confusion). In order to establish the connection between the canonical ensemble and thermodynamics, the Helmholtz free energy $F(\beta, \alpha, N)$ is defined as

$$F(\beta, \alpha, N) = -k_B T \ln Z(\beta, \alpha, N). \tag{5.23}$$

Note that, since $N \gg 1$, (5.8) may be approximated by

$$Z(\beta, \alpha, N) = \int_0^\infty dE \omega(E, \alpha, N) e^{-\beta E}$$

$$\simeq \frac{N}{k_B} \int_0^\infty de\, s'(e, \alpha) e^{N(-\beta e + s(e, \alpha)/k_B) + O(\ln N)}, \tag{5.24}$$

where $e = E/N$ and $s(e, \alpha)$ are the energy per particle and the entropy per particle, respectively, and (4.39) has been taken into account (recall the comment made on (4.37)). The exponent of the last factor of (5.24) has a maximum at $\tilde{e}$, and so the partition function may be approximated by (see Appendix B)

$$Z(\beta, \alpha, N) \simeq \frac{N}{k_B} s'(\tilde{e}, \alpha) \sqrt{\frac{2\pi}{N|g''(\tilde{e})|}} e^{N(-\beta\tilde{e}+s(\tilde{e},\alpha)/k_B)+O(\ln N)}, \tag{5.25}$$

where $g(e) \equiv -\beta e + s(e, \alpha)/k_B$.

Therefore

$$F(\beta, \alpha, N) \simeq N(\tilde{e} - Ts(\tilde{e}, \alpha)) + O(\ln N), \tag{5.26}$$

and so the Helmholtz free energy per particle in the thermodynamic limit reads

$$f(\beta, \alpha) = \text{TL}\left[\frac{1}{N} F(\beta, \alpha, N)\right] = \tilde{e} - Ts(\tilde{e}, \alpha), \tag{5.27}$$

corresponding to (2.26) with $\tilde{e} = \tilde{e}(\beta, \alpha)$, which is the value of the energy that maximizes the function $g(e)$. The equivalence between the microcanonical ensemble (entropy) and the canonical ensemble (Helmholtz free energy) occurs at the maximum of the probability density, in which the energy per particle is $\tilde{e}$. The condition for the maximum of $g(e)$ reads

$$\frac{1}{T} = \left(\frac{\partial s(e, \alpha)}{\partial e}\right)_{e=\tilde{e}}, \tag{5.28}$$

i.e., the temperatures of the system and of the external world are equal at the maximum.

5.4 Classical Ideal Gas

As a simple application, consider an ideal gas of N identical particles of mass m in a closed region R of volume V at the temperature $T = 1/k_B\beta$. Due to the identical nature of the particles, the one-particle partition function is the same for all of them, and so from (5.13) it follows that the partition function of the gas is given by

$$Z(\beta, V, N) = [Z(\beta, V, 1)]^N, \tag{5.29}$$

where $Z(\beta, V, 1)$ is the one-particle partition function, namely

$$Z(\beta, V, 1) = \frac{1}{h^3} \int d\mathbf{r} \int d\mathbf{p} e^{-\beta[\mathbf{p}^2/2m + \phi_R(\mathbf{r})]} = \frac{V}{\Lambda^3}, \tag{5.30}$$

and

$$\Lambda = \frac{h}{\sqrt{2\pi m k_B T}}, \tag{5.31}$$

is the thermal de Broglie wavelength associated to the particle. This example shows that, in a classical ideal gas, in order to determine the thermodynamic properties in the canonical ensemble (the Helmholtz free energy), one only needs to compute the Gaussian integral in (5.30). In the microcanonical ensemble the volume of the 3 N-dimensional hypersphere was required to determine the entropy $S(E, V, N)$. Thus, the simplification obtained with the canonical ensemble is evident.

The Helmholtz free energy (5.23) then reads

$$F(\beta, V, N) = -k_B T \ln Z(\beta, V, N) = -N k_B T \ln\left(\frac{V}{\Lambda^3}\right), \qquad (5.32)$$

which is not an extensive variable in the thermodynamic limit. The reason behind this anomaly has again to do with the fact that the particles of the gas have been taken to be distinguishable (Gibbs paradox). Note that, as in the microcanonical ensemble, the paradox may be solved in an ad hoc manner if the canonical ensemble (5.11) and (5.12) is written as

$$\rho(q, p) = \frac{1}{N! h^{3N}} \frac{1}{Z^*(\beta, V, N)} e^{-\beta H_N(q, p; V)}, \qquad (5.33)$$

with

$$Z^*(\beta, V, N) = \frac{1}{N! h^{3N}} \int dq \int dp \, e^{-\beta H_N(q, p; V)}, \qquad (5.34)$$

which does not alter the average values taken over the ensemble.

From (5.34) it follows that the partition function of the ideal gas becomes

$$Z^*(\beta, V, N) = \frac{1}{N!} [Z(\beta, V, 1)]^N, \qquad (5.35)$$

so that upon using the Stirling's approximation, it follows that the Helmholtz free energy, defined through $Z^*(\beta, V, N)$, is given by

$$F(\beta, V, N) = -k_B T \ln Z^*(\beta, V, N)$$
$$= N k_B T \left(\ln\left(\frac{N \Lambda^3}{V}\right) - 1\right) + O(\ln N), \qquad (5.36)$$

and the Helmholtz free energy per particle $f(T, \rho)$ is finite in the thermodynamic limit

$$f(T, \rho) = \mathrm{TL}\left[\frac{F(\beta, V, N)}{N}\right] = k_B T \left(\ln\left(\rho \Lambda^3\right) - 1\right), \qquad (5.37)$$

where $\rho = N/V < \infty$.

5.5   Ideal Gas in an External Potential

Consider an ideal gas of N particles of mass m contained in a closed region R of volume V at temperature T in an external potential $\phi(\mathbf{r})$. The one-particle partition function is

$$
\begin{aligned}
Z(\beta, V, 1) &= \frac{1}{h^3} \int_R d\mathbf{r} \int d\mathbf{p}\, e^{-\beta[p^2/2m + \phi_R(\mathbf{r}) + \phi(\mathbf{r})]} \\
&= \left(\frac{V}{\Lambda^3}\right) \frac{1}{V} \int_R d\mathbf{r}\, e^{-\beta\phi(\mathbf{r})},
\end{aligned}
\tag{5.38}
$$

and that of the gas is given by

$$
Z^*(\beta, V, N) = \frac{1}{N!} \left(\frac{V}{\Lambda^3}\right)^N \left(\frac{1}{V} \int_R d\mathbf{r}\, e^{-\beta\phi(\mathbf{r})}\right)^N .
\tag{5.39}
$$

Therefore, the Helmholtz free energy $F(\beta, V, N) = -k_B T \ln Z*(\beta, V, N)$ reads

$$
\begin{aligned}
\beta F(\beta, V, N) &= -\ln\left(\frac{1}{N!}\left(\frac{V}{\Lambda^3}\right)^N\right) - N\ln\left(\frac{1}{V}\int_R d\mathbf{r}\, e^{-\beta\phi(\mathbf{r})}\right) \\
&\equiv \beta F[\phi].
\end{aligned}
\tag{5.40}
$$

Note that in order to determine the Helmholtz free energy one has to know the value of the external potential $\phi(\mathbf{r})$ at all points of the Euclidean space that are contained in the region R, i.e., $F(\beta, V, N)$ is a functional of $\phi(\mathbf{r})$, which is denoted by $F[\phi]$ (see Appendix C). On the other hand, the external potential induces a non-uniform local density of particles $\rho_1(\mathbf{r})$. It can be easily seen that

$$
\rho_1(\mathbf{r}) = \left\langle \sum_{j=1}^{N} \delta(\mathbf{r} - \mathbf{r}_j) \right\rangle = N \frac{e^{-\beta\phi(\mathbf{r})}}{\int_R d\mathbf{r}' e^{-\beta\phi(\mathbf{r}')}},
\tag{5.41}
$$

where the average value has been determined with the canonical ensemble. In the case of a uniform gravitation field $\phi(\mathbf{r}) = mgz$, (5.41) yields the well-known barometric formula.

As the first functional derivative of $F[\phi]$ with respect to the external potential, $\delta F[\phi]/\delta\phi(\mathbf{r})$ is defined by (see Appendix C)

$$
\delta F[\phi] \equiv F[\phi + \delta\phi] - F[\phi] = \int_R d\mathbf{r}\, \frac{\delta F[\phi]}{\delta\phi(\mathbf{r})} \delta\phi(\mathbf{r}),
\tag{5.42}
$$

from (5.40) one has

$$
\begin{aligned}
\beta \delta F[\phi] &= -N \ln \left(\frac{\int_R d\mathbf{r}\, e^{-\beta[\phi(\mathbf{r}) + \delta\phi(\mathbf{r})]}}{\int_R d\mathbf{r}'\, e^{-\beta\phi(\mathbf{r}')}} \right) \\
&= -N \ln \left(\frac{\int_R d\mathbf{r}\, e^{-\beta\phi(\mathbf{r})}(1 - \beta\delta\phi(\mathbf{r}) + \cdots)}{\int_R d\mathbf{r}'\, e^{-\beta\phi(\mathbf{r}')}} \right) \\
&= -N \ln \left(1 - \beta \frac{\int_R d\mathbf{r}\, e^{-\beta\phi(\mathbf{r})}\delta\phi(\mathbf{r})}{\int_R d\mathbf{r}'\, e^{-\beta\phi(\mathbf{r}')}} + \cdots \right) \\
&= N\beta \frac{\int_R d\mathbf{r}\, e^{-\beta\phi(\mathbf{r})}\delta\phi(\mathbf{r})}{\int_R d\mathbf{r}'\, e^{-\beta\phi(\mathbf{r}')}} + \cdots
\end{aligned}
\tag{5.43}
$$

or

$$
\beta \delta F[\phi] = \beta \int_R d\mathbf{r}\, \rho_1(\mathbf{r})\delta\phi(\mathbf{r}) + \cdots
$$

i.e.,

$$
\frac{\delta F[\phi]}{\delta\phi(\mathbf{r})} = \rho_1(\mathbf{r}),
\tag{5.44}
$$

where use has been made of (5.41) and (5.42). Therefore, the functional derivative of the Helmholtz free energy with respect to the external potential is the local density of particles.

The concept of Legendre transform may now be generalized to the case of functionals, by introducing the intrinsic Helmholtz free energy functional $\mathcal{F}[\rho_1]$:

$$
\mathcal{F}[\rho_1] = F[\phi] - \int_R d\mathbf{r}\,\phi(\mathbf{r})\rho_1(\mathbf{r}),
\tag{5.45}
$$

i.e., in the new functional $\mathcal{F}[\rho_1]$ the local density of particles is the independent variable. From (5.41) it follows that

$$
\phi(\mathbf{r}) = k_B T \left(-\ln \rho_1(\mathbf{r}) + \ln N - \ln \int_R d\mathbf{r}'\, e^{-\beta\phi(\mathbf{r}')} \right).
\tag{5.46}
$$

Upon substitution of (5.46) into (5.45) and taking into account that

$$
\int_R d\mathbf{r}\,\rho_1(\mathbf{r}) = N
$$

one obtains from (5.40)

$$\mathcal{F}[\rho_1] = k_B T \int_R d\mathbf{r} \rho_1(\mathbf{r}) \left(\ln\left(\rho_1(\mathbf{r})\Lambda^3\right) - 1 \right) + O(\ln N), \tag{5.47}$$

where the Stirling's approximation has been used. In the thermodynamic limit (5.47) reduces to

$$\mathcal{F}[\rho_1] = k_B T \int d\mathbf{r} \rho_1(\mathbf{r}) \left(\ln\left(\rho_1(\mathbf{r})\Lambda^3\right) - 1 \right). \tag{5.48}$$

Note that in a homogeneous system ($\phi(\mathbf{r}) = 0$), $\rho_1(\mathbf{r}) = \rho$, the intrinsic Helmholtz free energy functional becomes a function of the density of the gas and (5.48) reduces to (5.37).

It must be pointed out that, in order to derive (5.48), the existence of an external potential inducing a non-uniform local density of particles has been assumed. This constitutes an example of a symmetrybreaking field considered in Chap. 3. As shown in that chapter, when taking first the thermodynamic limit and subsequently the zero-field limit, it is possible to find phases that do not have the same symmetry as the Hamiltonian. In particular, in the study of non-uniform phases (a solid) or inhomogeneous systems (an interface), (5.48) is the ideal part of the intrinsic Helmholtz free energy functional.

5.6 Equipartition Theorem

In the previous chapter the equipartition theorem in the microcanonical ensemble, according to which the average kinetic energy per degree of freedom is equal to $k_B T/2$, has been derived. In order to obtain this theorem in the canonical ensemble, consider a system of N particles whose Hamiltonian (see (4.60)) is the sum of the kinetic energy H_N^{id}, of the potential energy H_N^{int}, and the interaction energy between the particles and the container H_N^R. Due to the additivity of the energy, the partition function may be factorized with the result where

$$Z(\beta, V, N) = \frac{1}{h^{3N}} \int d\mathbf{r}^N \int d\mathbf{p}^N e^{-\beta(H_N id + H_N^{int} + H_N^R)}$$
$$= Z^{id}(\beta, V, N) Z^{int}(\beta, V, N), \tag{5.49}$$

where

$$Z^{id}(\beta, V, N) = [Z(\beta, V, 1)]^N, \tag{5.50}$$

is the ideal partition function (5.29) and (5.30), and

$$Z^{int}(\beta, V, N) = \frac{1}{V^N} \int d\mathbf{r}^N e^{-\beta(H_N^{int} + H_N^R)} \tag{5.51}$$

is the partition function of the interactions. The average kinetic energy of the system is given by

$$
\begin{aligned}
\langle H_N^{id} \rangle &= \frac{1}{h^{3N} Z(\beta, V, N)} \int d\mathbf{r}^N \int d\mathbf{p}^N H_N^{id} e^{-\beta(H_N^{id}+H_N^{int}+H_N^R)} \\
&= -\frac{1}{\beta h^{3N} Z(\beta, V, N)} \left(\frac{\partial}{\partial \lambda} \int d\mathbf{r}^N \int \mathbf{p} \mathbf{p}^N e^{-\beta(\lambda H_N^{id}+H_N^{int}+H_N^R)} \right)_{\lambda=1} \\
&= -\frac{1}{\beta Z(\beta, V, N)} \left(\frac{\partial}{\partial \lambda} Z^{id}(\lambda\beta, V, N) Z^{int}(\beta, V, N) \right)_{\lambda=1},
\end{aligned}
$$

so that from (5.49) one has

$$
\langle H_N^{id} \rangle = -\frac{1}{\beta} \left(\frac{\partial}{\partial \lambda} \ln Z^{id}(\lambda\beta, V, N) \right)_{\lambda=1}, \tag{5.52}
$$

or, alternatively, from (5.50)

$$
\langle H_N^{id} \rangle = -\frac{N}{\beta} \left(\frac{\partial}{\partial \lambda} \ln Z(\lambda\beta, V, 1) \right)_{\lambda=1}. \tag{5.53}
$$

The derivative in (5.53) may be readily evaluated since $Z(\beta, V, 1) = V/\Lambda^3$ leading to the equipartition theorem:

$$
\langle H_N^{id} \rangle = \frac{3}{2} N k_B T. \tag{5.54}
$$

Note that this result is a consequence of the factorization of the partition functions (5.49) and (5.50).

Another interesting example is that of a system of N harmonic oscillators whose Hamiltonian is given by the Einstein model (4.31). Since the energy is additive (ideal system), the partition function may be factorized and the average energy, in analogy with (5.53), is determined as

$$
\langle H_N \rangle = -\frac{N}{\beta} \left(\frac{\partial}{\partial \lambda} \ln Z(\lambda\beta, \omega, 1) \right)_{\lambda=1}. \tag{5.55}
$$

In this case, the partition function of a single harmonic oscillator is given by

$$
\begin{aligned}
Z(\beta, \omega, 1) &= \frac{1}{h^3} \int d\mathbf{r} \int d\mathbf{p} e^{-\beta(\mathbf{p}^2/2m+m\omega^2(\mathbf{r}-\mathbf{R})^2/2)} \\
&= \frac{1}{\Lambda^3} \left(\frac{2\pi}{\beta m \omega^2} \right)^{3/2}, \tag{5.56}
\end{aligned}
$$

and so, from (5.55) and (5.56) one finds

$$\langle H_N \rangle = 3Nk_BT, \tag{5.57}$$

which shows that the average energy of each one-dimensional oscillator is k_BT. As was already pointed out in the previous chapter, the equipartition theorem is a result of classical statistical physics which has no counterpart in quantum statistical physics. In the following sections it is shown that, when applied to radiation or to a harmonic solid, the equipartition theorem is at odds with experimental results. Therefore, these systems should be treated within quantum statistical physics.

5.7 Classical Theory of Radiation

Consider Maxwell's equations for the electromagnetic field in vacuum in a closed region R of volume V at the absolute temperature T. If $\mathbf{E} \equiv \mathbf{E}(\mathbf{r}, t)$ and $\mathbf{B} \equiv \mathbf{B}(\mathbf{r}, t)$ are the electric and magnetic fields, respectively, these equations read

$$\nabla \times \mathbf{E} = -\frac{1}{c}\frac{\partial \mathbf{B}}{\partial t}, \; \nabla \times \mathbf{B} = \frac{1}{c}\frac{\partial \mathbf{E}}{\partial t}, \; \nabla \cdot \mathbf{E} = 0, \; \nabla \cdot \mathbf{B} = 0. \tag{5.58}$$

where c is the velocity of light in vacuum.

Since $\mathbf{E}$ and $\mathbf{B}$ are related with the vector potential $\mathbf{A} \equiv \mathbf{A}(\mathbf{r}, t)$ through the expressions:

$$\mathbf{B} = \nabla \times \mathbf{A}, \; \mathbf{E} = -\frac{1}{c}\frac{\partial \mathbf{A}}{\partial t}, \tag{5.59}$$

Maxwell's Equations (5.58) may be condensed into two equations, namely the D' Alembert equation or wave equation

$$\nabla^2 \mathbf{A} - \frac{1}{c^2}\frac{\partial^2 \mathbf{A}}{\partial t^2} = 0, \tag{5.60}$$

and the transversality condition:

$$\nabla \cdot \mathbf{A} = 0. \tag{5.61}$$

As usual, consider plane-wave type solutions to (5.60) and (5.61), i.e.,

$$\hat{\varepsilon}_\alpha(\mathbf{k})e^{i(\mathbf{k}\cdot\mathbf{r}-\omega t)}, \tag{5.62}$$

where $\hat{\varepsilon}_\alpha(\mathbf{k})$ is the polarization vector, which is a real unit vector whose direction depends on $\mathbf{k}$. The transversality condition then reads

$$\mathbf{k} \cdot \hat{\varepsilon}_\alpha(\mathbf{k}) = 0, \tag{5.63}$$

which indicates that, for a fixed wave vector $\mathbf{k}$, there are two independent polarization states, since $\hat{\varepsilon}_\alpha(\mathbf{k})$ and $\mathbf{k}$ are orthogonal. On the other hand, upon substitution of (5.62) into (5.60) one finds:

$$\omega = kc, \tag{5.64}$$

which is the dispersion relation, with $k = |\mathbf{k}|$.

The general solution to the wave equation is a real linear superposition of plane waves (5.62) with coefficients $c_{\mathbf{k},\alpha}$, i.e.,

$$\mathbf{A} = \sqrt{\frac{4\pi}{V}} \sum_{\mathbf{k}} \sum_{\alpha} \left\{ c_{\mathbf{k},\alpha} \hat{\varepsilon}_\alpha(\mathbf{k}) e^{i(\mathbf{k}\cdot\mathbf{r}-\omega t)} + c^*_{\mathbf{k},\alpha} \hat{\varepsilon}_\alpha(\mathbf{k}) e^{-i(\mathbf{k}\cdot\mathbf{r}-\omega t)} \right\}, \tag{5.65}$$

and so from (5.59) the fields $\mathbf{E}$ and $\mathbf{B}$ and the Hamiltonian of the electromagnetic field may be determined, namely

$$H = \frac{1}{8\pi} \int_R d\mathbf{r} \left(|\mathbf{E}|^2 + |\mathbf{B}|^2 \right). \tag{5.66}$$

If $\mathbf{A}$ is periodic of period L ($L^3 = V$) in the three directions of space (see below) and one takes into account the fact that

$$\frac{1}{V} \int_R d\mathbf{r} \left\{ \hat{\varepsilon}_\alpha(\mathbf{k}) \cdot \hat{\varepsilon}_{\alpha'}(\mathbf{k}') \right\} e^{i(\mathbf{k}-\mathbf{k}')\cdot\mathbf{r}} = \delta_{\alpha\alpha'} \delta_{\mathbf{k}\mathbf{k}'},$$

then (5.66) becomes

$$H = \sum_{\mathbf{k}} \sum_{\alpha} 2 \left(\frac{\omega}{c} \right)^2 c_{\mathbf{k},\alpha}(t) c^*_{\mathbf{k},\alpha}(t), \tag{5.67}$$

where

$$c_{\mathbf{k},\alpha}(t) = c_{\mathbf{k},\alpha} e^{-i\omega t}. \tag{5.68}$$

Introducing the variables

$$P_{\mathbf{k},\alpha}(t) = -\frac{i\omega}{c} \left\{ c_{\mathbf{k},\alpha}(t) - c^*_{\mathbf{k},\alpha}(t) \right\}, \tag{5.69}$$

and

$$Q_{\mathbf{k},\alpha}(t) = \frac{1}{c} \left\{ c_{\mathbf{k},\alpha}(t) + c^*_{\mathbf{k},\alpha}(t) \right\}, \tag{5.70}$$

the energy of the electromagnetic field is given by

$$H = \sum_{k} \sum_{\alpha} \frac{1}{2} \{ P_{k,\alpha}^2(t) + \omega^2 Q_{k,\alpha}^2(t) \}, \tag{5.71}$$

which is the Hamiltonian of a system of uncoupled harmonic oscillators, since $P_{k,\alpha}$ and $Q_{k,\alpha}$ are seen to be canonical variables, i.e.,

$$\dot{P}_{k,\alpha} = -\frac{\partial H}{\partial Q_{k,\alpha}}, \quad \dot{Q}_{k,\alpha} = \frac{\partial H}{\partial P_{k,\alpha}}. \tag{5.72}$$

Up to this stage, the result (5.71) is independent of the nature of the statistics. If the radiation is treated classically, the average energy of each oscillator is, because of the equipartition theorem, $k_B T$. In order to determine the number of oscillators, assume that the region R is a cubic box whose side is L and that the vector potential is periodic, of period L, in the three directions of space. The number of oscillators for which the modulus of the wave vector k lies between k and $k + dk$ is, therefore,

$$\rho(k)dk = \frac{V}{8\pi^3} 4\pi k^2 dk = \frac{V}{2\pi^2} k^2 dk, \tag{5.73}$$

(note that, due to the periodicity condition, this result coincides with the one derived for the number of quantum states for a free particle (1.66)).

It is important to realize that, for each oscillator with wave vector $\mathbf{k}$, there are two independent polarization states, as has been deduced from the transversality condition. Starting from the dispersion relation (5.64) and from (5.73) one finds that the total number of oscillators with angular frequency between ω and $\omega + d\omega$ is given by

$$\rho(\omega)d\omega = 2\frac{V}{2\pi^2} \frac{1}{c^3} \omega^2 d\omega = \frac{V}{\pi^2 c^3} \omega^2 d\omega. \tag{5.74}$$

The average energy of the radiation in this interval of frequencies is, therefore,

$$e(\omega)d\omega = \frac{V}{\pi^2 c^3} k_B T \omega^2 d\omega, \tag{5.75}$$

and so the heat capacity of the radiation C_V is given by

$$C_V = \frac{\partial E}{\partial T} = \frac{\partial}{\partial T} \int_0^\infty d\omega \, e(\omega) = k_B \frac{V}{\pi^2 c^3} \int_0^\infty d\omega \omega^2 = \infty, \tag{5.76}$$

an anomaly known as the ultraviolet catastrophe. This result, as will be shown later in Sect. 5.14, is a consequence of the application of the equipartition theorem.

5.8 Classical Theory of Solids

The crystalline structure of a solid of N atoms, whose position vectors are $\mathbf{r}_j$ ($j = 1, 2, \ldots, N$), is determined by its potential energy $H_N^{\text{int}}(\mathbf{r}^N)$, which is a sum of pair potentials (1.4). From a classical point of view, the state of minimum energy of the solid corresponds to a configuration in which the atoms are localized at the equilibrium positions $\mathbf{R}_j$ ($j = 1, 2, \ldots N$) of a Bravais lattice. This static image of the structure of a solid is inadequate because the atoms experience deviations with respect to their equilibrium positions. If $\mathbf{u}_j = \mathbf{u}_j(\mathbf{R}_j) = \mathbf{r}_j - \mathbf{R}_j$ denotes the displacement vector of atom j with respect to its equilibrium position, the potential energy may then be written as

$$H_N^{\text{int}}(\mathbf{r}^N) = \frac{1}{2} \sum_{i=1}^{N} \sum_{i \neq j=1}^{N} V\left(|\mathbf{R}_i - \mathbf{R}_j + \mathbf{u}_i - \mathbf{u}_j|\right). \tag{5.77}$$

When the relative displacements of the atoms $\mathbf{u}_i - \mathbf{u}_j$ are small compared with the lattice constant, the potential energy (5.77) may be expanded in a Taylor series up to second order (an approximation called the "harmonic approximation"), namely

$$\begin{aligned} H_N^{\text{int}}(\mathbf{r}^N) = {} & \frac{1}{2} \sum_{i=1}^{N} \sum_{i \neq j=1}^{N} V\left(|\mathbf{R}_i - \mathbf{R}_j|\right) \\ & + \frac{1}{4} \sum_{i=1}^{N} \sum_{i \neq j=1}^{B} (\mathbf{u}_i - \mathbf{u}_j) \cdot \frac{\partial^2 V\left(|\mathbf{R}_i - \mathbf{R}_j|\right)}{\partial \mathbf{R}_i \partial \mathbf{R}_j} \cdot (\mathbf{u}_i - \mathbf{u}_j), \end{aligned}$$

which does not contain the linear term because the equilibrium configuration is a minimum of the potential energy and which may be rewritten as

$$H_N^{\text{int}}(\mathbf{r}^N) = H_N^{\text{int}}(\mathbf{R}^N) + \frac{1}{4} \sum_{i=1}^{N} \sum_{i \neq j=1}^{N} (\mathbf{u}_i - \mathbf{u}_j) \cdot \mathcal{U}(\mathbf{R}_i - \mathbf{R}_j) \cdot (\mathbf{u}_i - \mathbf{u}_j), \tag{5.78}$$

where $H_N^{\text{int}}(\mathbf{R}^N)$ is the potential energy when the particles are at their equilibrium positions, and $\mathcal{U}(\mathbf{R})$ is the second-order tensor,

$$\mathcal{U}^{\mu\nu}(\mathbf{R}) = \frac{\partial^2 V(|\mathbf{R}|)}{\partial R^\mu \partial R^\nu}, \tag{5.79}$$

with R^μ the Cartesian components of $\mathbf{R}$.

After development of the products in the harmonic term of (5.78) one finds

$$H_N^{\text{int}}(\mathbf{r}^N) = H_N^{\text{int}}(\mathbf{R}^N) + \frac{1}{2} \sum_{i=1}^{N} \sum_{j=1}^{N} \mathbf{u}_i \cdot \mathcal{D}(\mathbf{R}_i - \mathbf{R}_j) \cdot \mathbf{u}_j, \tag{5.80}$$

where the displacement tensor $\mathcal{D}(\mathbf{R})$ has been defined through (note that in (5.80) the double sum includes the terms $i = j$, which do not appear in (5.78))

$$\mathcal{D}(\mathbf{R}) = \delta_{\mathbf{R},0} \sum_{\mathbf{R}'} \mathcal{U}(\mathbf{R}') - \mathcal{U}(\mathbf{R}), \tag{5.81}$$

which has the following properties:

$$\mathcal{D}(\mathbf{R}) = \mathcal{D}(-\mathbf{R}), \quad \mathcal{D}^{\mu\nu}(\mathbf{R}) = \mathcal{D}^{\nu\mu}(\mathbf{R}), \quad \sum_{\mathbf{R}} \mathcal{D}(\mathbf{R}) = 0. \tag{5.82}$$

If one adds to (5.80) the kinetic energy of the atoms, of mass m, the corresponding Hamilton's equations read (note that the mechanical state of the system is described by the momenta $\mathbf{p}_i$ and by the displacement vectors $\mathbf{u}_i$)

$$\dot{\mathbf{u}}_i = \frac{\mathbf{p}_i}{m}, \quad \dot{\mathbf{p}}_i = -\sum_{j=1}^{N} \mathcal{D}(\mathbf{R}_i - \mathbf{R}_j) \cdot \mathbf{u}_j, \tag{5.83}$$

or, alternatively,

$$m\,\ddot{\mathbf{u}}_i = -\sum_{j=1}^{N} \mathcal{D}(\mathbf{R}_i - \mathbf{R}_j) \cdot \mathbf{u}_j, \tag{5.84}$$

which are the equations of motion of the displacement vectors $\mathbf{u}_i$.

It is worth noting the analogy between the Hamiltonian (5.67) of the electromagnetic field in vacuum and the harmonic term in the potential energy of the solid (5.80), which are both positive definite quadratic forms. In the solid, since the displacement tensor is real and symmetric, it is always possible to find a coordinate transformation in which the transformed tensor is diagonal. In this coordinate system, the Hamiltonian is the sum of the potential energy at the equilibrium positions and of a set of $3N$ one-dimensional uncoupled harmonic oscillators (normal modes). The average energy of the solid is thus given by

$$E = H_N^{\text{int}}(\mathbf{R}^N) + 3Nk_BT, \tag{5.85}$$

since the average energy of each normal mode is k_BT. Hence, the heat capacity of the solid C_V turns out to be

$$C_V = \frac{\partial E}{\partial T} = 3Nk_B, \tag{5.86}$$

which is independent of the temperature (Dulong–Petit law). This result is a good approximation at high temperatures, but the experiments indicate that at low temperatures $C_V \sim T^3$. Note that the possible deviations from the Dulong–Petit law when anharmonic terms are included should be relevant at high temperatures, since one would expect that under such conditions the relative displacements of the atoms should increase. From the previous argument it follows that the Dulong–Petit law is incorrect at low temperatures because of the application of the equipartition theorem and is not due to the harmonic approximation considered for the atomic oscillations.

5.9 Quantum Canonical Ensemble

In the previous two sections, it has been shown that the application of the equipartition theorem to the energy of the electromagnetic radiation in vacuum and to the solid in the harmonic approximation leads to results that are completely at odds with the experimental findings. Therefore, in this section the quantum statistical treatment of a system that exchanges energy with the external world, i.e., the quantum canonical ensemble, is discussed.

The derivation of the canonical ensemble from the microcanonical ensemble in classical statistical physics was based on the fact that the energy was additive and that the system had much fewer degrees of freedom than the external world. In this way, from the probability density (4.1), which is uniform on the energy surface, the probability density (5.11) was obtained in which the weight of the different mechanical states is proportional to $e^{-\beta H_N(q,p;\alpha)}$, where $H_N(q, p; \alpha)$ is the Hamiltonian of the system and $T = 1/k_B\beta$ the temperature of the external world. Since the concepts that allow one to derive one ensemble from the other are also valid in quantum mechanics, from the quantum microcanonical ensemble, in which all the quantum states of the Hamiltonian operator (4.85) whose energies $E_n^{(N)}(\alpha)$ lie in the interval between E and $E + \Delta E$ have the same probability (4.89), one obtains the quantum canonical ensemble, in which the density operator is proportional to $e^{-\beta \hat{H}_N(\alpha)}$, where $\hat{H}_N(\alpha)$ is the Hamiltonian operator. Hence, one postulates that the density operator $\hat{\rho}$ of a system of N particles that exchanges energy with the external world at the temperature $T = 1/k_B\beta$ is diagonal in the energy representation and that its elements are given by

$$\rho_{mn} = \langle \Psi_m | \hat{\rho} | \Psi_n \rangle = \frac{\delta_{mn}}{Z(\beta, \alpha, N)} e^{-\beta E_n^{(N)}(\alpha)}, \tag{5.87}$$

where n indicates the quantum state that corresponds to the eigenvector $|\Psi_n\rangle$ and whose energy is $E_n^{(N)}(\alpha)$. The normalization condition (3.30) of the density operator implies that

$$Z(\beta, \alpha, N) = \sum_n e^{-\beta E_n^{(N)}(\alpha)}, \tag{5.88}$$

where the sum extends over all quantum states of the eigenvalue Eq. (4.85). Equations (5.87) and (5.88) define the quantum canonical ensemble, where $Z(\beta, \alpha, N)$

is the quantum partition function. Note the analogy between (5.12) and (5.88). In the first case, one integrates over all the mechanical states in the phase space of the system, whereas in the second the sum is over all the quantum states of the eigenvalue equation of the Hamiltonian operator. Note also that the partition function (5.88) is the Laplace transform of the density of quantum states (4.87), namely

$$Z(\beta, \alpha, N) = \int_0^\infty dE \omega(E, \alpha, N) e^{-\beta E}$$

$$= \sum_n \int_0^\infty dE \delta(E - E_n^{(N)}(\alpha)) e^{-\beta E}$$

$$= \sum_n e^{-\beta E_n^{(N)}(\alpha)}. \tag{5.89}$$

As pointed out in Sect. 4.9, the convolution law for the density of quantum states cannot be used to determine the thermodynamic properties of an ideal quantum gas from $\omega(E, V, 1)$. For this reason, the partition function (5.89) of a quantum ideal gas does not factorize as the product of one-particle partition functions, and hence there is no equipartition theorem in quantum statistical physics (see Sect. 5.6).

Note, finally, that the expressions for the average energy (5.15), the energy fluctuation (5.17), and the equation of state (5.21) of a classical system may be used for a quantum system provided the partition function is determined with (5.88). For instance, consider the average value of the Hamiltonian operator which, according to (3.28), with $\hat{H}_N \equiv \hat{H}_N(\alpha)$, reads

$$\langle \hat{H}_N \rangle = \sum_n \langle \Psi_n | \hat{H}_N \hat{\rho} | \Psi_n \rangle = \sum_n \sum_m \langle \Psi_n | \hat{H}_N | \Psi_n \rangle \langle \Psi_m | \hat{\rho} | \Psi_n \rangle$$

$$= \frac{1}{Z(\beta, \alpha, N)} \sum_n E_n^{(N)}(\alpha) e^{-\beta E_n^{(N)}(\alpha)}$$

$$= -\frac{1}{Z(\beta, \alpha, N)} \frac{\partial}{\partial \beta} \sum_n e^{-\beta E_n^{(N)}(\alpha)}$$

$$= -\frac{1}{Z(\beta, \alpha, N)} \left(\frac{\partial}{\partial \beta} \right) Z(\beta, \alpha, N), \tag{5.90}$$

which coincides with (5.15).

Similarly, the Helmholtz free energy of a quantum system is defined by (5.23) when the partition function is determined using (5.88).

5.10   Ideal Quantum Systems

Consider a system of N particles defined by a Hamiltonian operator $\hat{H}_N(\alpha)$ which is the sum of one-particle operators (see Sect. 1.9), i.e.,

$$\hat{H}_N(\alpha) = \sum_{j=1}^{N} \hat{H}_1^j(\alpha). \tag{5.91}$$

If the particles are identical, the operators $\hat{H}_1^j(\alpha) \equiv \hat{H}_1(\alpha)$ have the same eigenvalue equation, which will be written as

$$\hat{H}_1(\alpha)|\phi_i\rangle = \varepsilon_i(\alpha)|\phi_i\rangle, \tag{5.92}$$

where it has been assumed that the spectrum of $\hat{H}_1(\alpha)$ is discrete. Once the eigenvalues $\varepsilon_i(\alpha)$ are known, the quantum state of the system is completely characterized when the occupation numbers of the particle quantum states are specified. For a given configuration $\{n_i\}$ the energy of the system is given by

$$E_{\{n_i\}}^{(N)}(\alpha) = \sum_i {}^* n_i \varepsilon_i(\alpha), \tag{5.93}$$

and so the partition function may be written as

$$Z(\beta, \alpha, N) = \sum_{\{n_i\}} {}^* e^{-\beta \sum_i n_i \varepsilon_i(\alpha)}, \tag{5.94}$$

that is a sum over all configurations $\{n_i\}$ which, due to the conservation of the number of particles, satisfy the following condition:

$$\sum_i n_i = N. \tag{5.95}$$

This restriction has been indicated with an asterisk in (5.93) and (5.94).

The evaluation of (5.94) is not simple due to (5.95). Further, one must also bear in mind that for bosons and fermions (indistinguishable particles) only those quantum states having a well-defined symmetry have to be considered. Indeed, while for bosons the wave function is symmetric under the permutation of any pair of particles, and the occupation numbers may take any integer value $0 \le n_i \le N$, for fermions the wave function is antisymmetric under the exchange of any two particles, and the occupation numbers may only take the values $n_i = 0, 1$ due to Pauli's exclusion principle.

As will be analyzed in the next section, there is, however, one especially interesting case in which (5.94) may be exactly determined in spite of the restriction (5.95).

5.11   Maxwell–Boltzmann Statistics

Consider a system of identical distinguishable particles, i.e., particles that may be numbered. Note that, in this case, the mere specification of the particle occupation numbers n_i is not enough to determine the quantum state of the system. For instance, let $n_1 = 1$, $n_2 = N - 1$, $n_i = 0$ $(i \geq 3)$. Since the particles are distinguishable, the quantum state of the system will only be completely specified once one knows which particle is in the energy state $\varepsilon_1(\alpha)$ (the first one, the second one, etc.). There are, therefore, N possible quantum states that correspond to this configuration. Since all these states have the same energy, the fact of considering distinguishable particles introduces a degeneration of the energy levels of the system when these are specified by the occupation numbers n_i. In general, given a configuration $\{n_i\}$ the degeneration of the quantum state of the system is given by

$$\frac{N!}{\Pi_i n_i!}, \tag{5.96}$$

and so the partition function turns out to be

$$Z(\beta, \alpha, N) = \sum_{\{n_i\}}{}^* \frac{N!}{\Pi_i n_i!} e^{-\beta \sum_i n_i \varepsilon_i(\alpha)}$$

$$= \sum_{\{n_i\}} \frac{N!}{n_1! \dots n_r! \dots} \left(e^{-\beta \varepsilon_1(\alpha)}\right)^{n_1} \dots \left(e^{-\beta \varepsilon_r(\alpha)}\right)^{n_r} \dots \tag{5.97}$$

It is seen that (5.97), when the numbers n_i satisfy the condition (5.95), is Newton's expansion of the multinomial:

$$Z(\beta, \alpha, N) = \left(e^{-\beta \varepsilon_1(\alpha)} + \dots + e^{-\beta \varepsilon_r(\alpha)} + \dots\right)^N = [Z(\beta, \alpha, 1)]^N, \tag{5.98}$$

where

$$Z(\beta, \alpha, 1) = \sum_r e^{-\beta \varepsilon_r(\alpha)} \tag{5.99}$$

is the one-particle partition function. Note that (5.98) is the quantum equivalent of the classical statistics of an ideal system (5.29). As will be analyzed in the next chapter, the Maxwell–Boltzmann statistics (5.98) is, under certain conditions, a good approximation to the quantum statistics (bosons and fermions), in which the particles are indistinguishable. In contrast with these quantum statistics, in the Maxwell–Boltzmann statistics the thermodynamic properties of an ideal system of N particles may be obtained from the one-particle partition function $Z(\beta, \alpha, 1)$.

The average value of the occupation number in a particle state r is given by

$$\langle n_r \rangle = \frac{1}{Z(\beta, \alpha, N)} \sum_{\{n_i\}}^* \frac{N!}{\Pi_i n_i!} n_r e^{-\beta \sum_i n_i \varepsilon_i(\alpha)}, \tag{5.100}$$

which may be written as

$$\langle n_r \rangle = \frac{N e^{-\beta \varepsilon_r(\alpha)}}{Z(\beta, \alpha, N)} \sum_{\{n_i\}}^* \frac{(N-1)!}{n_1! \ldots (n_r - 1)! \ldots}$$
$$\times \left(e^{-\beta \varepsilon_1(\alpha)} \right)^{n_1} \ldots \left(e^{-\beta \varepsilon_r(\alpha)} \right)^{n_r - 1} \ldots$$
$$= \frac{N e^{-\beta \varepsilon_r(\alpha)}}{Z(\beta, \alpha, N)} [Z(\beta, \alpha, 1)]^{N-1}, \tag{5.101}$$

where, once more, use has been made of Newton's expansion of the multinomial. According to (5.98), one finally obtains

$$\langle n_r \rangle = N \frac{e^{-\beta \varepsilon_r(\alpha)}}{Z(\beta, \alpha, 1)}, \tag{5.102}$$

which is known as the Maxwell–Boltzmann statistics. Note that the average occupation number of a particle quantum state is proportional to the Boltzmann factor $e^{-\beta \varepsilon_r(\alpha)}$.

5.12 Maxwell–Boltzmann's Ideal Gas

In order to determine the thermodynamic properties of an ideal quantum gas in the Maxwell–Boltzmann statistics, one has to evaluate the partition function (5.99) of a free particle. As shown in Chap. 1, the one-particle quantum states are specified by three quantum numbers, n_x, n_y, and n_z, that may take the values $0, \pm1, \pm2, \ldots$ The energy of a particle of mass m contained in a cubic box of side L in a quantum state n_x, n_y, n_z is given by

$$\varepsilon_{n_x, n_y, n_z}(V) = \frac{2\pi^2 \hbar^2}{m V^{2/3}} \left(n_x^2 + n_y^2 + n_z^2 \right), \tag{5.103}$$

where the volume of the box, $V = L^3$, is the external parameter. The partition function of a free particle is, therefore,

$$Z(\beta, V, 1) = \sum_{n_x = -\infty}^{\infty} \sum_{n_y = -\infty}^{\infty} \sum_{n_z = -\infty}^{\infty} e^{-\beta 2\pi^2 \hbar^2 \left(n_x^2 + n_y^2 + n_z^2 \right)/m V^{2/3}}. \tag{5.104}$$

The expression in (5.104) may be evaluated approximately when the volume of the box is large. To that end, note that the components of the wave vector k_x, k_y, k_z

are quantized, namely $k_x = 2\pi n_x/L$, $k_y = 2\pi n_y/L$, $k_z = 2\pi n_z/L$. These components vary, in modulus, by $2\pi/L$ when each of the quantum numbers changes by one unit. If L is large, this variation tends to zero, and the components of the wave vector may be considered as continuous variables so that one has

$$Z(\beta, V, 1) = \frac{V}{8\pi^3} \int\limits_{-\infty}^{\infty} dk_x \int\limits_{-\infty}^{\infty} dk_y \int\limits_{-\infty}^{\infty} dk_z e^{-\beta\hbar^2(k_x^2+k_y^2+k_z^2)/2m}, \tag{5.105}$$

and the Gaussian integrals may be readily evaluated to obtain

$$Z(\beta, V, 1) = \frac{V}{\Lambda^3}, \tag{5.106}$$

where Λ is the thermal de Broglie wavelength (5.31). Note that in classical statistical physics, Planck's constant is arbitrarily introduced in the determination of the one-particle partition function. In this way, the classical partition function of a free particle (5.30) coincides with (5.106). In the latter case, however, Planck's constant comes from the solution of Schrödinger's equation for the free particle in the box.

It is easy to show, using (5.98) and (5.106), that in the Maxwell–Boltzmann statistics the equation of state of the ideal gas is $p = \rho k_B T$ and that the average energy per particle is $e = 3k_B T/2$, in agreement with the equipartition theorem. On the other hand, the Helmholtz free energy (5.23) is not extensive. Such a result is logical since in the Maxwell–Boltzmann statistics particles are considered to be distinguishable. Therefore, to avoid this anomaly, it is necessary to introduce the Gibbs factor $1/N!$ when determining the partition function of the gas, namely $Z^*(\beta, V, N) = [Z(\beta, V, 1)]^N/N!$. This correction does not alter the equation of state and the energy equation of the gas.

5.13 Brillouin's Paramagnetism

Consider a system of N distinguishable particles of mass m, charge e, and total angular momentum $\hat{\mathbf{j}}$ in a magnetic field $\mathbf{B}$. These particles might be, for instance, atoms placed on the Bravais lattice of a solid, so that the Maxwell–Boltzmann statistics may be applied to such a system. To simplify the calculations, it will be assumed that the system is ideal, while the kinetic energy of the atoms will not be considered.

As was pointed out in Chap. 1, the interaction energy of the total angular momentum $\hat{\mathbf{j}}$ with the field $\mathbf{B}$ is one of the $2j + 1$ values given by the expression

$$\varepsilon_m(B) = -g\mu_B m B, \quad (m = -j, -j+1, \ldots, j-1, j), \tag{5.107}$$

where g is the Landé factor, $\mu_B = e\hbar/2mc$ is Bohr's magneton, and $B = |\mathbf{B}|$. The partition function of an atom

$$Z(\beta, B, 1) = \sum_{m=-j}^{j} e^{\beta g \mu_B m B} = \sum_{m=-j}^{i} e^{mx/j} \tag{5.108}$$

is the sum of the terms of a geometric series with $x = \beta_g \mu_B B j$, i.e.,

$$Z(\beta, B, 1) = e^{-x} \frac{e^{(2j+1)x/j} - 1}{e^{x/j} - 1} = \frac{\sinh\left(\left(1 + \frac{1}{2j}\right)x\right)}{\sinh\left(\frac{x}{2j}\right)}. \tag{5.109}$$

Since $Z(\beta, B, N) = [Z(\beta, B, 1)]^N$, the total magnetic moment of the system turns out to be

$$M = N k_B T \frac{\partial \ln Z(\beta, B, 1)}{\partial B} = N g \mu_B j B_j(x), \tag{5.110}$$

where $B_j(x)$ is the Brillouin function defined as

$$B_j(x) = \left(1 + \frac{1}{2j}\right) \coth\left(\left(1 + \frac{1}{2j}\right)x\right) - \frac{1}{2j} \coth\left(\frac{x}{2j}\right). \tag{5.111}$$

In Fig. 5.1 the Brillouin function is represented for $j = 1/2$, in which case $B_{1/2}(x) = 2 \coth(2x) - \coth x = \tanh x$, $j = 1$ and $j = 3/2$. Note that $B_j(x) \simeq 1$ $(x \gg 1)$ and that all the curves pass through the origin. The latter feature shows that, in the absence of a magnetic field, the magnetic moment of the system is zero (paramagnetism).

These limiting cases may be derived straightforwardly from (5.108). Consider, first, that $x \gg 1$. The partition function of an atom may then be written as

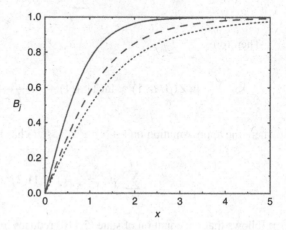

Fig. 5.1 Brillouin function $B_j(x)$ for $j = 1/2$ (*continuous line*), $j = 1$ (*broken line*), and $j = 3/2$ (*dotted line*)

$$Z(\beta, B, 1) = \sum_{m=-j}^{j} e^{mx/j} = e^x\left(1 + e^{-x/j} + e^{-2x/j} + \cdots\right) \simeq e^x, \qquad (5.112)$$

or, alternatively,

$$\ln Z(\beta, B, 1) \simeq x, \qquad (5.113)$$

so that the equation of state (5.110) becomes

$$M \simeq N g\mu_B j \quad (x \gg 1). \qquad (5.114)$$

This result leads to the conclusion that if the magnetic field is intense or the temperature is low $(g\mu_B B \gg k_B T)$ all the magnetic moments will orient in the direction of the field.

When $x \ll 1$, the exponential in (5.108) may be expanded up to second order to obtain

$$Z(\beta, B, 1) = \sum_{m=-j}^{j} e^{mx/j} = \sum_{m=-j}^{j} \left(1 + \frac{x}{j}m + \frac{1}{2}\frac{x^2}{j^2}m^2 + \cdots\right)$$

$$= (2j + 1)\left(1 + \frac{1}{2(2j + 1)}\frac{x^2}{j^2} \sum_{m=-j}^{j} m^2 + \cdots\right), \qquad (5.115)$$

since

$$\sum_{m=-j}^{j} m = 0.$$

Therefore,

$$\ln Z(\beta, B, 1) \simeq \ln(2j + 1) + \frac{1}{2(2j + 1)}\frac{x^2}{j^2} \sum_{m=-j}^{j} m^2, \qquad (5.116)$$

where the approximation $\ln(1 + x) \simeq x (x \ll 1)$ has been used. Since

$$\sum_{m=-j}^{j} m^2 = \frac{1}{3}j(j + 1)(2j + 1),$$

it follows that the equation of state (5.110) reduces in this case to

$$M \simeq \frac{1}{3}N(g\mu_B)^2 \frac{j(j + 1)}{k_B T} B \quad (x \ll 1), \qquad (5.117)$$

which can be cast in the form

$$m_V = \chi B \tag{5.118}$$

where $m_V = M/V$ is the magnetic moment per unit volume,

$$\chi = \frac{1}{3}\rho(g\mu_B)^2 \frac{j(j+1)}{k_B T} \tag{5.119}$$

is the magnetic susceptibility, which is inversely proportional to the absolute temperature (Curie's law), and $\rho = N/V$ is the density. An immediate conclusion of the theory of Brillouin is that since an ideal system of magnetic moments is paramagnetic, a necessary condition in order for the ferromagnetic phase to exist is that there must be interactions between the magnetic moments (see Chap. 9).

5.14 Photon Gas

When analyzing the classical radiation it was shown that the energy of the electromagnetic field in vacuum may be written as an infinite sum of terms each of which is the Hamiltonian of a one-dimensional harmonic oscillator. When these oscillators are treated classically, the application of the equipartition theorem leads to an infinite heat capacity of the radiation. Consider now the quantum treatment of these oscillators. As was discussed in Chap. 1, the energy of an oscillator of angular frequency ω_i may take the following values:

$$\varepsilon_i(\omega_i) = n_i \hbar \omega_i, \quad (n_i = 0, 1, 2, \ldots), \tag{5.120}$$

where n_i is the number of photons in the state of frequency ω_i, (see Sect. 1.8), and the zero-point energy has not been accounted for. Note that in order to specify the quantum state of the radiation, one has to fix an infinite set of variables n_i and that the energy of this state is given by

$$E_{\{n_i\}}(\{\omega_i\}) = \sum_i n_i \hbar \omega_i. \tag{5.121}$$

The fundamental difference between (5.93) and (5.121) is that, in the latter, the numbers n_i do not have to comply with the restriction (5.95); in other words, there is no conservation of the number of photons. The quantum partition function of the electromagnetic radiation in the vacuum contained in a volume V at the absolute temperature T is, therefore,

$$Z(\beta, V) = \sum_{\{n_i\}} e^{-\beta \Sigma_i n_i \hbar \omega_i} = \prod_i \sum_{n_i=0}^{\infty} e^{-\beta n_i \hbar \omega_i} = \prod_i \frac{1}{1 - e^{-\beta \hbar \omega_i}}, \tag{5.122}$$

or, alternatively,

$$\ln Z(\beta, V) = -\sum_i \ln\left(1 - e^{-\beta \hbar \omega_i}\right), \tag{5.123}$$

where the sum in (5.123) extends to all photon states.

It is convenient to point out at this stage that in the quantum theory of radiation it is shown that the photon is a boson of spin 1 and zero rest mass, so that its energy is $\varepsilon = pc = \hbar k c = \hbar \omega$. Although in principle the photon has three possible spin states, only two of them are independent, corresponding to the two states of polarization of the electromagnetic wave. The sum over the photon states in (5.123) extends to the oscillator states with any polarization (5.74), so that in the limit $V \to \infty$, one has

$$\ln Z(\beta, V) = -\frac{V}{\pi^2 c^3} \int_0^{\infty} d\omega \, \omega^2 \ln\left(1 - e^{-\beta \omega}\right). \tag{5.124}$$

Integrating by parts in (5.124) leads to

$$\ln Z(\beta, V) = \frac{2V}{\pi^2 c^3} \left(\frac{k_B T}{\hbar}\right)^3 \zeta(4), \tag{5.125}$$

where

$$\zeta(n) = \frac{1}{\Gamma(n)} \int_0^{\infty} dx \frac{x^{n-1}}{e^x - 1}, \tag{5.126}$$

is the Riemann function $\left(\zeta(4) = \pi^4/90\right)$. From (5.125) it follows that the equation of state and the energy equation of the electromagnetic radiation are given by

$$p = \frac{\pi^2}{45 \hbar^3 c^3} (k_B T)^4, \tag{5.127}$$

and

$$E = \frac{\pi^2 V}{15 \hbar^3 c^3} (k_B T)^4, \tag{5.128}$$

which implies that the heat capacity C_V is finite.

One may compare the classical and quantum treatments of the radiation by computing the average number of photons in a photon state, namely

$$\langle n_r \rangle = \frac{1}{Z(\beta, V)} {\sum_{\{n_i\}}}' n_r e^{-\beta \sum_i n_i \hbar \omega_i}. \qquad (5.129)$$

Due to the factorization of the different terms in (5.129) and (5.122), one has

$$\langle n_r \rangle = \left(1 - e^{-\beta \hbar \omega_r}\right) \sum_{n_r=0}^{\infty} n_r e^{-\beta \hbar \omega_r n_r}, \qquad (5.130)$$

and, since the first term in the series (5.130) is zero, the sum may be evaluated in the following way:

$$S \equiv \sum_{n_r=1}^{\infty} n_r e^{-\beta \hbar \omega_r n_r} = \sum_{m_r=0}^{\infty} (m_r + 1) e^{-\beta \hbar \omega_r (m_r+1)}$$

$$= e^{-\beta \hbar \omega_r} \left(S + \sum_{m_r=0}^{\infty} e^{-\beta \hbar \omega_r m_r} \right)$$

$$= e^{-\beta \hbar \omega_r} \left(S + \frac{1}{1 - e^{-\beta \hbar \omega_r}} \right), \qquad (5.131)$$

i.e.,

$$S = \frac{1}{\left(e^{\beta \hbar \omega_r} - 1\right)\left(1 - e^{-\beta \hbar \omega_r}\right)},$$

i.e.,

$$\langle n_r \rangle = \frac{1}{e^{\beta \hbar \omega_r} - 1}, \qquad (5.132)$$

which is known as Planck's statistics.

In the limit $V \to \infty$, one may determine from (5.132) the average energy of the radiation in the frequency interval between ω and $\omega + d\omega$. To that end, one has to take the product of the number of photon states in that interval (5.74), $V \omega^2 d\omega / \pi^2 c^3$ times the energy of the photon, $\hbar \omega$ times the average number of photons in the interval (5.132), to obtain

$$e(\omega)d\omega = \frac{V}{\pi^2 c^3} \omega^2 d\omega \hbar \omega \frac{1}{e^{\beta \hbar \omega} - 1}. \qquad (5.133)$$

Integration of (5.133) over the whole interval of frequencies leads to (5.128). Note, however, that in the limit of low frequencies or high temperatures ($\beta \hbar \omega \ll 1$) the fraction in (5.133) may be approximated by $1/\beta \hbar \omega$, and so in this limit

Fig. 5.2 Energy density of the radiation $e_0\,(x) = x^3/(e^x - 1)$ as a function of the variable $x = \beta\hbar\omega$ (*continuous line*). The curve has a maximum at $x = 2.821$. The *broken line* is the classical result $e_0\,(x) \simeq x^2$

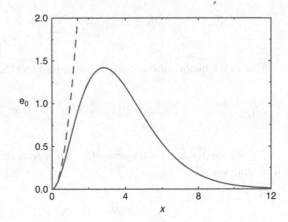

$$e(\omega)d\omega \simeq \frac{V}{\pi^2 c^3}k_B T\omega^2 d\omega \quad (\beta\hbar\omega \ll 1), \tag{5.134}$$

which results from multiplying the number of photon states in the interval by the classical average energy of the oscillator $k_B T$ (see (5.75)). This equation shows that the classical result is only a low-frequency approximation to the Planck statistics so that if extrapolated to all frequencies it leads to an infinite heat capacity of the radiation.

It is seen that (5.133) may be expressed in terms of the variable $x = \beta\hbar\omega$ as

$$e_0(x)dx \equiv \frac{\pi^2 c^3}{V}\frac{\hbar^3}{(k_B T)^4}e(\omega)d\omega = \frac{x^3}{e^x - 1}dx, \tag{5.135}$$

which is represented in Fig. 5.2. The function $e_0(x)$ has a maximum at $x = 2.821$, which implies that the angular frequency at which $e(\omega)$ reaches its maximum grows proportionally to the temperature (Wien's displacement law).

5.15   Phonon Gas

In the Hamiltonian of a solid in the harmonic approximation, the term corresponding to the deviations of the atoms from their equilibrium positions is a positive definite quadratic form. As is well known, the time evolution of the $3\,N$ deviations may be described by means of a set of $3\,N$ normal modes that satisfy equations of motion of the harmonic oscillator type. When these oscillators are treated classically, upon application of the equipartition theorem one obtains the Dulong–Petit law. The quantum study of the harmonic oscillations in a solid is, therefore, similar to the one just carried out for the electromagnetic radiation. Once more, the energy of each oscillator is given by

$$\varepsilon_i(\omega_i) = n_i \hbar \omega_i, \quad (n_i = 0, 1, 2, \ldots), \tag{5.136}$$

where n_i is the number of phonons in the state of frequency ω_i. The partition function of the phonon gas is then

$$Z(\beta, V, 3N) = \sum_{\{n_i\}} e^{-\beta \sum_{i=1}^{3N} n_i \hbar \omega_i} = \prod_{i=1}^{3N} \sum_{n_i=0}^{\infty} e^{-\beta n_i \hbar \omega_i}$$

$$= \prod_{i=1}^{3N} \frac{1}{1 - e^{-\beta \hbar \omega_i}}, \tag{5.137}$$

where neither the potential energy at the mechanical equilibrium nor (as in the case of the photon gas) the zero-point energy of the oscillators has been considered.

There are two fundamental differences between the photon gas and the phonon gas. The first one has to do with the fact that in the solid the number of oscillators is $3N$, while that of the electromagnetic radiation is infinite. The second one is that in the solid the exact form of the dispersion relation $\omega_i = \omega_i(k)$, where k is the modulus of the wave vector, is not known exactly.

The average energy of the phonon gas is given by

$$E = \frac{\partial}{\partial \beta} \sum_{i=1}^{3N} \ln\left(1 - e^{-\beta \hbar \omega_i}\right) = \sum_{i=1}^{3N} \frac{\hbar \omega_i}{e^{\beta \hbar \omega_i} - 1}, \tag{5.138}$$

which in the high-temperature limit $\beta \hbar \omega_i \ll 1$ may be approximated by

$$E \simeq \sum_{i=1}^{3N} \frac{\hbar \omega_i}{(1 + \beta \hbar \omega_i + \cdots) - 1} = 3N k_B T. \tag{5.139}$$

This is the result that one would obtain using the equipartition theorem and which leads to the Dulong–Petit law (5.86).

A simple approximation of (5.138) may be obtained if one assumes that all the oscillators have the same frequency ω_E (quantum Einstein model), so that

$$E = 3N \frac{\hbar \omega_E}{e^{\beta \hbar \omega_E} - 1}, \tag{5.140}$$

and, therefore,

$$C_V = 3N k_B \left(\frac{\Theta_E}{T}\right)^2 \frac{e^{\Theta_E/T}}{\left(e^{\Theta_E/T} - 1\right)^2}, \tag{5.141}$$

where the Einstein temperature has been defined as $\Theta_E = \hbar \omega_E / k_B$. Since Θ_E is unknown, it may be taken as a free parameter in (5.141) so that this expression fits

Fig. 5.3 Specific heat at constant volume, c_V/k_B, of a crystalline lattice as a function of $x = T/\Theta_E$ (Einstein's theory, *dotted line*) and $x = T/\Theta_D$ (Debye's theory, *continuous line*). *The broken line* is the classical result

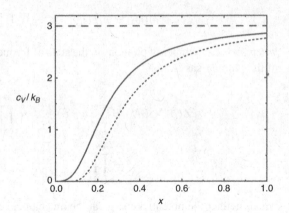

the experimental data. The results obtained in this fashion indicate that, although the heat capacity of the solid in the Einstein model has a qualitatively correct behavior, it decreases at low temperatures as

$$C_V \simeq 3Nk_B \left(\frac{\Theta_E}{T}\right)^2 e^{-\Theta_E/T}, \qquad (5.142)$$

when $T \ll \Theta_E$, whereas the experimental results indicate that in this limit $C_V \sim T^3$ (Fig. 5.3).

Finally, in order to establish the Debye theory, consider again the equations of motion of the displacement vectors $\mathbf{u}_i$ in the harmonic approximation (5.84). If one seeks for plane-wave solutions, namely

$$\hat{\varepsilon}_\alpha(\mathbf{k}) e^{i(\mathbf{k}\cdot\mathbf{R}_i - \omega_\alpha t)}, \qquad (5.143)$$

where $\hat{\varepsilon}_\alpha(\mathbf{k})$ is the polarization vector, which is a unit vector whose direction depends on $\mathbf{k}$, upon substitution of (5.143) into (5.84) one finds

$$\tilde{\mathcal{D}}(\mathbf{k}) \cdot \hat{\varepsilon}_\alpha(\mathbf{k}) = m\omega_\alpha^2(\mathbf{k})\hat{\varepsilon}_\alpha(\mathbf{k}), \qquad (5.144)$$

where

$$\tilde{\mathcal{D}}(\mathbf{k}) = \sum_{\mathbf{R}} \mathcal{D}(\mathbf{R}) e^{-i\mathbf{k}\cdot\mathbf{R}} = \tilde{\mathcal{D}}(-\mathbf{k}). \qquad (5.145)$$

Note that (5.144) is the eigenvalue equation of the operator $\tilde{\mathcal{D}}(\mathbf{k})$, whose eigenvectors are $\hat{\varepsilon}_\alpha(\mathbf{k})$ and its eigenvalues $m\omega_\alpha^2(\mathbf{k})$. From (5.145) it follows that $\tilde{\mathcal{D}}(\mathbf{k})$ is a real tensor, even in $\mathbf{k}$. The tensor is, moreover, symmetric since $\mathcal{D}^{\mu\nu}(\mathbf{R}) = \mathcal{D}^{\nu\mu}(\mathbf{R})$. For a fixed wave vector $\mathbf{k}$, let $\hat{\varepsilon}_\alpha(\mathbf{k})(\alpha = 1, 2, 3)$ be a set of orthonormal eigenvectors:

$$\hat{\varepsilon}_\alpha(\mathbf{k}) \cdot \hat{\varepsilon}_{\alpha'}(\mathbf{k}) = \delta_{\alpha,\alpha'}. \tag{5.146}$$

Note that, in contrast with the case of the electromagnetic radiation, in a solid the transversality condition does not hold so that, once the wave vector $\mathbf{k}$ is fixed, there are three independent states of polarization.

In order to obtain an adequate description of the heat capacity of the solid at low temperatures, consider the tensor $\tilde{\mathcal{D}}(\mathbf{k})$. According to (5.138) the most important contribution when $T \to 0\,(\beta \to \infty)$ corresponds to the low frequencies a or, as shown in what follows, to the small wave vectors. Therefore, (5.145) may be approximated by

$$\tilde{\mathcal{D}}(\mathbf{k}) = \sum_{\mathbf{R}} \mathcal{D}(\mathbf{R})\left(1 - i\mathbf{k} \cdot \mathbf{R} - \frac{1}{2}(\mathbf{k} \cdot \mathbf{R})^2 + \cdots\right)$$

$$= -\frac{k^2}{2} \sum_{\mathbf{R}} \mathcal{D}(\mathbf{R})(\hat{\mathbf{k}} \cdot \mathbf{R})^2, \tag{5.147}$$

where $\hat{\mathbf{k}} = \mathbf{k}/k$ is the unit vector in the direction of $\mathbf{k}$. The first two terms of the expansion (5.147) are zero, the first one because of (5.82) and the second one due to the fact that

$$\sum_{\mathbf{R}} \mathcal{D}(\mathbf{R}) \cdot \mathbf{R} = 0, \tag{5.148}$$

since $\mathcal{D}(\mathbf{R}) = \mathcal{D}(-\mathbf{R})$.

With this approximation, the eigenvalues of (5.144) may be written as

$$\omega_\alpha(\mathbf{k}) = c_\alpha(\hat{\mathbf{k}})k, \tag{5.149}$$

where $c_\alpha(\hat{\mathbf{k}})$ are the square roots of the eigenvalues of the tensor

$$-\frac{1}{2m} \sum_{\mathbf{R}} \mathcal{D}(\mathbf{R})(\hat{\mathbf{k}} \cdot \mathbf{R})^2. \tag{5.150}$$

In an isotropic medium it is always possible to choose one of the directions of the vectors $\hat{\varepsilon}_\alpha(\mathbf{k})$ in the direction of $\mathbf{k}$ (longitudinal mode), so that the other two polarization vectors are orthogonal to $\mathbf{k}$ (transverse modes). Note that the dispersion relation (5.149), which is valid when $\mathbf{k} \to 0$, is analogous to one of the electromagnetic radiation in the vacuum (5.64). In the case of the radiation, c is the velocity of light, while in (5.149) $c_\alpha(\hat{\mathbf{k}})$ is the velocity of the longitudinal or transversal sound wave.

Assume that, as in the case of radiation, periodic conditions are imposed to (5.143) in a cubic box of side L, so that the number of phonon states whose modulus of the wave vector lies between k and $k + dk$ is (5.73). For the longitudinal wave, the number of phonon states with angular frequency lying between ω and $\omega + d\omega$ is, therefore,

$$\rho_l(\omega)d\omega = \frac{V}{2\pi^2}\frac{1}{c_l^3}\omega^2 d\omega, \tag{5.151}$$

while for the transverse wave it is

$$\rho_t(\omega)d\omega = 2\frac{V}{2\pi^2}\frac{1}{c_t^3}\omega^2 d\omega, \tag{5.152}$$

where c_l and c_t are the velocities of the longitudinal and transversal sound waves. The total number of phonon states is given by

$$\rho(\omega)d\omega = \frac{V}{2\pi^2}\frac{1}{c_l^3}\omega^2 d\omega + 2\frac{V}{2\pi^2}\frac{1}{c_t^3}\omega^2 d\omega = 3\frac{V}{2\pi^2}\frac{1}{\bar{c}^3}\omega^2 d\omega, \tag{5.153}$$

where $\bar{c}$ has been defined as

$$\frac{3}{\bar{c}^3} \equiv \frac{1}{c_l^3} + \frac{2}{c_t^3}. \tag{5.154}$$

Since there are $3N$ phonon states, one must impose an upper frequency limit, ω_D, referred to as the Debye frequency, such that

$$3\frac{V}{2\pi^2}\frac{1}{\bar{c}^3}\int_0^{\omega_D} d\omega\,\omega^2 = 3N, \tag{5.155}$$

i.e.,

$$\omega_D^3 = 6\pi^2 \rho \bar{c}^3, \tag{5.156}$$

where $\rho = N/V$ is the average density of the solid. The average energy of the phonon gas is, according to (5.138),

$$E = \frac{3V}{2\pi^2\bar{c}^3}\int_0^{\omega_D} d\omega\,\omega^2\frac{\hbar\omega}{e^{\beta\hbar\omega} - 1}, \tag{5.157}$$

and hence

$$C_V = \frac{3Vk_B^4}{2\pi^2\bar{c}^3\hbar^3}T^3\int_0^{\beta\hbar\omega_D} dx\,x^4\frac{e^x}{(e^x - 1)^2}, \tag{5.158}$$

Table 5.1 Debye temperature Θ_D in Kelvin for some elements

	Θ_D		Θ_D
Li	400	Al	394
Na	150	Si	625
K	100	As	285
Cu	315	Fe	420
Ag	215	Ni	375
Au	170	Pt	230

Source N. W. Ashcroft and N. D. Mermin, *Solid State Physics*, Saunders, Philadelphia (1976)

or

$$C_V = 9Nk_B\left(\frac{T}{\Theta_D}\right)^3 \int\limits_0^{\Theta_D/T} dx\, x^4 \frac{e^x}{(e^x - 1)^2}, \tag{5.159}$$

where the Debye temperature, Θ_D, has been defined as $k_B\Theta_D = \hbar\omega_D$.

At low temperatures ($T \ll \Theta_D$) the upper limit in the integral may be replaced by infinity and since

$$\int\limits_0^\infty dx\, x^4 \frac{e^x}{(e^x - 1)^2} = \frac{4\pi^4}{15},$$

one finally finds for the specific heat at constant volume $c_V = C_V/N$

$$c_V = \frac{12\pi^4}{5} k_B \left(\frac{T}{\Theta_D}\right)^3. \tag{5.160}$$

This equation has an adjustable parameter, the Debye temperature, in a similar way as in the Einstein model and the law $c_V \sim T^3$ is the one obtained in the experiments (see Fig. 5.3). A typical Debye temperature is of the order of $10^2\, K$ (Table 5.1). Note that if $T \ll \Theta_D$ quantum effects are important, while if $T \gg \Theta_D$ the classical result (the Dulong–Petit law) is obtained.

Further Reading

1. K. Huang, *Statistical Mechanics* (Wiley, New York, 1987). *Contains a good first introduction to quantum statistical physics.*
2. L.D. Landau, E.M. Lifshitz, *Statistical Physics*, 3rd ed. (Pergamon Press, Oxford, 1989). *More difficult to read but covers a very large number of topics.*

3. L.E. Reichl, *A Modern Course in Statistical Physics,* 2nd ed. (Wiley, New York, 1998). *Contains an alternative derivation of the Gibbs ensembles.*

4. N.W. Ashcroft, N.D. Mermin, *Solid State Physics* (Saunders, Philadelphia, 1976). *Contains a detailed study of the phonon gas, whose notation was followed here.*

Chapter 6
Grand Canonical Ensemble

Abstract In this chapter, the grand canonical ensemble describing a system that exchanges energy and particles with the external world is introduced. The density fluctuations at the critical point and the ideal quantum boson and fermion gases are presented as key applications of this ensemble.

6.1 Classical Grand Canonical Ensemble

When two closed systems of volumes V_1 and V_2, which together constitute an isolated system of energy E and volume $V = V_1 + V_2$, exchange energy in an additive way, $E = E_1 + E_2$, the probability density of the energy of a subsystem is, according to (5.1),

$$\rho_1(E_1) = \frac{\omega_1(E_1, V_1, N_1)\omega_2(E - E_1, V_2, N_2)}{\omega(E, V, N)}, \tag{6.1}$$

where the numbers of particles, N_1 and N_2 ($N = N_1 + N_2$), are constant for each subsystem.

If the subsystems may also exchange particles, wherein what follows they will be considered to be all of the same species, the question immediately arises as to what the probability density, denoted by $\rho_1(E_1, N_1)$, for the subsystem to have an energy E_1 and a number of particles N_1 will be. Note that in (6.1) N_1 and N_2 are numbers that do not indicate which particles belong to one or the other subsystem. From a classical point of view, the mere specification of these numbers is not enough to determine the state of each subsystem, i.e., if the N particles of the total system are numbered, which is always possible in classical mechanics, the state of the subsystem will only be completely determined when one knows which particles belong to it. Since one wants to determine the probability density that the subsystem has any N_1 particles and (6.1) is the same for each group of N_1 particles, $\rho_1(E_1, N_1)$ is the expression (6.1) times the number of possible ways to distribute N particles among the subsystems in groups of N_1 and N_2 particles (since the groups are mutually exclusive), namely

© The Author(s), under exclusive license to Springer Nature Switzerland AG 2021
M. Baus and C. F. Tejero, *Equilibrium Statistical Physics*,
https://doi.org/10.1007/978-3-030-75432-7_6

$$\rho_1(E_1, N_1) = \frac{N!}{N_1! N_2!} \frac{\omega_1(E_1, V_1, N_1)\omega_2(E - E_1, V_2, N - N_1)}{\omega(E, V, N)}, \qquad (6.2)$$

in which the combinatorial factor corresponds to the degeneration of the states of the subsystems due to their characterization through the numbers N_1 and N_2 (note that this kind of combinatorial factor already appeared in Chap. 5 in the study of the Maxwell–Boltzmann statistics). The above equation may be written as

$$\rho_1(E_1, N_1) = \frac{\omega_1^*(E_1, V_1, N_1)\omega_2^*(E - E_1, V_2, N - N_1)}{\omega^*(E, V, N)}, \qquad (6.3)$$

or, alternatively,

$$\rho_1(E_1, N_1) = \omega_1^*(E_1, V_1, N_1) \frac{\Omega_2^*(E - E_1, V_2, N - N_1; \Delta E)}{\Omega^*(E, V, N; \Delta E)}, \qquad (6.4)$$

where

$$\omega^*(E, V, N) \equiv \frac{1}{N!}\omega(E, V, N), \qquad (6.5)$$

$$\Omega^*(E, V, N; \Delta E) \equiv \omega^*(E, V, N)\Delta E, \qquad (6.6)$$

and it has been assumed that $E - E_1 \gg \Delta E$.

Note that (6.1) and (6.3) are similar, with the difference that in the latter the functions $\omega(E, V, N)$ are divided by $N!$, following (6.5). Recall that both in the microcanonical and in the canonical ensembles the factor $1/N!$ was introduced in an ad hoc manner to avoid the Gibbs paradox. In the case of a classical subsystem that exchanges particles, this factor arises naturally for the reasons stated above. The normalization condition for $\rho_1(E_1, N_1)$ reads

$$\sum_{N_1=0}^{N} \int_0^E dE_1 \rho_1(E_1, N_1) = 1, \qquad (6.7)$$

which expresses the fact that the probability that the subsystem has any energy and any number of particles is one.

Note that since the probability density (6.3) is the product of a monotonously increasing function, $\omega_1^*(E_1, V_1, N_1)$, and a monotonously decreasing function, $\omega_2^*(E - E_1, V_2, N - N_1)$ of N_1 and E_1, $\rho_1(E_1, N_1)$ has a maximum at $\tilde{N}_1$ and $\tilde{E}_1$. When $N \gg 1$, $1 \ll \tilde{N}_1 \ll N$, and $\tilde{E}_1 \ll E$, which is the normal situation in an experiment in which the system (subsystem 1) is macroscopic and much smaller than the external world, one should expect that wherever $\rho_1(E_1, N_1)$ differs appreciably from zero, the energy E_2 and the number of particles N_2 be such that $E_2 \gg E_1$ and $N_2 \gg N_1$.

The probability density (6.4) may be approximated by taking the Taylor expansion of the logarithm of $\Omega_2^*(E - E_1, V_2, N - N_1; \Delta E)$ around the point $E_2 = E$ and $N_2 = N$, namely

$$\ln\left(\frac{\Omega_2^*(E - E_1, V_2, N - N_1; \Delta E)}{\Omega_2^*(E, V_2, N; \Delta E)}\right) = -\beta_2 E_1 + \beta_2 \mu_2 N_1 + \cdots, \qquad (6.8)$$

with

$$\beta_2 \mu_2 \equiv -\left(\frac{\partial \ln \Omega_2^*(E_2, V_2, N_2; \Delta E)}{\partial N_2}\right)_{E_2=E, N_2=N}, \qquad (6.9)$$

where $T_2 = 1/k_B \beta_2$ (see (5.3)) and μ_2 are the temperature and the chemical potential of the external world, respectively. Note that (6.9), which defines the chemical potential in classical statistical physics, may be written from (4.40) (recall that if the external parameter is the volume and the particles may be permuted, the function $\Omega(E, V, N; \Delta E)$ has to be replaced by $\Omega^*(E, V, N; \Delta E)$) as

$$\mu = -T \frac{\partial S}{\partial N}, \qquad (6.10)$$

which is precisely (2.19).

With the approximation (6.8), the density probability (6.3) leads to

$$\rho_1(E_1, N_1) = \frac{\Omega_2^*(E, V_2, N; \Delta E)}{\Omega^*(E, V, N; \Delta E)} \omega_1^*(E_1, V_1, N_1) e^{-\beta_2 E_1} z_2^{N_1}, \qquad (6.11)$$

where the fugacity z has been defined as

$$z = e^{\beta \mu}. \qquad (6.12)$$

If the normalization condition (6.7) is taken into account, one obtains the following expression,

$$\rho_1(E_1, N_1) = \frac{1}{Q(\beta_2, V_1, z_2)} \omega_1^*(E_1, V_1, N_1) e^{-\beta_2 E_1} z_2^{N_1}, \qquad (6.13)$$

where

$$Q(\beta_2, V_1, z_2) = \sum_{N_1=0}^{N} z_2^{N_1} \int_0^E dE_1 \omega_1^*(E_1, V_1, N_1) e^{-\beta_2 E_1}, \qquad (6.14)$$

is the classical grand partition function of the system. Observe that, except for β_2 and z_2, all the variables appearing in (6.13) and (6.14) correspond to the system. In what follows, to simplify the notation, the subscripts will be dropped to write

$$\rho(E, N) = \frac{1}{Q(\beta, V, z)} \omega^*(E, V, N) e^{-\beta E} z^N, \tag{6.15}$$

with

$$Q(\beta, V, z) = \sum_{N=0}^{\infty} z^N \int_0^{\infty} dE \, \omega^*(E, V, N) e^{-\beta E}, \tag{6.16}$$

where the variables E, V, and N correspond to the system and the parameters β and z to the external world. Note that, due to the pronounced maximum of $\rho(E, N)$, in the grand partition function (6.16) the upper limits in both the sum and the integral have been replaced by infinity. The ensemble described by (6.15) and (6.16) is called the grand canonical ensemble, which is the ensemble that represents a system in thermal contact (that exchanges energy) and chemical contact (that exchanges particles) with another system (the external world) which has a much greater number of degrees of freedom.

Note that the classical grand partition function may be written as

$$Q(\beta, V, z) = \sum_{N=0}^{\infty} z^N Z^*(\beta, V, N), \tag{6.17}$$

which is a series in the fugacity whose coefficients are the partition functions of N particles (with $Z(\beta, V, 0) \equiv 1$):

$$Z^*(\beta, V, N) = \int_0^{\infty} dE \, \omega^*(E, V, N) e^{-\beta E}. \tag{6.18}$$

This is indeed (5.8) in which the phase volume $\omega(E, V, N)$ has been replaced by $\omega^*(E, V, N)$. Note that in the grand canonical ensemble the factor $1/N!$ appears because the probability density $\rho(E, N)$ refers to any number of particles N.

Consider now the probability density in the phase space (4.46) of a subsystem that exchanges energy with another subsystem. If one assumes that the subsystems may also exchange particles, the probability density $\rho_1(q_1, p_1, N_1)$ for the subsystem 1 to be in a state (q_1, p_1) and containing any Ni particles is, by the same kind of arguments leading to (6.3) from (6.1), the probability density (4.46) multiplied by the combinatorial factor $N!/N_1!N_2!$ (once more, the Hamiltonians are denoted by *H1 and H2* in order to simplify the notation), namely

$$\rho_1(q_1, p_1, N_1) = \frac{N!}{N_1! N_2!} \frac{1}{h^{3N_1}} \frac{\omega_2(E - H_1, V_2, N - N_1)}{\omega(E, V, N)}$$

$$= \frac{1}{N_1! h^{3N_1}} \frac{\Omega_2^*(E - H_1, V_2, N - N_1; \Delta E)}{\Omega^*(E, V, N; \Delta E)}, \qquad (6.19)$$

which may be approximated using a Taylor series expansion as

$$\ln\left(\frac{\Omega_2^*(E - H_1, V_2, N - N_1; \Delta E)}{\Omega_2^*(E, V_2, N; \Delta E)}\right) = -\beta_2 H_1 + \beta_2 \mu_2 N_1 + \cdots, \qquad (6.20)$$

i.e., a result analogous to (6.8). Upon substitution of (6.20) into (6.19), imposing the normalization condition

$$\sum_{N_1=0}^{N} \int dq_1 \int dp_1 \rho_1(q_1, p_1, N_1) = 1, \qquad (6.21)$$

and, finally, dropping the subscripts, all of them corresponding to the variables of the system, except for the temperature T_2 (or the parameter β_2) and the chemical potential μ_2 (or the fugacity z_2) of the external world, one has

$$\rho(q, p, N) = \frac{1}{N! h^{3N}} \frac{1}{Q(\beta, V, z)} z^N e^{-\beta H_N(q, p; V)}, \qquad (6.22)$$

with

$$Q(\beta, V, z) = \sum_{N=0}^{\infty} z^N \frac{1}{N! h^{3N}} \int dq \int dp \, e^{-\beta H_N(q, p; V)}, \qquad (6.23)$$

where the integral in (6.23) extends over the whole phase space of the system without restriction, and the upper limit of the sum has been replaced by infinity. Note that (6.23) may be written as (6.17) with

$$Z^*(\beta, V, N) = \frac{1}{N! h^{3N}} \int dq \int dp \, e^{-\beta H_N(q, p; V)}, \qquad (6.24)$$

which, in contrast to (5.12), includes the factor $1/N!$ (see also (5.34)).

6.2 Mean Values and Fluctuations

In a system described by the grand canonical ensemble, the energy and the number of particles are random variables whose probability density is (6.15). The Hamiltonian of a system of N particles, $H_N \equiv H_N(q, p; V)$, is a function of the mechanical state (q, p) whose probability density is (6.22). Therefore,

$$\langle H_N \rangle = \frac{1}{Q(\beta, V, z)} \sum_{N=0}^{\infty} z^N \frac{1}{N! h^{3N}} \int dq \int dp H_N e^{-\beta H_N}$$

$$= \frac{1}{Q(\beta, V, z)} \sum_{N=0}^{\infty} z^N \frac{1}{N! h^{3N}} \left(-\frac{\partial}{\partial \beta} \right) \int dq \int dp e^{-\beta H_N}, \tag{6.25}$$

i.e.,

$$\langle H_N \rangle = -\frac{1}{Q(\beta, V, z)} \left(\frac{\partial}{\partial \beta} \right) Q(\beta, V, z), \tag{6.26}$$

which is known as the energy equation.

Similarly, the mean quadratic value of H_N is obtained as follows:

$$\langle H_N^2 \rangle = \frac{1}{Q(\beta, V, z)} \left(\frac{\partial^2}{\partial \beta^2} \right) Q(\beta, V, z), \tag{6.27}$$

The fluctuation of the energy of the system is given by

$$\langle H_N^2 \rangle - \langle H_N \rangle^2 = \frac{\partial}{\partial \beta} \left(\frac{1}{Q(\beta, V, z)} \frac{\partial Q(\beta, V, z)}{\partial \beta} \right), \tag{6.28}$$

i.e.,

$$\langle H_N^2 \rangle - \langle H_N \rangle^2 = -\frac{\partial}{\partial \beta} \langle H_N \rangle, \tag{6.29}$$

which is analogous to (5.18). Note, however, that the average values in the grand canonical ensemble are functions of β, V, and z, while the average values in the canonical ensemble are functions of β, V, and N. This point is addressed in more detail below.

Since in the grand canonical ensemble the number of particles is a random variable, its average value is given by

$$\langle N \rangle = \frac{1}{Q(\beta, V, z)} \sum_{N=0}^{\infty} N z^N \frac{1}{N! h^{3N}} \int dq \int dp e^{-\beta H_N}$$

$$= \frac{1}{Q(\beta, V, z)} \left(z \frac{\partial}{\partial z} \right) \sum_{N=0}^{\infty} z^N \frac{1}{N! h^{3N}} \int dq \int dp e^{-\beta H_N}, \tag{6.30}$$

i.e.,

$$\langle N \rangle = \frac{1}{Q(\beta, V, z)} \left(z \frac{\partial}{\partial z} \right) Q(\beta, V, z). \tag{6.31}$$

In a similar way one has

$$\langle N^2 \rangle = \frac{1}{Q(\beta, V, z)} \left(z \frac{\partial}{\partial z} \right) \left(z \frac{\partial}{\partial z} \right) Q(\beta, V, z), \tag{6.32}$$

and the fluctuation in the number of particles turns out to be

$$\langle N^2 \rangle - \langle N \rangle^2 = z \frac{\partial}{\partial z} \left(\frac{1}{Q(\beta, V, z)} z \frac{\partial Q(\beta, V, z)}{\partial z} \right), \tag{6.33}$$

i.e.,

$$\langle N^2 \rangle - \langle N \rangle^2 = z \frac{\partial}{\partial z} \langle N \rangle. \tag{6.34}$$

The pressure, which is the intensive variable corresponding to the external parameter V, is obtained as the average value of the dynamical function $-\partial H_N / \partial V$ with the average performed with the probability density (6.22):

$$p = -\frac{1}{Q(\beta, V, z)} \sum_{N=0}^{\infty} z^N \frac{1}{N! h^{3N}} \int dq \int dp \frac{\partial H_N}{\partial V} e^{-\beta H_N}$$

$$= \frac{1}{\beta Q(\beta, V, z)} \sum_{N=0}^{\infty} z^N \frac{1}{N! h^{3N}} \frac{\partial}{\partial V} \int dq \int dp \, e^{-\beta H_N}, \tag{6.35}$$

or, alternatively,

$$p = \frac{1}{\beta Q(\beta, V, z)} \frac{\partial Q(\beta, V, z)}{\partial V}. \tag{6.36}$$

Note that (6.36), $p = p(\beta, V, z)$, is not the equation of state of the fluid $p = p(\beta, V, \langle N \rangle)$. Since in the grand canonical ensemble the fugacity is an independent variable, all the average values are parametric functions of z. In order to obtain, for instance, the equation of state $p = p(\beta, V, \langle N \rangle)$, one has to eliminate z from the equation for the average number of particles (6.31), i.e., $z = z(\beta, V, \langle N \rangle)$, and then substitute this result in (6.36), $p = p(\beta, V, z(\beta, V, \langle N \rangle))$. This is not a simple process as will be shown throughout this chapter.

6.3 Grand Potential

As indicated at the beginning of this chapter, the probability density (6.15) has a pronounced maximum for a value of the energy $\tilde{E}$ and the number of particles $\tilde{N}$. The conditions for the extremum of $\rho(E, N)$ are given by

$$\left(\frac{\partial \ln \Omega^*(E, V, N; \Delta E)}{\partial E}\right)_{E=\tilde{E}, N=\tilde{N}} - \beta = 0, \tag{6.37}$$

and

$$\left(\frac{\partial \ln \Omega^*(E, V, N; \Delta E)}{\partial N}\right)_{E=\tilde{E}, N=\tilde{N}} + \beta \mu = 0, \tag{6.38}$$

which express, according to (4.55) and (6.9), the thermal equilibrium (equality of temperatures) and the chemical equilibrium (equality of chemical potentials) between the system and the external world. These equilibrium conditions take place at the state of maximum probability, and so, as a consequence of the existence of this maximum, the series in the fugacity (6.17) may be approximated by the dominant term, namely

$$Q(\beta, V, z) \simeq z^{\tilde{N}} Z^*(\beta, V, \tilde{N}), \tag{6.39}$$

where $\tilde{N}(\beta, V, z)$ and $\tilde{E}(\beta, V, z)$ are the solution of the system of Eqs. (6.37) and (6.38). Multiplying by $k_B T$ and taking the logarithm, one has

$$k_B T \ln Q(\beta, V, z) \simeq \tilde{N} \mu + k_B T \ln Z^*(\beta, V, \tilde{N}) = \tilde{N} \mu - F(\beta, V, \tilde{N}), \tag{6.40}$$

where $F(\beta, V, \tilde{N})$ is the Helmholtz free energy. If one defines the Landau free energy or grand potential $\Omega(\beta, V, z)$ as

$$\Omega(\beta, V, z) = -k_B T \ln Q(\beta, V, z), \tag{6.41}$$

from (6.40) one obtains

$$\Omega(\beta, V, z) \simeq F(\beta, V, \tilde{N}) - \tilde{N} \mu, \tag{6.42}$$

which is the thermodynamic relation (2.36) evaluated at $\tilde{N}$. In the thermodynamic limit

$$\text{TL}\left[\frac{\Omega(\beta, V, z)}{\langle N \rangle}\right] = f(\beta, \rho) - \mu, \tag{6.43}$$

where $f(\beta, \rho)$ is the Helmholtz free energy per particle, and $\rho = \langle N \rangle / V$ is the average density (note that due to the maximum of probability $\langle N \rangle = \tilde{N}$). From (6.36) it follows that since p and β are intensive variables, in the thermodynamic limit $\ln Q(\beta, V, z)$ has to be a homogeneous function of degree one in V, i.e.,

$$\frac{p}{k_B T} = \text{TL}\left[\frac{\ln Q(\beta, V, z)}{V}\right], \tag{6.44}$$

or, alternatively,

$$\mathrm{TL}\left[\frac{\Omega(\beta, V, z)}{\langle N \rangle}\right] = -\frac{p}{\rho}, \tag{6.45}$$

which is (2.39). Note that the fact that in statistical physics the independent variables are β and z (instead of the thermodynamic variables T and μ) should lead to no confusion.

6.4 Classical Ideal Gas

As a simple application, consider the classical ideal gas in the grand canonical ensemble. The grand partition function of the gas is given by

$$Q(\beta, V, z) = \sum_{N=0}^{\infty} z^N \frac{1}{N!}[Z(\beta, V, 1)]^N = e^{zZ(\beta, V, 1)}, \tag{6.46}$$

where use has been made of (5.35). Therefore,

$$\ln Q(\beta, V, z) = zZ(\beta, V, 1) = z\frac{V}{\Lambda^3}, \tag{6.47}$$

because, according to (5.30), $Z(\beta, V, 1) = V/\Lambda^3$.

The equation for the average number of particles (6.31) reads

$$\langle N \rangle = z\frac{V}{\Lambda^3}, \tag{6.48}$$

from which it readily follows that $z = \rho\Lambda^3$, where $\rho = \langle N \rangle/V$ is the average density of the gas and hence

$$\mu = k_B T \ln(\rho\Lambda^3). \tag{6.49}$$

The solution to the system of parametric equations is in this case trivial, and so substitution of the value of the fugacity $z = \rho\Lambda^3$ into (6.26), (6.29), (6.34), and (6.36) leads to

$$\langle H_N \rangle = \frac{3}{2}\langle N \rangle k_B T, \tag{6.50}$$

$$\langle H_N^2 \rangle - \langle H_N \rangle^2 = \frac{15}{4}\langle N \rangle(k_B T)^2, \tag{6.51}$$

$$\langle N^2 \rangle - \langle N \rangle^2 = \langle N \rangle, \tag{6.52}$$

and

$$p = \rho k_B T . \tag{6.53}$$

Note that (6.50) is a generalization of the equipartition theorem, where N is replaced by $\langle N \rangle$. The energy fluctuation (6.51) is greater in this case than the one corresponding to the canonical ensemble, $3N(k_B T)^2/2$ assuming that $\langle N \rangle = N$, since when the number of particles fluctuates (something that does not occur in the canonical ensemble) these fluctuations also contribute to the energy fluctuation. Note that (6.51) and (6.52) show that the relative fluctuations

$$\frac{\langle H_N^2 \rangle - \langle H_N \rangle^2}{\langle H_N \rangle^2} = \frac{5}{3\langle N \rangle}, \quad \frac{\langle N^2 \rangle - \langle N \rangle^2}{\langle N \rangle^2} = \frac{1}{\langle N \rangle}, \tag{6.54}$$

tend to zero in the thermodynamic limit. On the other hand, (6.53) is, as is well known, the equation of state of the classical ideal gas.

6.5 Classical Ideal Gas in an External Potential

Consider now a classical ideal gas in a one-body external potential $\phi(\mathbf{r})$. From (6.46) and (5.38) the grand partition function $Q(\beta, V, z)$ reads

$$\ln Q(\beta, V, z) = z \frac{1}{\Lambda^3} \int_R d\mathbf{r}\, e^{-\beta \phi(\mathbf{r})}, \tag{6.55}$$

and, hence, the grand potential $\Omega(\beta, V, z) = -k_B T \ln Q(\beta, V, z)$ can be written as

$$\beta \Omega(\beta, V, z) = -\frac{1}{\Lambda^3} \int_R d\mathbf{r}\, e^{\beta u(\mathbf{r})} \equiv \beta \Omega[u], \tag{6.56}$$

which is a functional, denoted by $\Omega[u]$, of $u(\mathbf{r}) = \mu - \phi(\mathbf{r})$, with $\mu = k_B T \ln z$ the chemical potential. The external potential $\phi(\mathbf{r})$ induces a non-uniform local density of particles $\rho_1(\mathbf{r})$:

$$\rho_1(\mathbf{r}) = \left\langle \sum_{i=1}^N \delta(\mathbf{r} - \mathbf{r}_i) \right\rangle = \frac{1}{\Lambda^3} e^{\beta u(\mathbf{r})}, \tag{6.57}$$

which is the barometric formula for the density distribution in the presence of an external field. It is seen from (6.57) that there is a one-to-one correspondence between $\rho_1(\mathbf{r})$ and $u(\mathbf{r})$.

From (6.56) one has

$$\beta \delta \Omega[u] = -\frac{1}{\Lambda^3} \int_R d\mathbf{r} \left(e^{\beta[u(\mathbf{r})+\delta u(\mathbf{r})]} - e^{\beta u(\mathbf{r})} \right)$$

$$= -\frac{1}{\Lambda^3} \int_R d\mathbf{r} e^{\beta u(\mathbf{r})} \beta \delta u(\mathbf{r}) + \cdots \tag{6.58}$$

i.e., the first functional derivative of the grand potential reads

$$\frac{\delta \Omega[u]}{\delta u(\mathbf{r})} = -\rho_1(\mathbf{r}). \tag{6.59}$$

From (6.57) it follows that

$$\delta \rho_1(\mathbf{r}) = \frac{1}{\Lambda^3} \left(e^{\beta[u(\mathbf{r})+\delta u(\mathbf{r})]} - e^{\beta u(\mathbf{r})} \right) = \beta \frac{1}{\Lambda^3} e^{\beta u(\mathbf{r})} \delta u(\mathbf{r}) + \cdots$$

$$= \beta \int_R d\mathbf{r}' \rho_1(\mathbf{r}) \delta(\mathbf{r} - \mathbf{r}') \delta u(\mathbf{r}') + \cdots \tag{6.60}$$

i.e.,

$$\frac{\delta \rho_1(\mathbf{r})}{\delta u(\mathbf{r}')} = \beta \rho_1(\mathbf{r}) \delta(\mathbf{r} - \mathbf{r}'), \tag{6.61}$$

yielding

$$\frac{\delta^2 \Omega[u]}{\delta u(\mathbf{r}) \, \delta u(\mathbf{r}')} = -\beta \rho_1(\mathbf{r}) \delta(\mathbf{r} - \mathbf{r}') < 0, \tag{6.62}$$

i.e., $\Omega[u]$ is a concave functional of $u(\mathbf{r})$.

Now define the intrinsic Helmholtz free energy functional $\mathcal{F}[\rho_1]$ as the Legendre transform of $\Omega[u]$:

$$\mathcal{F}[\rho_1] = \Omega[u] + \int_R d\mathbf{r} \, u(\mathbf{r}) \rho_1(\mathbf{r}), \tag{6.63}$$

where $u(\mathbf{r})$ in (6.63) is the solution of (6.57), i.e.,

$$u(\mathbf{r}) = k_B T \ln \left(\rho_1(\mathbf{r}) \Lambda^3 \right). \tag{6.64}$$

An elementary calculation leads to

$$\mathcal{F}[\rho_1] = k_B T \int_R d\mathbf{r}\, \rho_1(\mathbf{r})\big(\ln\big(\rho_1(\mathbf{r})\Lambda^3\big) - 1\big). \tag{6.65}$$

As will be shown in the next chapter, the equilibrium local density $\rho_1(\mathbf{r})$ of an open system in an external potential $\phi(\mathbf{r})$ can be obtained by a variational principle. For an ideal gas the demonstration is simple. Indeed, define the functional

$$A[\overline{\rho}_1] \equiv \mathcal{F}[\overline{\rho}_1] - \int_R d\mathbf{r}\, u(\mathbf{r})\overline{\rho}_1(\mathbf{r}), \tag{6.66}$$

and observe that when $\overline{\rho}_1(\mathbf{r}) = \rho_1(\mathbf{r})$, (6.63) and (6.66) yield $A[\rho_1] = \Omega[u]$, i.e., the functional reduces to the grand potential. The first functional derivative of (6.66) at constant $u(\mathbf{r})$ is

$$\frac{\delta A[\overline{\rho}_1]}{\delta\overline{\rho}_1(\mathbf{r})} = k_B T\, \ln\big(\overline{\rho}_1(\mathbf{r})\Lambda^3\big) - u(\mathbf{r}), \tag{6.67}$$

which, after (6.64), vanishes at the equilibrium local density $\rho_1(\mathbf{r})$. Moreover, the second functional derivative is given by

$$\frac{\delta^2 A[\overline{\rho}_1]}{\delta\overline{\rho}_1(\mathbf{r})\delta\overline{\rho}_1(\mathbf{r}')} = k_B T\frac{\delta(\mathbf{r} - \mathbf{r}')}{\overline{\rho}_1(\mathbf{r})}, \tag{6.68}$$

i.e., $A[\overline{\rho}_1]$ is a convex functional of $\overline{\rho}_1(\mathbf{r})$. In summary, the equilibrium local density $\rho_1(\mathbf{r})$ minimizes the functional $A[\overline{\rho}_1]$ for a fixed external potential $u(\mathbf{r})$. In the next chapter it will be shown that this variational principle also holds for interacting systems.

6.6   Two-Particle Distribution Function

Consider a system of interacting particles in a one-particle external potential $\phi(\mathbf{r})$. The Hamiltonian is

$$H_N(\mathbf{r}^N, \mathbf{p}^N) = \sum_{j=1}^N \frac{\mathbf{p}_j^2}{2m} + U_N(\mathbf{r}^N)$$

$$= \sum_{j=1}^N \frac{\mathbf{p}_j^2}{2m} + H_N^{int}(\mathbf{r}^N) + \sum_{j=1}^N \phi(\mathbf{r}_j), \tag{6.69}$$

which is the sum of the kinetic energy, the interatomic potential energy, and the energy of the one-particle external potential. Observe that in (6.69) the interatomic potential energy may consist of pair, triple, etc., interactions. The grand partition function is

$$Q(\beta, V, z) = \sum_{N=0}^{\infty} z^N \frac{1}{N! \Lambda^{3N}} \int_R d\mathbf{r}^N \exp[-\beta U_N(\mathbf{r}^N)], \qquad (6.70)$$

where the elementary integration over the momenta has been performed. For simplicity, denote by $\widehat{\rho}_1(\mathbf{r})$ the dynamical function that represents the local density of particles (4.69), i.e.,

$$\widehat{\rho}_1(\mathbf{r}) = \sum_{j=1}^{N} \delta(\mathbf{r} - \mathbf{r}_j). \qquad (6.71)$$

Since $z = e^{\beta \mu}$ one finds

$$N\mu - \sum_{j=1}^{N} \phi(\mathbf{r}_j) = \int_R d\mathbf{r} \sum_{j=1}^{N} \delta(\mathbf{r} - \mathbf{r}_j)(\mu - \phi(\mathbf{r}))$$

$$= \int_R d\mathbf{r}\, u(\mathbf{r})\widehat{\rho}_1(\mathbf{r}), \qquad (6.72)$$

and the grand potential $\Omega(\beta, V, z) = -k_B T \ln Q(\beta, V, z) \equiv \Omega[u]$, which is a functional of $u(\mathbf{r}) = \mu - \phi(\mathbf{r})$, can be written as

$$e^{-\beta \Omega[u]} = \sum_{N=0}^{\infty} \frac{1}{N! \Lambda^{3N}} \int_R d\mathbf{r}^N I_N(\mathbf{r}^N; [u]), \qquad (6.73)$$

where

$$I_N(\mathbf{r}^N; [u]) \equiv \exp\left[-\beta\left\{H_N^{int}(\mathbf{r}^N) - \int_R d\mathbf{r}\, u(\mathbf{r})\widehat{\rho}_1(\mathbf{r})\right\}\right]. \qquad (6.74)$$

This result is a straightforward extension of that obtained for the ideal gas. Observe, however, that for interacting systems the two first functional derivatives of $\Omega[u]$ yield different distribution functions. The first functional derivative of (6.73) is

$$\frac{\delta \Omega[u]}{\delta u(\mathbf{r})} = -e^{\beta \Omega[u]} \sum_{N=0}^{\infty} \frac{1}{N! \Lambda^{3N}} \int_R d\mathbf{r}^N \widehat{\rho}_1(\mathbf{r}) I_N(\mathbf{r}^N; [u]), \qquad (6.75)$$

i.e.,

$$\frac{\delta\Omega[u]}{\delta u(\mathbf{r})} = -\left\langle\hat{\rho}_1(\mathbf{r})\right\rangle = -\rho_1(\mathbf{r}), \tag{6.76}$$

where $\rho_1(\mathbf{r})$ is the non-uniform local density of particles (one-particle distribution function) induced by the external potential. A further differentiation of (6.75) leads to

$$\frac{\delta^2\Omega[u]}{\delta u(\mathbf{r})\delta u(\mathbf{r}')} = -\beta e^{\beta\Omega[u]} \sum_{N=0}^{\infty} \frac{1}{N!\Lambda^{3N}} \int_R d\mathbf{r}^N \hat{\rho}_1(\mathbf{r})\hat{\rho}_1(\mathbf{r}') I_N(\mathbf{r}^N;[u])$$
$$+ \beta\frac{\delta\Omega[u]}{\delta u(\mathbf{r})}\frac{\delta\Omega[u]}{\delta u(\mathbf{r}')} \tag{6.77}$$

i.e.,

$$\frac{\delta^2\Omega[u]}{\delta u(\mathbf{r})\delta u(\mathbf{r}')} = -\beta\left(\left\langle\hat{\rho}_1(\mathbf{r})\hat{\rho}_1(\mathbf{r}')\right\rangle - \left\langle\hat{\rho}_1(\mathbf{r})\right\rangle\left\langle\hat{\rho}_1(\mathbf{r}')\right\rangle\right) < 0, \tag{6.78}$$

i.e., $\Omega[u]$ is a concave functional of $u(\mathbf{r})$.

From (6.71) one has

$$\left\langle\hat{\rho}_1(\mathbf{r})\hat{\rho}_1(\mathbf{r}')\right\rangle = \left\langle\sum_{j=1}^{N}\delta(\mathbf{r}-\mathbf{r}_j)\sum_{i=1}^{N}\delta(\mathbf{r}'-\mathbf{r}_i)\right\rangle$$
$$= \delta(\mathbf{r}-\mathbf{r}')\left\langle\sum_{j=1}^{N}\delta(\mathbf{r}-\mathbf{r}_j)\right\rangle$$
$$+ \left\langle\sum_{j=1}^{N}\sum_{j\neq i=1}^{N}\delta(\mathbf{r}-\mathbf{r}_j)\delta(\mathbf{r}'-\mathbf{r}_i)\right\rangle, \tag{6.79}$$

where the contributions $i=j$ and $i\neq j$ have been separated, yielding

$$\left\langle\hat{\rho}_1(\mathbf{r})\hat{\rho}_1(\mathbf{r}')\right\rangle = \rho_1(\mathbf{r})\,\delta(\mathbf{r}-\mathbf{r}') + \rho_2(\mathbf{r},\mathbf{r}'), \tag{6.80}$$

with $\rho_2(\mathbf{r},\mathbf{r}')$ the two-particle distribution function defined as

$$\rho_2(\mathbf{r},\mathbf{r}') = \left\langle\sum_{j=1}^{N}\sum_{j\neq i=1}^{N}\delta(\mathbf{r}-\mathbf{r}_j)\delta(\mathbf{r}'-\mathbf{r}_i)\right\rangle. \tag{6.81}$$

From (6.76), (6.78), and (6.80) one obtains

$$\frac{\delta \rho_1(\mathbf{r})}{\delta u(\mathbf{r}')} = \beta \left[\rho_2(\mathbf{r}, \mathbf{r}') + \rho_1(\mathbf{r}) \, \delta(\mathbf{r} - \mathbf{r}') - \rho_1(\mathbf{r}) \rho_1(\mathbf{r}') \right]. \tag{6.82}$$

It is seen, by comparison of (6.82) and (6.61), that for an ideal gas $\rho_2(\mathbf{r}, \mathbf{r}') = \rho_1(\mathbf{r}) \rho_1(\mathbf{r}')$.

6.7 Density Fluctuations

The fluctuation in the number of particles in the grand canonical ensemble is given by (6.34) and, as has been indicated in Sect. 6.2, it is a function of β, V, and z. This fluctuation, as will be shown shortly, is related to the isothermal compressibility coefficient of the system. To that end, consider the identities:

$$\left(\frac{\partial \langle N \rangle}{\partial z} \right)_{\beta,V} = \left(\frac{\partial \langle N \rangle}{\partial p} \right)_{\beta,V} \left(\frac{\partial p}{\partial z} \right)_{\beta,V}, \tag{6.83}$$

and

$$\left(\frac{\partial \langle N \rangle}{\partial p} \right)_{\beta,V} = -\left(\frac{\partial \langle N \rangle}{\partial V} \right)_{\beta,p} \left(\frac{\partial V}{\partial p} \right)_{\beta,\langle N \rangle}, \tag{6.84}$$

where, to be more precise, the variables that remain constant in each partial derivative have been indicated by a subscript. Note that if $\langle N \rangle = \langle N \rangle (\beta, V, p)$, since β and p are intensive variables, in the thermodynamic limit $\langle N \rangle$ has to be a homogeneous function of degree one in V so that

$$\mathrm{TL} \left[\left(\frac{\partial \langle N \rangle}{\partial V} \right)_{\beta,p} \right] = \mathrm{TL} \left[\frac{\langle N \rangle}{V} \right] = \rho, \tag{6.85}$$

where ρ is the average density of particles. On the other hand, from (6.31) and (6.36) it follows that

$$\begin{aligned}
z \left(\frac{\partial p}{\partial z} \right)_{\beta,V} &= z \left(\frac{\partial}{\partial z} \right)_{\beta,V} k_B T \left(\frac{\partial \ln Q(\beta, V, z)}{\partial V} \right)_{\beta,z} \\
&= k_B T \left(\frac{\partial}{\partial V} \right)_{\beta,z} z \left(\frac{\partial \ln Q(\beta, V, z)}{\partial z} \right)_{\beta,V} \\
&= k_B T \left(\frac{\partial \langle N \rangle}{\partial V} \right)_{\beta,z},
\end{aligned} \tag{6.86}$$

and since β and z are intensive variables

$$\mathrm{TL}\left[z\left(\frac{\partial p}{\partial z}\right)_{\beta,V}\right] = k_B T \ \mathrm{TL}\left[\frac{\langle N \rangle}{V}\right] = k_B T \rho \,. \tag{6.87}$$

Using (6.83) and (6.87), it is found that (6.34) may be expressed as

$$\mathrm{TL}\left[\frac{\langle N^2 \rangle - \langle N \rangle^2}{\langle N \rangle}\right] = \rho k_B T \chi_T \,, \tag{6.88}$$

where the isothermal compressibility coefficient χ_T is defined by

$$\chi_T \equiv -\frac{1}{V}\left(\frac{\partial V}{\partial p}\right)_{\beta,\langle N \rangle} \,. \tag{6.89}$$

Equation (6.88) expresses the fluctuation in the number of particles with respect to the average value as a function of the temperature, the average density, and the isothermal compressibility coefficient of the system. Note that in the thermodynamic limit $\chi_T = \chi_T(T, \rho)$, and hence if the system is a fluid, either a vapor or a liquid phase, where χ_T is finite

$$\mathrm{TL}\left[\frac{\langle N^2 \rangle - \langle N \rangle^2}{\langle N \rangle}\right]_{T,\rho} < \infty \,. \tag{6.90}$$

An important exception to this result takes place at the liquid–vapor critical point, since the critical isotherm T_c has an inflexion point (see Chap. 9) at $\rho = \rho_c$, where p_c is the critical density, i.e., $\chi_T(T_c, \rho_c) = \infty$. Therefore, at the critical point,

$$\mathrm{TL}\left[\frac{\langle N^2 \rangle - \langle N \rangle^2}{\langle N \rangle}\right]_{T_c,\rho_c} = \infty, \tag{6.91}$$

and the density fluctuations become observable in a phenomenon called "critical opalescence" (i.e., during the dispersion of light, the resulting fluctuations of the index of refraction of the fluid make it to acquire a milky appearance).

6.8   Correlations at the Critical Point

The density fluctuations of a fluid may also be analyzed taking a different approach. Consider again the dynamical function (6.71):

$$\hat{\rho}_1(\mathbf{r}) = \sum_{j=1}^{N} \delta(\mathbf{r} - \mathbf{r}_j) \,. \tag{6.92}$$

Assume that the fluid is formed by N particles contained in a closed region R of volume V at temperature T. Let $R' \subset R$ be an open region of volume V'. The dynamical function number of particles contained in R', $N(q, p; V')$, is, according to (6.92):

$$N(q, p; V') = \int_{R'} d\mathbf{r} \sum_{j=1}^{N} \delta(\mathbf{r} - \mathbf{r}_j), \qquad (6.93)$$

whose average value in the grand canonical ensemble turns out to be

$$\langle N \rangle_{V'} = \int_{R'} d\mathbf{r} \left\langle \sum_{j=1}^{N} \delta(\mathbf{r} - \mathbf{r}_j) \right\rangle = \int_{R'} d\mathbf{r} \rho_1(\mathbf{r}), \qquad (6.94)$$

where

$$\rho_1(\mathbf{r}) = \left\langle \sum_{j=1}^{N} \delta(\mathbf{r} - \mathbf{r}_j) \right\rangle, \qquad (6.95)$$

is the local density of particles.

In order to determine the fluctuation in the number of particles in R', one has to compute the square of (6.93) and take the average over the grand canonical ensemble, namely

$$\langle N^2 \rangle_{V'} = \int_{R'} d\mathbf{r} \int_{R'} d\mathbf{r}' \left\langle \sum_{j=1}^{N} \sum_{i=1}^{N} \delta(\mathbf{r} - \mathbf{r}_j) \delta(\mathbf{r}' - \mathbf{r}_i) \right\rangle \qquad (6.96)$$

or, using (6.80)

$$\langle N^2 \rangle_{V'} = \int_{R'} d\mathbf{r}\, \rho_1(\mathbf{r}) + \int_{R'} d\mathbf{r} \int_{R'} d\mathbf{r}'\, \rho_2(\mathbf{r}, \mathbf{r}'). \qquad (6.97)$$

In the thermodynamic limit $V \to \infty$, $N \to \infty$ with $\rho = N/V < \infty$, the fluid is invariant under any translation or rotation, so that for every vector $\mathbf{a}$ one has

$$\rho_1(\mathbf{r}) = \rho_1(\mathbf{r} + \mathbf{a}),$$

and

$$\rho_2(\mathbf{r}, \mathbf{r}') = \rho_2(\mathbf{r} + \mathbf{a}, \mathbf{r}' + \mathbf{a}).$$

From the first equation it follows that the local density of particles is uniform (independent of $\mathbf{r}$) $\rho_1(\mathbf{r}) = \rho$ and from the second equation it follows that the two-particle distribution function only depends on the relative position $\mathbf{r} - \mathbf{r}'$. Due to the invariance under rotations, $\rho_2(\mathbf{r}, \mathbf{r}')$ has to be a function of the scalar $|\mathbf{r} - \mathbf{r}'|$, i.e., $\rho_2(\mathbf{r}, \mathbf{r}') = \rho_2(|\mathbf{r} - \mathbf{r}'|)$. Therefore, (6.97) may be written as

$$\langle N^2 \rangle_{V'} - \langle N \rangle^2_{V'} = \rho V' + \rho^2 \int_{R'} d\mathbf{r} \int_{R'} d\mathbf{r}' h(|\mathbf{r} - \mathbf{r}'|), \tag{6.98}$$

where the total correlation function $h(r)$ has been defined by

$$\rho^2 h(r) \equiv \rho_2(r) - \rho^2. \tag{6.99}$$

Finally, in the thermodynamic limit $\langle N \rangle_{V'}/V' \to \rho$ and from (6.98) it follows that

$$\mathrm{TL}\left[\frac{\langle N^2 \rangle - \langle N \rangle^2}{\langle N \rangle} \right] = 1 + \rho \int d\mathbf{r} h(r; T, \rho), \tag{6.100}$$

where the integral extends over the whole Euclidean space, and the change in notation from $h(r)$ to $h(r; T,\rho)$ is analyzed in the following. From the comparison of (6.100) with (6.88) one has

$$1 + \rho \int d\mathbf{r}\, h(r; T, \rho) = \rho k_B T \chi_T(T, \rho), \tag{6.101}$$

which is called the compressibility equation and is one of the most important equations in equilibrium statistical physics. Before examining the interest of (6.101), one must provide an interpretation for the total correlation function $h(r; T,\rho)$.

Note that from the definition of the two-particle distribution function (6.81), it follows that this is the marginal probability density of finding any two particles of the fluid at points $\mathbf{r}$ and $\mathbf{r}'$ of the Euclidean space, since the average contains the product $\delta(\mathbf{r} - \mathbf{r}_j)\delta(\mathbf{r}' - \mathbf{r}_i)$, which implies that the variables $\mathbf{r}_j$ and $\mathbf{r}_i$ are not integrated over in the statistical average (see Appendix B). In (6.81) the integration is over the position of the remaining particles of the fluid, as well as over the momenta of all the particles including the j-th and the i-th. On the other hand, the fluid is in equilibrium with the external world of temperature T and chemical potential μ, since the average is over the grand canonical ensemble. Since in the thermodynamic limit $\mu = \mu(T, \rho)$, the two-particle distribution function in (6.81) should be written in this limit as $\rho_2(|\mathbf{r} - \mathbf{r}'|; T, \rho)$. This is the notation that has been used in (6.101). Although this function is extremely complicated one may still make some simple considerations about it.

In the first place, note that in the thermodynamic limit the individual probability density of finding any one particle at a point $\mathbf{r}$ of the Euclidean space, a quantity denoted by $\rho_1(\mathbf{r})$, is a constant (it cannot depend on $\mathbf{r}$). Hence, in the thermodynamic limit it follows that $\rho_1(\mathbf{r}) = \rho$, which is the average density of particles. As seen in Sect. 6.6, in an ideal fluid $\rho_2(|\mathbf{r} - \mathbf{r}'|; T, \rho) = \rho^2$, which expresses that the marginal probability density of finding any two particles of the fluid at $\mathbf{r}$ and $\mathbf{r}'$ is the product of the individual probability densities, i.e., in ideal systems the coordinates of the particles are statistically independent random variables. In an ideal fluid, therefore, $h(r; T, \rho) = 0$ and from (6.101) it follows that $\chi_T(T, \rho) = (\rho k_B T)^{-1}$, which is a result that may be easily derived from the equation of state $p = \rho k_B T$.

From the statistical independence in an ideal fluid one concludes that in a fluid with interactions $\rho_2(|\mathbf{r} - \mathbf{r}'|; T, \rho) \neq \rho^2$, i.e., the interaction creates a correlation between the particles. One would expect that $\rho_2(|\mathbf{r} - \mathbf{r}'|; T, \rho) \rightarrow \rho^2$ when $|\mathbf{r} - \mathbf{r}'| \rightarrow \infty$, because in statistical physics one considers short-range interactions (see Chap. 3), and thus there should be no correlations between the particles if they are very far away from each other. Therefore, one defines the total correlation function $h(r; T, \rho)$ through (6.99), which satisfies the condition

$$\lim_{r \to \infty} h(r; T, \rho) = 0. \tag{6.102}$$

Another point to bear in mind is that, although the interactions in a fluid create correlations between the particles, in order to derive (6.101) no specific form for the interaction potential has been required and so the compressibility equation is valid for all classical fluids, with the only restriction that the interaction potential has to be of short range. Note further that in its derivation it has been assumed that the density of the fluid ρ is uniform, which does not allow the use of the compressibility equation in the liquid–vapor coexistence region (see Chap. 11).

Once these considerations have been made, the importance of the compressibility equation should be clear. Note that when the fluid is in the vapor phase or in the liquid phase $\chi_T(T, \rho) \neq \infty$ and, as a consequence, the integral of the total correlation function is finite. Because the particles have a finite size (see Chap. 7), $h(r; T, \rho) = -1 \, (r < \sigma)$, where σ is the diameter of the particle, as follows from (6.99), since the probability of finding two particles at a distance less than the diameter is zero $\rho_2(r; T, \rho) = 0 \, (r < \sigma)$. Hence

$$\int d\mathbf{r}\, h(r; T, \rho) = 4\pi \int_0^{\infty} dr\, r^2 h(r; T, \rho) < \infty, \tag{6.103}$$

which implies that $h(r; T, \rho)$ is short-ranged, in the sense that it must tend to zero at least as fast as r^{-4} to avoid the divergence of the integral at the upper limit.

On the contrary, at the critical point, $\chi_T(T_c, \rho_c) = \infty$, the integral of the correlation function diverges at the upper limit:

$$\int d\mathbf{r}\, h(r; T_c, \rho_c) = 4\pi \int_0^\infty dr\, r^2 h(r; T_c, \rho_c) = \infty, \tag{6.104}$$

which implies that $h(r; T_c, \rho_c)$ is of infinite range (for instance, a function of the type of r^{-1}). Therefore, at the liquid–vapor critical point the correlations are of infinite range even if the interaction potential is of short range. This result is fundamental in the study of the thermodynamic properties of a fluid in the vicinity of the critical point, usually referred to as "critical phenomena" (see Chap. 10).

6.9   Quantum Grand Canonical Ensemble

As has been analyzed in the classical case, the probability density in the phase space of a system that exchanges energy and particles with the external world, of temperature T and fugacity z, is proportional to the factor $z^N e^{-\beta H_N(q,p;V)}$, where $H_N(q,p;V)$ is the Hamiltonian of a system of N particles contained in a closed region R of volume V. For the derivation of the grand canonical ensemble from (6.1), the additivity of the energy and of the number of particles has been taken into account as well as the fact that the system is much smaller than the external world. Since these concepts are also valid in quantum mechanics, it is postulated that the density operator of a quantum system of N particles, whose Hamiltonian operator is $\hat{H}_N(V)$, is proportional to $z^{\hat{N}} e^{-\beta \hat{H}_N(V)}$, where $\hat{N}$ is the number of particles operator, which commutes with $\hat{H}_N(V)$. It is assumed, therefore, that the density operator is diagonal in the energy representation and that its elements are given by

$$\rho_{mn} = \langle \Psi_m | \hat{\rho} | \Psi_n \rangle = \frac{\delta_{mn}}{Q(\beta, V, z)} z^N e^{-\beta E_n^{(N)}(V)}, \tag{6.105}$$

where n denotes the quantum state of a system of N particles which corresponds to the eigenvector $|\Psi_n\rangle$ and whose energy is $E_n^{(N)}(V)$. From the normalization condition of the density operator (3.30) it follows that the quantum grand partition function $Q(\beta, V, z)$ in (6.105) is given by

$$Q(\beta, V, z) = \sum_{N=0}^\infty z^N \sum_n e^{-\beta E_n^{(N)}(V)} = \sum_{N=0}^\infty z^N Z(\beta, V, N), \tag{6.106}$$

where $Z(\beta, V,N)$ is the quantum partition function (5.88), whose sum extends to all quantum states of the eigenvalue equation of the Hamiltonian operator (4.85).

The expressions for the average values and the fluctuations derived for a classical system (see Sect. 6.2) may also be applied to a quantum system, although in this case the grand partition function must be evaluated using (6.106). Note that since the quantum partition function contains only those states whose wave function is either symmetric or antisymmetric, there is no Gibbs paradox in quantum systems.

6.10 Bose–Einstein and Fermi–Dirac Statistics

Consider an ideal gas of N particles of mass m contained in a closed region R of volume V at temperature T. The partition function of the gas is, according to (5.94), given by

$$Z(\beta, V, N) = \sum_{\{n_i\}} {}^* e^{-\beta \sum_i n_i \varepsilon_i(V)}, \tag{6.107}$$

where the energies $\varepsilon_i(V)$ of the particle quantum states are given by (1.63), namely

$$\varepsilon_i(V) \equiv \frac{2\pi^2 \hbar^2}{mV^{2/3}} (n_x^2 + n_y^2 + n_z^2), \quad (n_x, n_y, n_z = 0, \pm 1, \ldots), \tag{6.108}$$

a result obtained by dividing the Euclidean space in cubic boxes of side L and volume $V = L^3$ and imposing periodicity, with period L, of the wave function in all three directions of space. In (6.107) the asterisk indicates that the occupation numbers of the particle quantum states n_i, which may take the values $n_i = 0, 1$(fermions) and $n_i = 0, 1, 2, \ldots$(bosons), must satisfy the condition

$$\sum_i n_i = N \tag{6.109}$$

and it is this restriction which does not allow one to determine the partition function explicitly.

The grand partition function of the gas is given by

$$Q(\beta, V, z) = \sum_{N=0}^{\infty} z^N Z(\beta, V, N) = \sum_{N=0}^{\infty} z^N \sum_{\{n_i\}} {}^* e^{-\beta \sum_i n_i \varepsilon_i(V)}, \tag{6.110}$$

which is a series in the fugacity z whose coefficients are the partition functions $Z(\beta, V, N)$. In principle it does not seem likely for one to be able to sum a series whose coefficients are unknown. Note, however, that the restriction (6.109) that appears in each of the coefficients of the series is irrelevant in the evaluation of $Q(\beta, V, z)$ since summing over N, which may take any value between zero and infinity, is equivalent to summing over all the occupation numbers without restriction. In this way, (6.110) may be written as

$$Q(\beta, V, z) = \sum_{\{n_i\}} z^{\sum_i n_i} e^{-\beta \sum_i n_i \varepsilon_i(V)}, \tag{6.111}$$

where the numbers n_i are now independent variables (which do not have to satisfy the condition (6.109)), and so the grand partition function may be factorized as follows:

$$Q(\beta, V, z) = \prod_i \sum_{n_i} \left(z e^{-\beta \varepsilon_i(V)}\right)^{n_i}, \tag{6.112}$$

where the product extends to all one-particle quantum states.

For fermions ($n_i = 0, 1$) from (6.112) it immediately follows that

$$Q(\beta, V, z) = \prod_i \left(1 + z e^{-\beta \varepsilon_i(V)}\right). \tag{6.113}$$

For bosons ($n_i = 0, 1, 2, \ldots$) the sum over the occupation numbers in (6.112) is, for each of the one-particle quantum states of energy $\varepsilon_i(V)$, that of a geometric progression of infinite terms and of ratio $z e^{-\beta \varepsilon_i(V)}$. According to (6.108), the energies of the one-particle quantum states are positive, except for that of the ground state which is zero. In order for the series to converge, one has to impose, therefore, that $z < 1$, which guarantees that the ratio is less than one. With this condition, the grand partition function for the boson gas may be easily determined yielding

$$Q(\beta, V, z) = \prod_i \left(\frac{1}{1 - z e^{-\beta \varepsilon_i(V)}}\right) \quad (z < 1). \tag{6.114}$$

If one takes the logarithm of expressions (6.113) and (6.114), the two resulting equations may be condensed into one that reads

$$\ln Q(\beta, V, z) = \theta \sum_i \ln\left(1 + \theta z e^{-\beta \varepsilon_i(V)}\right), \tag{6.115}$$

where $\theta = 1$ for fermions and $\theta = -1$ for bosons, in which case the fugacity has to be less than one. From (6.115) and (6.31) it follows that the average number of particles of the gas is given by

$$\langle N \rangle = \sum_i \frac{z e^{-\beta \varepsilon_i(V)}}{1 + \theta z e^{-\beta \varepsilon_i(V)}}, \tag{6.116}$$

which is a sum that extends over all the one-particle quantum states, so that the average number of particles $\langle n_i \rangle$ in a quantum state i of energy $\varepsilon_i(V)$ turns out to be

$$\langle n_i \rangle = \frac{z e^{-\beta \varepsilon_i(V)}}{1 + \theta z e^{-\beta \varepsilon_i(V)}}, \tag{6.117}$$

which particularized to $\theta = -1$ and $\theta = 1$ gives rise to the Bose–Einstein and to the Fermi–Dirac statistics, respectively.

Note that from (6.115) and (6.36) it follows that the pressure of the gas is given by

$$p = -\sum_i \langle n_i \rangle \frac{\partial \varepsilon_i(V)}{\partial V}, \tag{6.118}$$

where use has been made of (6.117). From (6.108) one has

$$\frac{\partial \varepsilon_i(V)}{\partial V} = -\frac{2}{3V} \varepsilon_i(V), \tag{6.119}$$

and hence

$$p = \frac{2}{3V} \sum_i \langle n_i \rangle \, \varepsilon_i(V) = \frac{2}{3V} \langle H_N \rangle, \tag{6.120}$$

a result which is independent of the statistics.

For the sake of comparison, consider an ideal gas in the Maxwell–Boltzmann statistics. The partition function of the gas is given by

$$Z^*(\beta, V, N) = \frac{1}{N!}[Z(\beta, V, 1)]^N, \tag{6.121}$$

where

$$Z(\beta, V, 1) = \sum_i e^{-\beta \varepsilon_i(V)}, \tag{6.122}$$

is the one-particle partition function. The grand partition function of the gas is, therefore,

$$Q(\beta, V, z) = \sum_{N=0}^{\infty} z^N \frac{1}{N!}[Z(\beta, V, 1)]^N = e^{zZ(\beta, V, 1)}, \tag{6.123}$$

from which it follows that

$$\langle N \rangle = zZ(\beta, V, 1) = z \sum_i e^{-\beta \varepsilon_i(V)}, \tag{6.124}$$

i.e., the average number of particles in a quantum state in the Maxwell–Boltzmann statistics turns out to be

$$\langle n_i \rangle = ze^{-\beta \varepsilon_i(V)}. \tag{6.125}$$

Note that if $z \ll 1$, the denominator of (6.117) may be approximated by one and the two quantum statistics reduce to (6.125). Therefore, one calls this approximation the classical limit according to which the quantum gases may be studied with the Maxwell–Boltzmann statistics. In the limit of infinite volume, the one-particle partition function is V/Λ^3, so that from (6.124) it follows that the classical limit $z \ll 1$ may be expressed as

$$\rho \Lambda^3 = \rho \left(\frac{h}{2\pi m k_B T} \right)^{3/2} \ll 1, \qquad (6.126)$$

where $\rho = \langle N \rangle / V$ is the average density of particles. Note that if $v = 1/\rho$ is the volume per particle of the gas and a is the side of the cube $a^3 = v$, the inequality (6.126) may be written as $a \gg \Lambda$, which expresses that the average distance between particles is much greater than the thermal de Broglie wavelength. In this case, quantum effects are not important, and the boson and the fermion gases may be treated as if they were a Maxwell–Boltzmann gas, whose study is rather simpler.

This condition is met, the greater is the temperature of the gas and the smaller is the density, although one must keep in mind that these two thermodynamic variables are combined in (6.126).

6.11 Virial Expansions in the Classical Limit

The Bose–Einstein and Fermi–Dirac statistics reduce to the Maxwell–Boltzmann statistics in the classical limit $z \ll 1$, when the denominator of (6.117) may be approximated by one. This corresponds to considering only the first term in the series expansion $(1 + x)^{-1} = 1 - x + x^2 - x^3 \ldots$, with $x = \theta z e^{-\beta \varepsilon_i(V)}$. When including more terms, one obtains a series in the fugacity. Consider, first, the equation of the average number of particles (6.116), which may be written as

$$\langle N \rangle = \sum_i z e^{-\beta \varepsilon_i(V)} \left(1 - \theta z e^{-\beta \varepsilon_i(V)} + \cdots \right)$$

$$= z \sum_i e^{-\beta \varepsilon_i(V)} - \theta z^2 \sum_i e^{-2\beta \varepsilon_i(V)} + \cdots . \qquad (6.127)$$

In the thermodynamic limit, the one-particle partition function is $Z(\beta, V, 1) = V/\Lambda^3$, and thus it follows that

$$\sum_i e^{-2\beta \varepsilon_i(V)} = \frac{1}{2^{3/2}} \frac{V}{\Lambda^3}, \qquad (6.128)$$

so that (6.127) reads

$$\rho \Lambda^3 = z - \theta \frac{1}{2^{3/2}} z^2 + \cdots \tag{6.129}$$

Since $z \ll 1$, this equation may be solved by iteration, i.e.,

$$z = \rho \Lambda^3 + \theta \frac{1}{2^{3/2}} z^2 + \cdots$$

$$= \rho \Lambda^3 + \theta \frac{1}{2^{3/2}} (\rho \Lambda^3)^2 + \cdots , \tag{6.130}$$

where in the term in z^2, the fugacity, has been replaced by its value at the previous order, $z = \rho \Lambda^3$. When considering more terms in the series of (6.127), the fugacity is given by a series in the density, whose first two terms are the ones obtained in (6.130). This type of expansion (see Chap. 7) appears in classical statistical physics when there exists interaction between the particles. In this way, it is shown that an ideal quantum gas has a similar behavior to the one of a classical gas with interactions.

If one proceeds in the same way as in (6.127) with the logarithm of the grand partition function (6.115), since $\ln(1 + x) = x - x^2/2 + x^3/3 + \ldots$, one has

$$\ln Q(\beta, V, z) = \theta \sum_i \left(\theta z e^{-\beta \varepsilon_i(V)} - \frac{1}{2} z^2 e^{-2\beta \varepsilon_i(V)} + \cdots \right)$$

$$= \sum_i \left(z e^{-\beta \varepsilon_i(V)} - \frac{1}{2} \theta z^2 e^{-2\beta \varepsilon_i(V)} + \cdots \right), \tag{6.131}$$

so that in the thermodynamic limit

$$\ln Q(\beta, V, z) = z \frac{V}{\Lambda^3} - \frac{1}{2} \theta z^2 \frac{1}{2^{3/2}} \frac{V}{\Lambda^3} + \cdots . \tag{6.132}$$

The equation for the pressure is given by

$$\frac{p \Lambda^3}{k_B T} = z - \theta \frac{1}{2^{5/2}} z^2 + \cdots , \tag{6.133}$$

so that upon substitution into (6.133) of the value of the fugacity derived from (6.130) one obtains the virial expansion of the pressure, namely

$$p = \rho k_B T \left(1 + \theta \frac{1}{2^{5/2}} \rho \Lambda^3 + \cdots \right). \tag{6.134}$$

Note that if one only considers the first term, the result is the equation of state for the classical ideal gas $p = \rho k_B T$. From the second term it follows that

$$B_2(T) = \theta \frac{1}{2^{5/2}} \Lambda^3, \tag{6.135}$$

where $B_2(T)$ is the second virial coefficient (see the next chapter), which is positive for fermions and negative for bosons. In classical statistical physics, the second virial coefficient is positive when there is a repulsive interaction potential between the particles and negative if the potential is attractive. Therefore, quantum effects induce a "non-ideal" behavior of the boson and fermion gases. This fact is also reflected in the London–Placzek relation (see R. Balescu, *Equilibrium and Nonequilibrium Statistical Mechanics*, J. Wiley, New (1975)) which expresses that the total correlation function (see (6.99)) of a quantum ideal gas is given by

$$h(r; T, \rho) = -\theta \left| \int d\mathbf{p} \, e^{i\mathbf{p}\cdot\mathbf{r}/\hbar} \frac{ze^{-\beta p^2/2m}}{1 + \theta z e^{-\beta p^2/2m}} \right|^2, \tag{6.136}$$

which is negative for fermions and positive for bosons (note that the fraction in (6.136) is the average number of particles in a one-particle quantum state of momentum $\mathbf{p}$). It has been already stated in Sect. 6.8 that for a classical ideal gas $h(r; T, \rho) = 0$. Since the Bose–Einstein and Fermi–Dirac statistics reduce to the Maxwell–Boltzmann statistics when $\theta = 0$, note that the absence of correlations in a classical ideal gas is also contained in the London–Placzek equation. Observe further that due to Pauli's exclusion principle, if a fermion is in a quantum state, it is not possible to find another fermion in the same state, something that creates a negative correlation $(\rho_2(r; T, \rho) < \rho^2)$ between fermions which, classically, may be interpreted as if the particles interacted through a repulsive interaction potential. Since for bosons, on the contrary, the correlation is positive $(\rho_2(r; T, \rho) > \rho^2)$, and particles may accumulate in the same quantum state, which in classical statistical physics may be interpreted as if there was an attractive interaction potential between bosons.

Since quantum effects in the limit $z \ll 1$ appear as small corrections to the classical ideal behavior, one would expect their importance to be greater in very dense systems or at low temperatures. These cases are analyzed in the following sections.

6.12 Boson Gas: Bose–Einstein Condensation

The average number of particles for a boson gas is, according to (6.116), given by

$$\langle N \rangle = \frac{z}{1-z} + {\sum_i}' \frac{ze^{-\beta \varepsilon_i(V)}}{1 - ze^{-\beta \varepsilon_i(V)}}, \tag{6.137}$$

where the sum extends to the quantum states of energy $\varepsilon_i(V) \neq 0$, while the term $z/(1-z)$ is the contribution of the ground state. Note that at zero temperature ($\beta = \infty$) this is the only nonvanishing term in (6.137), which may be easily interpreted since, as the energy of the gas has to be a minimum, all the particles accumulate in the ground state. When the temperature is increased, at constant average density, the quantum states of energy $\varepsilon_i(V) \neq 0$ begin to be populated with particles that at $T = 0$ were in the state of minimum energy. In order to examine what happens at a temperature $T \neq 0$ one has to bear in mind that, since $\varepsilon_i(V) \sim V^{-2/3}$, in the thermodynamic limit these quantum states form a continuum and the sum in (6.137) may be replaced by the integral of the function, multiplied by the density of quantum states (1.67), namely

$$\langle N \rangle = \frac{z}{1-z} + 2\pi V \left(\frac{2m}{h^2}\right)^{3/2} \int_0^\infty d\varepsilon \sqrt{\varepsilon} \frac{ze^{-\beta\varepsilon}}{1 - ze^{-\beta\varepsilon}}. \tag{6.138}$$

In (6.138) the lower limit in the integral, ε_1, has been replaced by zero since $\varepsilon_1 \sim V^{-2/3}$, an approximation which is exact in the thermodynamic limit. Since $z < 1$, this integral may be written as a series in the fugacity. To that end, expanding $(1-x)^{-1} = \sum_{q=0}^\infty x^q$ and integrating with respect to ε, one obtains

$$\langle N \rangle = \frac{z}{1-z} + \frac{V}{\Lambda^3} \sum_{q=1}^\infty \frac{z^q}{q^{3/2}} = \frac{z}{1-z} + \frac{V}{\Lambda^3} g_{3/2}(z), \tag{6.139}$$

where

$$g_l(z) = \sum_{q=1}^\infty \frac{z^q}{q^l}. \tag{6.140}$$

Assume that the conditions of density and temperature are such that $z \ll 1$ (classical limit). If the fugacity is increased, $z < 1$, this also increases the two contributions to $\langle N \rangle$ in (6.139), because the functions $z/(1-z)$ and $g_{3/2}(z)$ grow monotonously with z. Note, however, that

$$\mathrm{TL}\left[\frac{1}{V}\frac{z}{1-z}\right] = 0 \quad (z < 1), \tag{6.141}$$

and so in the thermodynamic limit (6.139) reads

$$\rho = \frac{1}{\Lambda^3} g_{3/2}(z) \quad (z < 1), \tag{6.142}$$

i.e., only the particles in the states $\varepsilon_i(V) \neq 0$ contribute to the average density ρ of the gas. Increasing z in (6.142) at constant average density decreases the temperature of the gas. Since $g_{3/2}(z)$ has as an upper bound the Riemann function $\zeta(3/2)$, where

$$\zeta(l) = \sum_{q=1}^\infty \frac{1}{q^l}, \tag{6.143}$$

from (6.142) it follows that the gas may be cooled up to a critical value of the temperature $T_c(p)$ defined by

$$\rho \Lambda_c^3(\rho) \equiv \zeta(3/2),$$ (6.144)

where $\zeta(3/2) \simeq 2.61238$, and

$$\Lambda_c(\rho) = \frac{h}{\sqrt{2\pi m k_B T_c(\rho)}}.$$ (6.145)

At temperatures below $T_c(\rho)$, the particles begin to accumulate in the ground state, and Eq. (6.142) must be modified. Indeed, if there is a nonzero fraction of particles in the ground state, i.e.,

$$\frac{z}{1-z} = \lambda\langle N\rangle,$$ (6.146)

where $0 < \lambda \leq 1$, one has

$$z = \frac{\lambda\langle N\rangle}{1+\lambda\langle N\rangle} = \frac{1}{1+\frac{1}{\lambda\langle N\rangle}},$$ (6.147)

so that, in the thermodynamic limit, (6.139) reads

$$\rho = \rho_0 + \frac{1}{\Lambda^3}\zeta(3/2) \quad (z \simeq 1),$$ (6.148)

where

$$\rho_0 = \mathrm{TL}\left[\frac{1}{V}\frac{z}{1-z}\right] \quad (z \simeq 1)$$ (6.149)

is the average density of particles in the ground state, and $z \simeq 1$ denotes the value of the fugacity of the gas (6.147). In (6.148) one has further accounted that

$$\mathrm{TL}\left[\sum_{q=1}^{\infty}\frac{1}{q^{3/2}}\left(\frac{1}{1+\frac{1}{\lambda\langle N\rangle}}\right)^q\right] = \zeta(3/2).$$ (6.150)

Equations (6.142) and (6.148) may, therefore, be written as

$$\rho = \begin{cases} \frac{1}{\Lambda^3}g_{3/2}(z), & T > T_c(\rho) \\ \rho_0 + \frac{1}{\Lambda^3}\zeta(3/2), & T < T_c(\rho) \end{cases}$$ (6.151)

where $T_c(\rho)$ is defined by (6.144) and (6.145).

If there exists a critical value of the temperature $T_c(\rho)$, finite and different from zero, below which a nonzero fraction of particles accumulates in the ground state, one says that the boson gas experiences a Bose–Einstein condensation at the temperature $T_c(\rho)$. If $T > T_c(\rho)$ the gas is in a disordered phase, characterized by the fact that there are no particles in the ground state. When $T < T_c(\rho)$ the ordered phase (particles in the ground state) and the disordered phase (particles in states $\varepsilon_i(V) \neq 0$) coexist. The transition may be characterized by an order parameter (the fraction of particles, N_0/N, in the state of minimum energy) which, according to (6.144) and (6.151), may be written as

$$\frac{\rho_0}{\rho} = 1 - \frac{\zeta(3/2)}{\rho\Lambda^3} = 1 - \frac{\Lambda_c^3(\rho)}{\Lambda^3},$$

or, alternatively,

$$\frac{\rho_0}{\rho} = 1 - \left(\frac{T}{T_c(\rho)}\right)^{3/2}, \tag{6.152}$$

which is a maximum at the absolute zero of temperature, decreases as T is increased and vanishes for $T = T_c(\rho)$. Note that an essential point in the derivation of (6.151) has been the application of the thermodynamic limit. In a finite system, (6.141) is not valid, and for any temperature there is a nonzero fraction of particles in the ground state, i.e., $T_c(\rho) = \infty$.

The first experimental observation of Bose–Einstein condensation took place in 1995, i.e., 61 years after its theoretical prediction. The greatest problems were associated with the difficulty of obtaining very low temperatures (close to the absolute zero) and finding a boson system with very weak interactions such that it would behave as an ideal gas (the system must not crystallize at such low temperatures nor be a liquid, such as helium). Through a combination of optical and evaporation techniques, in 1995 temperatures of the order of 1 nK were obtained, which allowed the observation of condensation in a system of 2000 [87]Rb atoms. Since at those temperatures the atoms tend to adhere to the walls of the container, it is necessary to confine them using magnetic fields and hence to study the Bose–Einstein condensation in an external potential, which modifies the critical behavior of the gas (see V. Bagnato, D. E. Pritchard and D. Kleppner, Phys. Rev. A **35**, 4354 (1987)). Thus, the order parameter N_0/N in a harmonic potential $m\omega_0^2/2$ of frequency ω_0 vanishes according to the following law:

$$\frac{N_0}{N} = 1 - \left(\frac{T}{T_0(N)}\right)^3,$$

when $T < T_0(N)$, where $T_0(N)$ is a critical temperature given by

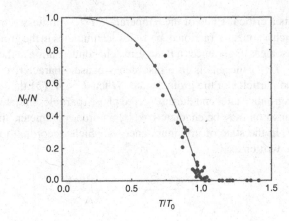

Fig. 6.1 Order parameter, N_0/N, as a function of T/T_0 in the Bose–Einstein condensation of an ideal boson gas in a harmonic potential whose critical temperature is $T_0 = T_0(N)$ (*continuous line*). The points refer to the experimental values in a system of forty thousand atoms of ^{87}Rb at 280 nK. *Source* J. R. Ensher, D. S. Jin, M. R. Matthews, C. E. Wieman and E. A. Cornell, Phys. Rev. Lett. **77**, 4984 (1996).

$$k_B T_0(N) = \hbar\omega_0 \left(\frac{N}{\zeta(3)}\right)^{1/3},$$

and $\zeta(3)$ is the Riemann function. In Fig. 6.1 the experimental values of the order parameter in the Bose–Einstein condensation of ^{87}Rb as a function of temperature are represented. Note that the results are accounted for by the theoretical predictions for an ideal boson gas in a harmonic potential. Afterward, condensation has also been observed in dilute gases of alkaline atoms. All these results represent the first macroscopic manifestations of quantum effects in an ideal gas with Bose–Einstein statistics.

6.12.1 Specific Heat

It is well known that in phase transitions (see Chap. 9) some thermodynamic variables are discontinuous or diverge at the transition. Note that the partition function of an ideal quantum gas (6.107) is a sum of exponentials, which are analytic functions of β, and hence of $T (T \neq 0)$. This sum extends to the occupation numbers n_i of the one-particle quantum states, with the restriction (6.109). If N is finite, the partition function is a sum of a finite number of terms and hence an analytic function of $T (T \neq 0)$. Only if $N \to \infty$ it is possible to predict a phase transition since, in this case, there are no analyticity theorems on the sum of an infinite number of analytic terms. As was already indicated in Chap. 3, only in the thermodynamic limit is it possible to demonstrate the existence of a phase transition.

In the grand canonical ensemble the number of particles is not an independent variable since it has been replaced by the fugacity. In this ensemble, therefore, the existence of a phase transition when $V \to \infty$ may only be shown when for a critical value of the fugacity, z_c, some thermodynamic variables are not analytic in z. In the Bose–Einstein condensation $z_c \to 1$ which, according to (6.146), corresponds to $\langle N \rangle \to \infty$.

In order to analyze the behavior of some thermodynamic variables at the transition, consider the expression for the logarithm of the grand partition function of the gas, namely

$$\ln Q(\beta, V, z) = -\ln(1 - z)$$
$$- 2\pi V \left(\frac{2m}{h^2}\right)^{3/2} \int_0^\infty d\varepsilon \sqrt{\varepsilon} \ln\left[1 - ze^{-\beta\varepsilon}\right], \tag{6.153}$$

where the sum over states $\varepsilon_i(V) \neq 0$ has been replaced by an integral in the limit of infinite volume. Expanding the logarithm in (6.153) in a series, $\ln(1 - x) = -\sum_{q=1}^\infty x^q/q$, and integrating with respect to ε, one has

$$\ln Q(\beta, V, z) = -\ln(1 - z) + \frac{V}{\Lambda^3} g_{5/2}(z). \tag{6.154}$$

From (6.26) it follows that the average energy per particle $e = \langle H_N \rangle / \langle N \rangle$ is given by

$$e = \frac{3}{2} k_B T \frac{1}{\rho \Lambda^3} g_{5/2}(z), \tag{6.155}$$

which when particularized to $z < 1$ and $z \simeq 1$ can be written as

$$e = \begin{cases} \frac{3}{2} k_B T \left(\frac{T}{T_c(\rho)}\right)^{3/2} \frac{g_{5/2}(z)}{\zeta(3/2)}, & T > T_c(\rho) \\ \frac{3}{2} k_B T \left(\frac{T}{T_c(\rho)}\right)^{3/2} \frac{\zeta(5/2)}{\zeta(3/2)}, & T < T_c(\rho) \end{cases} \tag{6.156}$$

where $\zeta(5/2) \simeq 1.34149$ and use has been made of (6.144). The average energy per particle is, therefore, a continuous function at $T = T_c(\rho)$, where it takes the value

$$(e)_{T=T_c(\rho)} = \frac{3}{2} k_B T_c(\rho) \frac{\zeta(5/2)}{\zeta(3/2)}. \tag{6.157}$$

From an experimental point of view, the specific heat $c_V = \partial e / \partial T$ is a more relevant thermodynamic variable than the energy since it can be experimentally measured. As shown in what follows, c_V is also continuous at $T = T_c(\rho)$. The method used in this derivation may be readily generalized to higher-order derivatives of the energy with respect to the temperature. In this way it is shown that the derivative $\partial c_V / \partial T$ is discontinuous at $T = T_c(\rho)$, which is a macroscopic manifestation of the Bose–Einstein condensation.

Consider, first, the specific heat c_V at a temperature below the critical temperature. From (6.156) one has

$$c_V = \frac{15}{4} k_B \left(\frac{T}{T_c(\rho)} \right)^{3/2} \frac{\zeta(5/2)}{\zeta(3/2)}, \tag{6.158}$$

for $T < T_c(\rho)$, so that when $T \to T_c(\rho) - 0$, i.e., when T approaches $T_c(\rho)$ from below then:

$$(c_V)_{T_c(\rho)-0} = \frac{15}{4} k_B \frac{\zeta(5/2)}{\zeta(3/2)}. \tag{6.159}$$

Note that, according to (6.151), when $T > T_c(\rho)$ the fugacity of the gas is a function of the temperature, and so after taking the derivative of the energy Eq. (6.156) the result is

$$c_V = \frac{15}{4} k_B \left(\frac{T}{T_c(\rho)} \right)^{3/2} \frac{g_{5/2}(z)}{\zeta(3/2)}$$
$$+ \frac{3}{2} k_B T \left(\frac{T}{T_c(\rho)} \right)^{3/2} \frac{1}{\zeta(3/2)} \frac{\partial g_{5/2}(z)}{\partial T}, \tag{6.160}$$

and since $z \partial g_q(z)/\partial z = g_{q-1}(z)$, the derivative of the last term of (6.160) reads

$$\frac{\partial g_{5/2}(z)}{\partial T} = \frac{\partial g_{5/2}(z)}{\partial z} \frac{\partial z}{\partial T} = \frac{g_{3/2}(z)}{z} \frac{\partial z}{\partial T}. \tag{6.161}$$

On the other hand, after derivation of (6.151) with respect to the temperature one has

$$-\frac{3}{2T} \rho \Lambda^3 = \frac{\partial g_{3/2}(z)}{\partial T} = \frac{g_{1/2}(z)}{z} \frac{\partial z}{\partial T}. \tag{6.162}$$

From (6.160) and (6.162) it follows finally that

$$c_V = \frac{15}{4} k_B \left(\frac{T}{T_c(\rho)} \right)^{3/2} \frac{g_{5/2}(z)}{\zeta(3/2)} - \frac{9}{4} k_B \frac{g_{3/2}(z)}{g_{1/2}(z)}, \tag{6.163}$$

where the fugacity is a function of $T/T_c(\rho)$ which is obtained from the solution of (6.144) and (6.151), i.e.,

$$g_{3/2}(z) = \zeta(3/2) \left(\frac{T_c(\rho)}{T} \right)^{3/2}. \tag{6.164}$$

At the transition $(z \to 1)$ $g_{1/2}(z)$ diverges, and the last term in (6.163) vanishes, i.e.,

$$(c_V)_{T_c(\rho)+0} = \frac{15}{4} k_B \frac{\zeta(5/2)}{\zeta(3/2)}, \tag{6.165}$$

and from (6.159) it follows that c_V is a continuous variable at $T = T_c(\rho)$.

By a similar argument, one may determine $\partial c_V / \partial T$ from (6.158) and (6.163), with the result

$$\left(\frac{\partial c_V}{\partial T}\right)_{T_c(\rho)-0} = \left(\frac{\partial c_V}{\partial T}\right)_{T_c(\rho)+0}$$
$$+ \frac{27}{8} \zeta(3/2) \frac{k_B}{T_c(\rho)} \lim_{z \to 1} \left(\frac{g_{3/2}(z)g_{-1/2}(z)}{\left\{g_{1/2}(z)\right\}^3}\right). \tag{6.166}$$

In order to obtain the limit appearing in (6.166), the following approximation for the functions $g_l(z)$ when $2l$ is an odd integer and $z \to 1$ $(z < 1)$ may be used (see J. E. Robinson, Phys. Rev. **83**, 678 (1951)):

$$g_l(z) = \Gamma(1-l)(-\ln z)^{l-1} + \sum_{s=0}^{\infty} \frac{(-1)^s}{s!} \zeta(l-s)(-\ln z)^s, \tag{6.167}$$

where $\Gamma(1-l)\Gamma(l) = \pi \operatorname{cosec}(\pi l)$. From (6.167) it follows that

$$\lim_{z \to 1} \left(\frac{g_{3/2}(z)g_{-1/2}(z)}{\left\{g_{1/2}(z)\right\}^3}\right) = \frac{\zeta(3/2)}{2\pi},$$

which leads to

$$\left(\frac{\partial c_V}{\partial T}\right)_{T_c(\rho)-0} - \left(\frac{\partial c_V}{\partial T}\right)_{T_c(\rho)+0} = \frac{27}{16\pi} \{\zeta(3/2)\}^2 \frac{k_B}{T_c(\rho)}, \tag{6.168}$$

i.e., the derivative of the specific heat is discontinuous at the transition, as shown in Fig. 6.2.

6.12.1.1 Equation of State

Consider now the equation of state (6.120). From (6.155) one has

$$p = \frac{k_B T}{\Lambda^3} g_{5/2}(z), \tag{6.169}$$

which may be written as

Fig. 6.2 Specific heat, c_V/k_B (*continuous line*), of an ideal boson gas as a function of T/T_c, where $T_c = T_c(\rho)$ is the critical temperature. The *broken line* is the classical result (Dulong–Petit law)

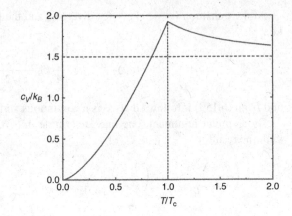

Fig. 6.2 Specific heat, c_V/k_B (*continuous line*), of an ideal boson gas as a function of T/T_c, where $T_c = T_c(\rho)$ is the critical temperature. The *broken line* is the classical result (Dulong–Petit law)

$$p = \begin{cases} \frac{k_BT}{\Lambda^3}g_{5/2}(z), & z < 1 \\ \frac{k_BT}{\Lambda^3}\zeta(5/2), & z \simeq 1 \end{cases} \tag{6.170}$$

Note that from the first of these equations it follows that $p = p(z, T)$. Since, according to (6.142), in the disordered phase $z = z(T, \rho)$, in such a phase the pressure $p = p(z(T, \rho), T) \equiv p(T, \rho)$ is a function of the density and of the temperature. From the second equation of (6.170), on the contrary, it is seen that when the ordered phase and the disordered phase coexist, $p = p(T)$. Observe that if the gas is compressed at constant temperature, according to (6.142) the transition from the disordered phase to the coexistence region takes place at a critical specific volume $v_c(T)$ defined by

$$v_c(T) \equiv \frac{\Lambda^3}{\zeta(3/2)}, \tag{6.171}$$

and so the equation of state is given by

$$p = \begin{cases} \frac{k_BT}{\Lambda^3}g_{5/2}(z), & v > v_c(T) \\ \frac{k_BT}{\Lambda^3}\zeta(5/2), & v < v_c(T) \end{cases} \tag{6.172}$$

The isotherms in the (v, p) plane of the Bose–Einstein gas are similar to those of a first-order transition. Comparing (6.172) with Fig. 2.10 it follows that the disordered phase of the Bose–Einstein gas at a temperature T_0 is the region $v > v_c(T_0)$ and that the binodal region is $v < v_c(T_0)$. The specific volumes of the binodals of Fig. 2.9 at the temperature T_0 are, therefore, $v_1 = v_c(T_0)$ and $v_2 = 0$, which is independent of temperature. Note that since $v_c(T) \sim T^{-3/2}$, when the temperature is increased, the width of the binodal region in the (v, T) plane decreases. Since the binodals coincide at the critical point (v_c, T_c) (see Fig. 2.15) this happens in the Bose–Einstein gas when $T_c = \infty$ and $v_c = 0$. Such a result shows, therefore, that it is not possible to

Fig. 6.3 Compressibility factor $Z = pv/k_B T$ of an ideal boson gas as a function of v/v_c, where $v_c = v_c(T)$ is the critical specific volume. *The broken line* is the classical result

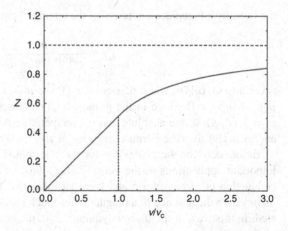

pass in a continuous way from the disordered phase to the completely ordered phase, in which all the particles are in the ground state.

In Fig. 6.3 the compressibility factor $Z = pv/k_B T$ of an ideal boson gas is represented as a function of $v/v_c(T)$, i.e.,

$$Z = \begin{cases} \frac{g_{5/2}(z)}{g_{3/2}(z)}, & v > v_c(T) \\ \frac{\xi(5/2)}{\xi(3/2)}\frac{v}{v_c(T)}, & v < v_c(T) \end{cases} \tag{6.173}$$

where, in the first of these equations, the fugacity is a function of $v/v_c(T)$ which is obtained by solving the following equation:

$$g_{3/2}(z) = \xi(3/2)\frac{v_c(T)}{v}. \tag{6.174}$$

6.13 Fermion Gas

Consider the properties of an ideal fermion gas at the absolute zero of temperature $\beta = \infty$. Since no two particles may be found in the same quantum state, due to Pauli's exclusion principle, the state of minimum energy of the gas is obtained when, starting from the ground state, the quantum states of lower energies are filled until all particles have been accommodated. Note that if in the expression for the average occupation number of a one-particle quantum state corresponding to the Fermi–Dirac statistics,

$$\langle n_i \rangle = \frac{ze^{-\beta\varepsilon_i(V)}}{1 + ze^{-\beta\varepsilon_i(V)}}, \tag{6.175}$$

the fugacity z is replaced in terms of the chemical potential μ, $z = e^{\beta\mu}$, namely

$$\langle n_i \rangle = \frac{1}{e^{\beta(\varepsilon_i(V)-\mu)} + 1}, \tag{6.176}$$

according to (6.176) when $\beta = \infty$, if $\varepsilon_i(V) < \mu$ then $\langle n_i \rangle = 1$, whereas if $\varepsilon_i(V) > \mu$ then $\langle n_i \rangle = 0$. Since in the grand canonical ensemble $z = z(T, \rho)$ and hence $\mu = \mu(T, \rho)$, at the absolute zero of temperature all those quantum states whose energy is less than the Fermi energy $\mu_F \equiv \mu(T = 0, \rho)$ are occupied.

Before deriving the expression for μ_F, one should point out that one of the most important applications of the Fermi–Dirac statistics is based on assuming that the conduction electrons in a metal form an ideal gas. If the spin of the electrons is included, in the absence of a magnetic field each translation state is doubly degenerated. In this way, in the thermodynamic limit, the equation for the average number of particles reads

$$\langle N \rangle = 4\pi V \left(\frac{2m}{h^2}\right)^{3/2} \int_0^\infty d\varepsilon \sqrt{\varepsilon} f(\varepsilon - \mu, T), \tag{6.177}$$

where use has been made of the density of quantum states (1.67), and the Fermi function $f(\varepsilon - \mu, T)$ has been defined as (Fig. 6.4)

$$f(\varepsilon - \mu, T) = \frac{1}{e^{\beta(\varepsilon-\mu)} + 1}. \tag{6.178}$$

Since at the absolute zero of temperature, the Fermi function reduces to a step function, one has

Fig. 6.4 Fermi function of an ideal fermion gas $f(\varepsilon - \mu, T)$ as a function of the energy ε at the absolute zero of temperature $f(\varepsilon - \mu_F, T = 0) = \Theta(\mu_F - \varepsilon)$ (*broken line*) and at a temperature $T \ll T_F$, where T_F is the Fermi temperature of the ideal fermion gas (*continuous line*)

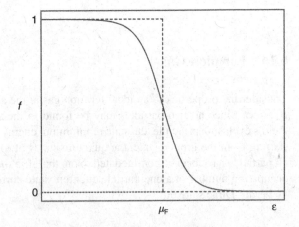

$$\langle N \rangle_{T=0} = 4\pi V \left(\frac{2m}{h^2} \right)^{3/2} \int\limits_0^\infty d\varepsilon \sqrt{\varepsilon} \Theta(\mu_F - \varepsilon), \tag{6.179}$$

whose integration is immediate, yielding

$$\mu_F = \frac{h^2}{2m} \left(\frac{3\rho}{8\pi} \right)^{2/3}, \tag{6.180}$$

where ρ is the average density.

Observe that even at $T = 0$ the average energy of the gas is not zero. Its value may be obtained from (6.26) or, alternatively, by multiplying the density of quantum states by the energy and by the average occupation number of a quantum state, namely

$$\langle H_N \rangle = 4\pi V \left(\frac{2m}{h^2} \right)^{3/2} \int\limits_0^\infty d\varepsilon \, \varepsilon^{3/2} f(\varepsilon - \mu, T), \tag{6.181}$$

which at the absolute zero of temperature may be written as

$$\langle H_N \rangle_{T=0} = 4\pi V \left(\frac{2m}{h^2} \right)^{3/2} \int\limits_0^\infty d\varepsilon \, \varepsilon^{3/2} \Theta(\mu_F - \varepsilon). \tag{6.182}$$

Once more, the integration in (6.182) is immediate, and the result may be written, using (6.180), in the following way:

$$e_0 = \frac{3}{5} \mu_F, \tag{6.183}$$

where e_0 is the average energy per particle. From (6.120) it follows that the pressure at the absolute zero of temperature is given by

$$p_0 = \frac{2}{5} \rho \, \mu_F. \tag{6.184}$$

In order to determine the thermodynamic properties of a fermion gas when $T \neq 0$, one may write (6.177) in terms of the variable $x = \beta(\varepsilon - \mu)$ as

$$\rho \Lambda^3 = \frac{4}{\sqrt{\pi}} \int\limits_{-\beta\mu}^\infty dx \frac{\sqrt{\beta\mu + x}}{e^x + 1}. \tag{6.185}$$

The integral I in (6.185) may be split as

$$
\begin{aligned}
I &= \int_{-\beta\mu}^{0} dx \, \frac{\sqrt{\beta\mu+x}}{e^{x}+1} + \int_{0}^{\infty} dx \, \frac{\sqrt{\beta\mu+x}}{e^{x}+1} \\
&= \int_{0}^{\beta\mu} dx \, \frac{\sqrt{\beta\mu-x}}{e^{-x}+1} + \int_{0}^{\infty} dx \, \frac{\sqrt{\beta\mu+x}}{e^{x}+1},
\end{aligned}
\tag{6.186}
$$

and since

$$
\frac{1}{e^{-x}+1} = 1 - \frac{1}{e^{x}+1},
\tag{6.187}
$$

one has

$$
\begin{aligned}
I &= \int_{0}^{\beta\mu} dx \sqrt{\beta\mu-x} - \int_{0}^{\beta\mu} dx \, \frac{\sqrt{\beta\mu-x}}{e^{x}+1} \\
&\quad + \int_{0}^{\infty} dx \, \frac{\sqrt{\beta\mu+x}}{e^{x}+1},
\end{aligned}
\tag{6.188}
$$

i.e.,

$$
I = \frac{2}{3}(\beta\mu)^{3/2} + I_1 + I_2,
\tag{6.189}
$$

where

$$
I_1 = \int_{0}^{\beta\mu} dx \, \frac{\sqrt{\beta\mu+x} - \sqrt{\beta\mu-x}}{e^{x}+1},
\tag{6.190}
$$

and

$$
I_2 = \int_{\beta\mu}^{\infty} dx \, \frac{\sqrt{\beta\mu+x}}{e^{x}+1}.
\tag{6.191}
$$

Assume now that $\beta\mu \gg 1$ (this hypothesis will be justified a posteriori), in which case the last integral in (6.189) is negligible. For instance, if $\beta\mu = 10^2$, one has $I_2 = 5.3 \times 10^{-43}$, while the first term of the right hand side of (6.189) is $\simeq 6 \times 10^2$. On the other hand, if one adds and subtracts to the numerator of I_1 the first term of the Taylor series expansion of $\sqrt{\beta\mu+x} - \sqrt{\beta\mu-x}$, one has

$$
I_1 = I_3 + I_4,
\tag{6.192}
$$

with

$$I_3 = \int\limits_0^{\beta\mu} \frac{dx}{e^x + 1} \left(\sqrt{\beta\mu + x} - \sqrt{\beta\mu - x} - \frac{x}{\sqrt{\beta\mu}} \right), \qquad (6.193)$$

and

$$I_4 = \frac{1}{\sqrt{\beta\mu}} \int\limits_0^{\beta\mu} dx \frac{x}{e^x + 1}. \qquad (6.194)$$

For $\beta\mu = 10^2$ the result is $I_3 = 7.1 \times 10^{-6}$, while the value of $\beta\mu$ in the upper limit of the integral in (6.194) may be replaced by infinity. Due to the fact that

$$\int\limits_0^\infty dx \frac{x}{e^x + 1} = \frac{\pi^2}{12}, \qquad (6.195)$$

it then follows that

$$I = \frac{2}{3}(\beta\mu)^{3/2} + \frac{\pi^2}{12} \frac{1}{\sqrt{\beta\mu}} + \cdots \qquad (6.196)$$

Observe that for $\beta\mu = 10^2$, $I_1 = 0.0822538$ while the last contribution to (6.196) is 0.0822467, which proves the validity of the approximation.

Using (6.196) the equation for the average number of particles (6.185) reads

$$\mu_F^{3/2} = \mu^{3/2}\left(1 + \frac{\pi^2}{8} \frac{1}{(\beta\mu)^2} + \cdots \right), \qquad (6.197)$$

where use has been made of the expression for the Fermi energy (6.180). This equation may be solved by iteration as

$$\mu = \mu_F\left(1 + \frac{\pi^2}{8} \frac{1}{(\beta\mu)^2} + \cdots \right)^{-2/3} \simeq \mu_F\left(1 + \frac{\pi^2}{8} \frac{1}{(\beta\mu_F)^2} + \cdots \right)^{-2/3}$$

$$\simeq \mu_F\left(1 - \frac{\pi^2}{12} \frac{1}{(\beta\mu_F)^2} + \cdots \right), \qquad (6.198)$$

since μ does not differ much from μ_F and $(1 + x)^{-q} \simeq 1 - qx(x \ll 1)$.

If the Fermi temperature T_F is defined as

$$\mu_F = k_B T_F, \qquad (6.199)$$

then (6.198) may be rewritten as

$$\mu = \mu_F \left(1 - \frac{\pi^2}{12} \left(\frac{T}{T_F} \right)^2 + \cdots \right), \tag{6.200}$$

which is an approximate expression for the chemical potential of the gas at the temperature T. Since the Fermi temperature of a typical metal is of the order of $T_F \simeq 10^4$ K (Table 6.1), (6.200) indicates that at room temperature, $T \simeq 10^2$ K, the temperature correction is of the order of 10^{-4}, which leads to the conclusion that, to a good approximation, the electron gas may be studied as if it was at the absolute zero of temperature (in this case $\beta\mu \simeq \beta\mu_F = 10^2$, which is the numerical value used in the previous approximations). Nevertheless, it is necessary to include this small correction in order to determine the specific heat of the gas. To that end, consider (6.181) which, with the change of variable $\beta(\varepsilon - \mu) = x$, reads

$$\beta e \rho \Lambda^3 = \frac{4}{\sqrt{\pi}} \int\limits_{-\beta\mu}^{\infty} dx \frac{(\beta\mu + x)^{3/2}}{e^x + 1}, \tag{6.201}$$

where e is the average energy per particle when $T \neq 0$. The integral in (6.201) may be approximated by the same method that was followed in the equation for the average number of particles with the result

$$\int\limits_{-\beta\mu}^{\infty} dx \frac{(\beta\mu + x)^{3/2}}{e^x + 1} = \frac{2}{5}(\beta\mu)^{5/2} + \frac{\pi^2}{4}\sqrt{\beta\mu} + \cdots. \tag{6.202}$$

Within this approximation (6.201) reads

$$e = e_0 \left(\frac{\mu}{\mu_F} \right)^{5/2} \left(1 + \frac{5\pi^2}{8} \frac{1}{(\beta\mu)^2} + \cdots \right), \tag{6.203}$$

which may be solved by iteration, if one takes into account (6.200), i.e.,

Table 6.1 Fermi temperature T_F in Kelvin for some metals

	$T_F \times 10^{-4}$		$T_F \times 10^{-4}$
Li	5.51	Al	13.6
Na	3.77	Fe	13.0
K	2.46	Mn	12.7
Cu	8.16	Zn	11.0
Ag	6.38	Sn	11.8
Au	6.42	Pb	11.0

Source N. W. Ashcroft and N. D. Mermin, *Solid State Physics*, Saunders, Philadelphia (1976)

$$e \simeq e_0 \left(1 - \frac{\pi^2}{12} \left(\frac{T}{T_F} \right)^2 + \cdots \right)^{5/2} \left(1 + \frac{5\pi^2}{8} \left(\frac{T}{T_F} \right)^2 + \cdots \right)$$

$$\simeq e_0 \left(1 - \frac{5\pi^2}{24} \left(\frac{T}{T_F} \right)^2 + \cdots \right) \left(1 + \frac{5\pi^2}{8} \left(\frac{T}{T_F} \right)^2 + \cdots \right)$$

$$= e_0 \left(1 + \frac{5\pi^2}{12} \left(\frac{T}{T_F} \right)^2 + \cdots \right), \tag{6.204}$$

from where it follows that the specific heat c_V is given by

$$c_V = e_0 \frac{5\pi^2}{6} \frac{T}{T_F^2} = \frac{3}{2} k_B \left(\frac{\pi^2}{3} \frac{T}{T_F} \right), \tag{6.205}$$

which tends to zero proportionally to T. Note that at room temperature, the specific heat of the electron gas is much smaller than that of the classical ideal gas, namely $3k_B/2$.

At low temperatures, the specific heat of the metal is the sum of the contributions of the electron gas (6.205) and of the lattice (5.160), namely.

$$c_V = k_B \left(v \frac{\pi^2}{2} \frac{T}{T_F} + \frac{12\pi^4}{5} \left(\frac{T}{\Theta_D} \right)^3 \right), \tag{6.206}$$

where v is the number of conduction electrons per atom. Note that if $T_F = 10^4$ K, $\Theta_D = 10^2$ K, and $v = 1$, defining a temperature T_0 by the equation

$$\frac{\pi^2}{2} \frac{T_0}{T_F} = \frac{12\pi^4}{5} \left(\frac{T_0}{\Theta_D} \right)^3,$$

one finds $T_0 = 1.45$ K. This result shows that, in order to observe experimentally the term $c_V \sim T$ due to the conduction electrons, the solid has to be cooled to very low temperatures. Note that (6.206) is the equation of a straight line in the $(T^2, c_V/T)$ plane. This is the behavior observed experimentally, as shown in Fig. 6.5.

Fig. 6.5 Experimental
values of c_V/T (in J/Kmol
K^2) of metallic Ag as a
function of T^2 (in K^2).
Source W. S. Corak, M. P.
Garfunkel, C. B.
Satterthwaite and A. Wexler,
Phys. Rev. **98**, 1699 (1955)

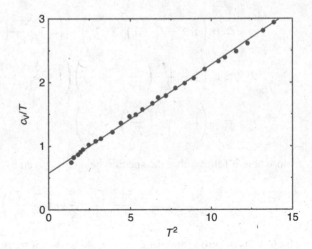

Further Reading

1. R. Balescu, *Equilibrium and Nonequilibrium Statistical Mechanics* (Wiley, New York, 1975). *Contains a derivation of the London–Placzek relation.*
2. N.W. Ashcroft, N.D. Mermin, *Solid State Physics* (Saunders, Philadelphia, 1976). *A classic introduction to the conduction electrons viewed as an ideal fermion gas.*
3. L.D. Landau, E.M. Lifshitz, *Statistical Physics,* 3rd ed. (Pergamon Press, Oxford, 1989). *Provides a compact treatment of the ideal quantum systems with many applications.*
4. R. Kubo, *Statistical Mechanics,* 8th ed. (North-Holland, Amsterdam, 1990). *Contains a series of lectures and problems with complete solutions.*
5. R. Balian, *From Microphysics to Macrophysics,* vol. 1 (Springer, Berlin, 1991). *A good balance between the basic principles and the applications.*
6. H.S. Robertson, *Statistical Thermophysics* (Prentice-Hall, New Jersey, 1993). *Introduces an alternative approach to the Gibbs ensembles.*
7. C. Townsend, W. Ketterle, S. Stiingari, Physics World **10**, 29 (1997). *A very readable account of the first observations of the Bose–Einstein condensation of ideal quantum gases.*

Part III
Non-ideal Systems

Chapter 7
Classical Systems with Interactions

Abstract In the ideal systems considered in the previous chapters the Hamiltonian $H_N(q, p; \alpha)$ and the Hamiltonian operator $\hat{H}_N(\alpha)$ are, respectively, a sum of one-particle dynamical functions and of one-particle operators (kinetic energy and harmonic oscillators). Real systems are characterized by the fact that, besides the kinetic energy, $H_N(q, p; \alpha)$ and $\hat{H}_N(\alpha)$ also include the potential energy, which describes how the particles, atoms, or molecules, interact with each other. As has been shown throughout the text, ideal quantum systems are more complex than classical ideal systems. This complexity is also greater in systems with interaction and so from here onward only classical systems will be considered. Since, in general, the interaction potential in real systems is not known exactly, it is rather common to introduce the so-called reference systems in which the potential energy of interaction is relatively simple, while their thermodynamic properties are nevertheless similar to those of real systems, and so they provide a qualitative description of the latter. Along this chapter some approximate methods for the determination of the free energy of an interacting system are considered. In the last section, a brief summary of the so-called numerical simulation methods is provided.

7.1 Thermodynamic Integration

Consider a real classical system of N particles in equilibrium with the external world at temperature T, whose Hamiltonian is $H = H_N(q, p; \alpha)$, where α denotes one or several external parameters (note that in this section the Hamiltonian is denoted by H instead of H_N). As analyzed in Chap. 5, the thermodynamic potential in the canonical ensemble is the Helmholtz free energy $F = F(\beta, \alpha, N)$, with $\beta = 1/k_B T$. Let $H^0 = H_N^0(q, p; \alpha)$ and $F^0 = F^0(\beta, \alpha, N)$ be the Hamiltonian and the Helmholtz free energy of a reference system whose thermodynamic properties are qualitatively similar to those of the real system. Consider a family of Hamiltonians $H(\lambda) = H_N(q, p; \alpha/\lambda)$ which depends on a parameter λ ($0 \leq \lambda \leq 1$), such that $H(\lambda = 1) = H$ and $H(\lambda = 0) = H^0$, like, for instance,

$$H(\lambda) = H^0 + \lambda(H - H^0). \tag{7.1}$$

© The Author(s), under exclusive license to Springer Nature Switzerland AG 2021
M. Baus and C. F. Tejero, *Equilibrium Statistical Physics*,
https://doi.org/10.1007/978-3-030-75432-7_7

The Helmholtz free energy $F(\lambda) = F(\beta, \alpha, N|\lambda)$ associated with the Hamiltonian $H(\lambda)$ is

$$e^{-\beta F(\lambda)} = \frac{1}{h^{3N}} \int dq \int dp e^{-\beta H(\lambda)}, \tag{7.2}$$

since $F(\lambda) = -k_B T \ln Z(\lambda)$, where $Z(\lambda) \equiv Z(\beta, \alpha, N|\lambda)$ is the partition function corresponding to the Hamiltonian $H(\lambda)$. Upon derivation of (7.2) with respect to λ one has

$$\frac{\partial F(\lambda)}{\partial \lambda} e^{-\beta F(\lambda)} = \frac{1}{h^{3N}} \int dq \int dp \frac{\partial H(\lambda)}{\partial \lambda} e^{-\beta H(\lambda)}, \tag{7.3}$$

or, alternatively,

$$\frac{\partial F(\lambda)}{\partial \lambda} = \left\langle \frac{\partial H(\lambda)}{\partial \lambda} \right\rangle_\lambda, \tag{7.4}$$

where

$$\langle a \rangle_\lambda = \frac{1}{h^{3N}} \frac{1}{Z(\lambda)} \int dq \int dp\, a(q, p) e^{-\beta H(\lambda)}, \tag{7.5}$$

is the average value of the dynamical function $a(q, p)$ in the canonical ensemble of Hamiltonian $H(\lambda)$.

Note that since $F(\lambda = 1) = F$ and $F(\lambda = 0) = F^0$, the identity

$$F = F^0 + \int_0^1 d\lambda \frac{\partial F(\lambda)}{\partial \lambda} \tag{7.6}$$

may be written, according to (7.4), as

$$F = F^0 + \int_0^1 d\lambda \left\langle \frac{\partial H(\lambda)}{\partial \lambda} \right\rangle_\lambda. \tag{7.7}$$

If this equation is particularized to the linear path (7.1), the result is

$$F = F^0 + \int_0^1 d\lambda \langle H - H^0 \rangle_\lambda, \tag{7.8}$$

and so from the mean value theorem,

$$\int_0^1 d\lambda \langle H - H^0 \rangle_\lambda = \langle H - H^0 \rangle_{\bar{\lambda}}, \tag{7.9}$$

it follows that

$$F = F^0 + \langle H - H^0 \rangle_{\bar{\lambda}}. \tag{7.10}$$

Observe that, although the Helmholtz free energy of the system may not be expressed as an average of a dynamical function, since it is a thermal variable, according to (7.10) the difference in free energy between the real system and the reference system, $\Delta F = F - F^\circ$, is equal to the average value of the difference in energy, $\Delta H = H - H^0$ in the canonical ensemble of Hamiltonian $H(\bar{\lambda})$, where $\bar{\lambda}$ is, in general, an unknown quantity. From (7.8) it follows that ΔF may also be determined as the integral over λ of $\langle \Delta H \rangle_\lambda$, an average value that may be computed numerically using simulation methods whose basic principles are considered at the end of this chapter.

7.2 Thermodynamic Perturbation Theory

Upon derivation of (7.3) with respect to λ and taking into account the definition of the average value of a dynamical function (7.5), one obtains

$$\frac{\partial^2 F(\lambda)}{\partial \lambda^2} = \left\langle \frac{\partial^2 H(\lambda)}{\partial \lambda^2} \right\rangle_\lambda - \beta \left\langle \left(\frac{\partial H(\lambda)}{\partial \lambda} \right)^2 \right\rangle_\lambda + \beta \left\langle \frac{\partial H(\lambda)}{\partial \lambda} \right\rangle_\lambda^2$$

$$= \left\langle \frac{\partial^2 H(\lambda)}{\partial \lambda^2} \right\rangle_\lambda - \beta \left\langle \left(\frac{\partial H(\lambda)}{\partial \lambda} - \left\langle \frac{\partial H(\lambda)}{\partial \lambda} \right\rangle_\lambda \right)^2 \right\rangle_\lambda, \tag{7.11}$$

and since

$$\langle (a - \langle a \rangle_\lambda)^2 \rangle_\lambda \geq 0 \tag{7.12}$$

it follows that

$$\frac{\partial^2 F(\lambda)}{\partial \lambda^2} \leq \left\langle \frac{\partial^2 H(\lambda)}{\partial \lambda^2} \right\rangle_\lambda. \tag{7.13}$$

From (7.4) and (7.11) it is straightforward to see that the identity

$$\frac{\partial F(\lambda)}{\partial \lambda} = \left[\frac{\partial F(\lambda)}{\partial \lambda}\right]_{\lambda=0} + \int_0^\lambda d\lambda' \frac{\partial^2 F(\lambda')}{\partial \lambda'^2}, \tag{7.14}$$

yields the inequality

$$\frac{\partial F(\lambda)}{\partial \lambda} \leq \left\langle \frac{\partial H(\lambda)}{\partial \lambda}\right\rangle_0 + \int_0^\lambda d\lambda' \left\langle \frac{\partial^2 H(\lambda')}{\partial \lambda'^2}\right\rangle_{\lambda'}, \tag{7.15}$$

where $\langle \cdots \rangle_0 = \langle \cdots \rangle_{\lambda=0}$. When particularizing this equation to the family of Hamiltonians (7.1) and subsequently integrating between the reference system and the real system, one has

$$F \leq F^0 + \langle H - H^0\rangle_0, \tag{7.16}$$

which is referred to as the Gibbs–Bogoliubov inequality. This equation expresses that the free energy difference, $F - F^0$, has as an upper bound the energy difference, $H - H^0$, averaged over the canonical ensemble of the reference system. An interesting feature of the Gibbs–Bogoliubov inequality is that the free energy of the real system F may be determined approximately by using a variational principle. Thus, if the Hamiltonian of the reference system depends on one (or several) parameter γ, $H^0 = H^0(\gamma)$, then $F^0 = F^0(\gamma)$ and the best estimate of F is obtained by minimizing the upper bound (7.16) with respect to γ, namely

$$F \simeq \min_\gamma \left[F^0(\gamma) + \langle H - H^0(\gamma)\rangle_0\right]. \tag{7.17}$$

An application of this method is provided in Sect. 8.4.

Sometimes the reference system is chosen using some physical criterion that fixes all its parameters, so it is not possible to apply the variational method (7.17). In such a case (7.11) may be written as

$$\frac{\partial^2 F(\lambda)}{\partial \lambda^2} \simeq \left\langle \frac{\partial^2 H(\lambda)}{\partial \lambda^2}\right\rangle_\lambda, \tag{7.18}$$

where it has been assumed that the fluctuations of $\partial H(\lambda)/\partial \lambda$ are small for every λ. With this approximation, when particularizing (7.15) to the family of Hamiltonians (7.1) and subsequently integrating between the reference system and the real system, one finds

$$F \simeq F^0 + \langle H - H^0\rangle_0, \tag{7.19}$$

which corresponds to the first order of a perturbative expansion of F in $\langle H - H^\circ \rangle_0$ around F^0.

An important application of (7.19) is the van der Waals theory. For its derivation, take the interaction potential H^{int} to be the sum of a repulsive part, H_R^{int}, and of an attractive part, H_A^{int}. If the Hamiltonian of the reference system is $H^0 = H^{\text{id}} + H_R^{\text{int}}$, i.e., the one of the real system without the attractive part of the potential, (7.19) is a first-order perturbative expansion in which H_A^{int} is the perturbation. In the van der Waals theory one then has

$$F \simeq F^0 + \langle H_A^{\text{int}} \rangle_0, \tag{7.20}$$

and so if H_A^{int} is the sum of pair potential interactions, namely

$$H_A^{\text{int}} = \frac{1}{2} \sum_{i=1}^{N} \sum_{i \neq j=1}^{N} V_A(|\mathbf{r}_i - \mathbf{r}_j|), \tag{7.21}$$

it follows that

$$\langle H_A^{\text{int}} \rangle_0 = \frac{1}{2} \int_R d\mathbf{r} \int_R d\mathbf{r}' \rho_2^0(\mathbf{r}, \mathbf{r}') V_A(|\mathbf{r} - \mathbf{r}'|), \tag{7.22}$$

where $\rho_2^0(\mathbf{r}, \mathbf{r}')$ is the two-particle distribution function (see (6.81)) of the reference system (note that, in this case, the average is over the canonical ensemble).

If as an additional approximation correlations are neglected in (7.22), i.e., if $\rho_2^0(\mathbf{r}, \mathbf{r}') = \rho_1^0(\mathbf{r})\rho_1^0(\mathbf{r}')$, where $\rho_1^0(\mathbf{r})$ is the local density of particles of the reference system, one has

$$\langle H_A^{\text{int}} \rangle_0 = \frac{1}{2} \int_R d\mathbf{r} \int_R d\mathbf{r}' \rho_1^0(\mathbf{r})\rho_1^0(\mathbf{r}') V_A(|\mathbf{r} - \mathbf{r}'|) = \frac{1}{2} \int_R d\mathbf{r} \rho_1^0(\mathbf{r})\phi_A(\mathbf{r}), \tag{7.23}$$

where

$$\phi_A(\mathbf{r}) = \int_R d\mathbf{r}' \rho_1^0(\mathbf{r}') V_A(|\mathbf{r} - \mathbf{r}'|) \tag{7.24}$$

is the mean potential at $\mathbf{r}$ due to the attractive part of the potential. That is the reason why (7.23) is also called a mean (potential) field approximation. Therefore, the Helmholtz free energy in the van der Waals theory is given by

$$F = F^0 + \frac{1}{2} \int_R d\mathbf{r} \int_R d\mathbf{r}' \rho_1^0(\mathbf{r})\rho_1^0(\mathbf{r}') V_A(|\mathbf{r} - \mathbf{r}'|). \tag{7.25}$$

It is important to point out that in order to apply the van der Waals theory (and, in general, in all the perturbative methods) it is necessary to know the thermodynamic and structural properties of the reference system, e.g., F^0 and $\rho_1^0(\mathbf{r})$ in the van der Waals theory. The determination of these properties is not simple and requires additional approximations, as will be analyzed in the next chapter.

7.3 Virial Expansions

The Helmholtz free energy per particle f of a fluid may be written as

$$f = f^{id} + f^{int},\qquad(7.26)$$

where

$$f^{id} = k_B T\left(\ln\left(\rho\Lambda^3\right) - 1\right)\qquad(7.27)$$

is the ideal part (5.37) and f^{int} is the contribution of the interactions. At low densities f^{iint} may be expanded (in the thermodynamic limit) in a Taylor series expansion in the density of particles $\rho = N/V$. Such a series is known as the virial expansion and the coefficients in the series as the virial coefficients. If $B_n^* = B_n^*(T)$ are the virial coefficients of βf^{int}, i.e.,

$$\beta f^{int} = \sum_{n=1}^{\infty} B_{n+1}^* \rho^n,\qquad(7.28)$$

from the thermodynamic relation,

$$p = \rho^2 \frac{\partial f}{\partial \rho},\qquad(7.29)$$

it follows that the virial expansion of the pressure p is given by

$$\beta p = \rho + \sum_{n=1}^{\infty} B_{n+1}\rho^{n+1}, \quad B_{n+1} = n B_{n+1}^*.\qquad(7.30)$$

On the other, hand, from (7.27), (7.28), and (7.30) it readily follows that the ideal contribution, μ^{id}, and the contribution of the interactions, μ^{int}, to the chemical potential,

$$\mu = f + \frac{p}{\rho},\qquad(7.31)$$

are, after (6.49), given by

$$\beta\mu^{\text{id}} = \ln(\rho\Lambda^3),$$

and

$$\beta\mu^{\text{int}} = \sum_{n=1}^{\infty} B_{n+1}^{**}\rho^n, \quad B_{n+1}^{**} = (n+1)B_{n+1}^*, \tag{7.32}$$

which is the virial expansion of μ^{int}.

Since the determination of the virial coefficients B_n is not simple, one often considers only the correction due to the second coefficient, which may be evaluated analytically for some interaction potentials. In this approximation (note that $B_2^{**} = 2B_2^* = 2B_2$),

$$\beta p^{\text{int}} \simeq B_2\rho^2, \quad \beta f^{\text{int}} \simeq B_2\rho, \quad \beta\mu^{\text{int}} \simeq 2B_2\rho. \tag{7.33}$$

In order to derive the analytical expression of B_2, consider again (7.8),

$$F = F^0 + \int_0^1 d\lambda \langle H - H^0 \rangle_\lambda, \tag{7.34}$$

particularized to the case in which H^0 is the kinetic energy of the system $H^0 = H^{\text{id}}$. One then has that the free energy of the interactions, $F^{\text{int}} = F - F^{\text{id}}$, is given by

$$F^{\text{int}} = \int_0^1 d\lambda \langle H^{\text{int}} \rangle_\lambda, \tag{7.35}$$

where $H^{\text{int}} = H - H^{\text{id}}$ is the potential energy. If H^{int} is a sum of pair potential interactions, i.e.,

$$H^{\text{int}} = \frac{1}{2}\sum_{i=1}^{N}\sum_{i\neq j=1}^{N} V(|\mathbf{r}_i - \mathbf{r}_j|), \tag{7.36}$$

then (7.35) may be written as

$$F^{\text{int}} = \frac{1}{2}\int_0^1 d\lambda \int_R d\mathbf{r} \int_R d\mathbf{r}'\rho_2(\mathbf{r}, \mathbf{r}'|\lambda)V(|\mathbf{r} - \mathbf{r}'|), \tag{7.37}$$

where $\rho_2(\mathbf{r}, \mathbf{r}'|\lambda)$ is the two-particle distribution function of the system whose Hamiltonian is $H(\lambda) = H^{id} + \lambda H^{int}$ (see (6.81)), namely

$$\rho_2(\mathbf{r}, \mathbf{r}'|\lambda) = \left\langle \sum_{i=1}^{N} \sum_{i \neq j=1}^{N} \delta(\mathbf{r} - \mathbf{r}_i)\delta(\mathbf{r}' - \mathbf{r}_j) \right\rangle_\lambda . \tag{7.38}$$

If $Z^{int}(\lambda)$ is the partition function of the interactions of the system with Hamiltonian $H(\lambda)$, after (5.51) one then has

$$\rho_2(\mathbf{r}, \mathbf{r}'|\lambda) = \frac{N(N-1)}{V^N Z^{int}(\lambda)} \int_R d\mathbf{r}^N \delta(\mathbf{r} - \mathbf{r}_1)\delta(\mathbf{r}' - \mathbf{r}_2)e^{-\beta\lambda H^{int}} . \tag{7.39}$$

Note that, since the particles are identical, the double summation in (7.38) has been expressed as $N(N-1)$ times the product of two Dirac delta functions, for instance $\delta(\mathbf{r} - \mathbf{r}_1)\delta(\mathbf{r}' - \mathbf{r}_2)$, which eliminates two integrations (there is, therefore, a term V^{-2}) and localizes the interaction potential $V(|\mathbf{r}_1 - \mathbf{r}_2|)$. Hence, (7.39) may be written as:

$$\rho_2(\mathbf{r}, \mathbf{r}'|\lambda) = \rho^2\left(1 - \frac{1}{N}\right)e^{-\beta\lambda V(|\mathbf{r}-\mathbf{r}'|)} + \cdots , \tag{7.40}$$

where the dots denote other terms that depend on the thermodynamic variables N, V, and β, on the parameter λ and on $\mathbf{r}$ and $\mathbf{r}'$ (in the thermodynamic limit such terms form a series in the density). Note that since in (7.37) one only wants to obtain the lowest order in the density, the terms represented by the dots in (7.40) may be neglected in first approximation. Since

$$\int_0^1 d\lambda V(|\mathbf{r} - \mathbf{r}'|)e^{-\beta\lambda V(|\mathbf{r}-\mathbf{r}'|)} = -\frac{1}{\beta}f(|\mathbf{r} - \mathbf{r}'|; T), \tag{7.41}$$

where

$$f(r; T) = e^{-\beta V(r)} - 1 \tag{7.42}$$

is referred to as the Mayer function, from (7.37), (7.40), and (7.41) it follows that

$$F^{int} = -\frac{1}{2\beta}\rho^2\left(1 - \frac{1}{N}\right)\int_R d\mathbf{r} \int_R d\mathbf{r}' f(|\mathbf{r} - \mathbf{r}'|; T) + \cdots . \tag{7.43}$$

The free energy per particle f^{int} in the thermodynamic limit then reads

$$\beta f^{\text{int}} = -\frac{1}{2}\rho \int d\mathbf{r} f(r; T) + \cdots, \tag{7.44}$$

which upon comparison with (7.33) allows one to obtain the expression for the second virial coefficient, namely

$$B_2 \equiv B_2(T) = -\frac{1}{2} \int d\mathbf{r} f(r; T). \tag{7.45}$$

Recall that, as stated in Chap. 3, in order to prove the existence of the thermodynamic limit, the Hamiltonian must fulfill the stability condition and the weak-decay condition. The latter implies that the pair potential interaction $V(r)$ has to decrease with the distance at least as $r^{-(d+\varepsilon)}$, where d is the dimensionality of space and ε is a positive constant. This condition is also required for the existence of the virial expansions. Indeed, consider (7.45) and let r_0 be the range of $V(r)$, i.e., $\beta V(r) \ll 1$ when $r > r_0$. The second virial coefficient may then be written as

$$B_2 = 2\pi \int_0^\infty dr r^2 \left(1 - e^{-\beta V(r)}\right) = 2\pi \int_0^{r_0} dr r^2 \left(1 - e^{-\beta V(r)}\right)$$

$$+ 2\pi \int_{r_0}^\infty dr r^2 \left(1 - e^{-\beta V(r)}\right). \tag{7.46}$$

In the last integral, the exponential may be expanded in a Taylor series yielding

$$\int_{r_0}^\infty dr r^2 \left(1 - e^{-\beta V(r)}\right) = \int_{r_0}^\infty dr r^2 \beta V(r) + \cdots, \tag{7.47}$$

and in order for the latter integral to be finite in the upper limit one must necessarily have $V(r) \sim r^{-(3+\varepsilon)}$. When this condition is fulfilled, one says that the interaction pair potential is short-ranged. Note that for long-ranged pair potentials B_2 is infinite (the same result holds then for all virial coefficients) and, therefore, no virial expansions exist. The reason is that, in these expansions, the thermodynamics of a system of N particles (the pressure) is divided into a sequence of terms involving one, two, three, etc., particles (the coefficients of the virial series); for instance, B_2 contains the interaction of only two particles. If the pair potential is long-ranged, this separation looses its meaning and hence all the terms in the series diverge. It is evident that this result does not imply, for instance, that there does not exist an equation of state, it only implies that the virial expansion of the pressure is inadequate (thus, in the Taylor series expansion of $e^{-x} - 1 = -x + x^2/2! + \cdots$ all the terms of the series diverge when $x \to \infty$, although the series is convergent). As also indicated in Chap. 3, an important example of a long-ranged pair potential is the Coulomb potential $V(r) = e^2/r$, where e is the electric charge of the particles. In systems of mobile charges or plasmas (note that a plasma should have at least two types of particles to maintain the overall electroneutrality), there exists a series expansion of f^{int} in the so-called plasma parameter whose coefficients are finite. The fundamental difference between a short-ranged potential and the Coulomb potential may be envisaged through the following argument. In a fluid whose pair potential is short-ranged, the relevant variables are the mass of the particles m, the density ρ, the temperature T, and the amplitude ε and the range r_0 of the potential. With these variables two dimensionless parameters may be formed: $\varepsilon/k_B T$ and ρr_0^3. If $\rho r_0^3 \ll 1$ the thermodynamic properties of the fluid may be expanded as a series of this parameter and one obtains the virial expansions. In a plasma one does not have the variables ε and r_0, but one has the electric charge e (although in a plasma the different components have different charges, it is usual to consider the so-called one-component plasma in which particles of charge e move in a background of continuous charge that maintains the overall electroneutrality). In this case, one may only form one dimensionless parameter with the variables m, ρ, T, and e, namely $\rho\lambda^3$, where λ is the Debye wavelength $\lambda = \sqrt{k_B T/4\pi e^2 \rho}$. If $\rho\lambda^3 \gg 1$ the thermodynamic properties of the plasma may be expanded in a series in the plasma parameter $\varepsilon = (\rho\lambda^3)^{-1} \ll 1$. The condition $\rho\lambda^3 \gg 1$ implies that there is a great number of particles in a Debye cube of side λ; i.e., charge electroneutrality is also maintained locally in such a way that the effective pair potential between the mobile charges is the Yukawa potential e^{-Kr}/r, with $K = 1/\lambda$, instead of the Coulomb potential (for more details, see Sect. 8.10.2 where the effective interaction between charged particles is analyzed). It is thus seen that the series expansions in $\varepsilon \sim \sqrt{\rho}$ are not analytic in the density and this is the reason why the coefficients of the virial series of a plasma are infinite.

7.4 Direct Correlation Function

In the next chapters, phases of matter (e.g., solids and liquid crystals) and inhomogeneous systems (e.g., a liquid–vapor interface) characterized by a non-uniform local density of particles $\rho_1(\mathbf{r})$ will be considered. As stated in Chap. 3, the standard method to produce an inhomogeneity in a system whose Hamiltonian is invariant under translation and rotation is to add to the Hamiltonian the contribution of a one-particle external potential $\phi(\mathbf{r})$, also referred to as the symmetry breaking field. As was discussed in Chap. 6 the grand potential becomes in this case a functional $\Omega[u]$ of $u(\mathbf{r}) = \mu - \phi(\mathbf{r})$, where p is the chemical potential. It has been also shown there that for an ideal gas there is a one-to-one correspondence between $\rho_1(\mathbf{r})$ and $u(\mathbf{r})$ and that $\Omega[u]$ is a concave functional of $u(\mathbf{r})$. These properties allowed to define the intrinsic Helmholtz free energy functional $\mathscr{F}[\rho_1]$ as the Legendre transform of $\Omega[u]$ and to formulate a variational principle in terms of the functional $A[\rho_1]$. The density functional approach thus focuses on functionals of $\rho_1 6.6(\mathbf{r})$ rather than $u(\mathbf{r})$.

For a system of interacting particles, it has been found in Sect. 6.6 that $\Omega[u]$ is also a concave functional of $u(\mathbf{r})$ and that the two first functional derivatives of $\Omega[u]$ are related to the local density of particles $\rho_1(\mathbf{r})$ and to the two-particle distribution function $\rho_2(\mathbf{r}, \mathbf{r}')$. Whereas it is clear that $\rho_1(\mathbf{r})$ is univocally determined by $u(\mathbf{r})$, one can prove the less obvious result that only one $u(\mathbf{r})$ can determine a specified $\rho_1(\mathbf{r})$. In this way, the equilibrium properties of inhomogeneous interacting systems can also be formulated in terms of the Legendre transform of $\Omega[u]$, i.e., from the intrinsic Helmholtz free energy functional $\mathscr{F}[\rho_1]$.

Consider an open system of interacting particles in a one-particle external potential $\phi(\mathbf{r})$. Since there is a one-to-one correspondence between $\rho_1(\mathbf{r})$ and $u(\mathbf{r})$, the intrinsic Helmlholtz free energy functional $\mathscr{F}[\rho_1]$ can be constructed as the Legendre transform of $\Omega[u]$:

$$\mathscr{F}[\rho_1] = \Omega[u] + \int_R d\mathbf{r}\, u(\mathbf{r})\rho_1(\mathbf{r}), \tag{7.48}$$

which, in what follows, will be split into the ideal part (6.65) and the contribution resulting from the interactions, i.e.,

$$\mathscr{F}[\rho_1] = k_B T \int_R d\mathbf{r}\,\rho_1(\mathbf{r})\big(\ln\big(\rho_1(\mathbf{r})\Lambda^3\big) - 1\big) + \mathscr{F}^{\text{int}}[\rho_1]. \tag{7.49}$$

Consider now the following definitions:

$$c_1(\mathbf{r}) \equiv -\beta \frac{\delta \mathscr{F}^{\text{int}}[\rho_1]}{\delta\rho_1(\mathbf{r})}, \quad c_2(\mathbf{r}, \mathbf{r}') \equiv -\beta \frac{\delta^2 \mathscr{F}^{\text{int}}[\rho_1]}{\delta\rho_1(\mathbf{r})\delta\rho_1(\mathbf{r}')}, \tag{7.50}$$

which are functionals of the local density, i.e., $c_1(\mathbf{r}) = c_1(\mathbf{r}, [\rho_1])$ and $c_2(\mathbf{r}, \mathbf{r}') = c_2(\mathbf{r}, \mathbf{r}', [\rho_1])$. In particular, $c_2(\mathbf{r}, \mathbf{r}')$ is known as the Ornstein–Zernike direct correlation function (or, simply, direct correlation function) which will henceforth be denoted by $c(\mathbf{r}, \mathbf{r}')$.

Using the well-known properties of the Legendre transform, from (7.48) and (6.76) one obtains

$$\frac{\delta\mathscr{F}[\rho_1]}{\delta\rho_1(\mathbf{r})} = u(\mathbf{r}). \tag{7.51}$$

The first functional derivative of (7.49) then yields, using (7.50),

$$k_B T \ln\big(\rho_1(\mathbf{r})\Lambda^3\big) = k_B T c_1(\mathbf{r}) + u(\mathbf{r}), \tag{7.52}$$

which can be reexpressed as

$$\rho_1(\mathbf{r}) = \frac{1}{\Lambda^3} e^{\beta[u(\mathbf{r}) + k_B T c_1(\mathbf{r})]}. \tag{7.53}$$

For an ideal gas $c_1(\mathbf{r}) = 0$, and (7.53) reduces to the barometric law (6.57). Thus, $-k_B T c_1(\mathbf{r})$ in (7.53) can be interpreted as an effective one-particle potential resulting from the interactions which, together with $u(\mathbf{r})$, determines $\rho_1(\mathbf{r})$.

A further differentiation of (7.52) leads to

$$k_B T \frac{\delta(\mathbf{r} - \mathbf{r}')}{\rho_1(\mathbf{r})} = k_B T c(\mathbf{r}, \mathbf{r}') + \frac{\delta u(\mathbf{r})}{\delta \rho_1(\mathbf{r}')}, \tag{7.54}$$

and using the identity

$$\int_R d\mathbf{r}'' \frac{\delta u(\mathbf{r})}{\delta \rho_1(\mathbf{r}'')} \frac{\delta \rho_1(\mathbf{r}'')}{\delta u(\mathbf{r}')} = \delta(\mathbf{r} - \mathbf{r}'), \tag{7.55}$$

from (6.82) and (7.54) one has

$$\rho_2(\mathbf{r}, \mathbf{r}') = \rho_1(\mathbf{r})\rho_1(\mathbf{r}') + \rho_1(\mathbf{r})\rho_1(\mathbf{r}')c(\mathbf{r}, \mathbf{r}')$$
$$+ \rho_1(\mathbf{r}) \int_R d\mathbf{r}'' c(\mathbf{r}, \mathbf{r}'') [\rho_2(\mathbf{r}'', \mathbf{r}') - \rho_1(\mathbf{r}'')\rho_1(\mathbf{r}')]. \tag{7.56}$$

By defining the pair correlation function $g(\mathbf{r}, \mathbf{r}')$ and the total correlation function $h(\mathbf{r}, \mathbf{r}')$ as

$$\rho_2(\mathbf{r}, \mathbf{r}') \equiv \rho_1(\mathbf{r})\rho_1(\mathbf{r}')g(\mathbf{r}, \mathbf{r}') \equiv \rho_1(\mathbf{r})\rho_1(\mathbf{r}')[h(\mathbf{r}, \mathbf{r}') + 1], \tag{7.57}$$

then (7.56) reads

$$h(\mathbf{r}, \mathbf{r}') = c(\mathbf{r}, \mathbf{r}') + \int_R d\mathbf{r}'' c(\mathbf{r}, \mathbf{r}'')\rho_1(\mathbf{r}'')h(\mathbf{r}'', \mathbf{r}'), \tag{7.58}$$

which is known as the Ornstein–Zernike equation. Observe that, as indicated in Sect. 6.8, in (7.58) a shorthand notation for the functionals $h(\mathbf{r}, \mathbf{r}')$ and $c(\mathbf{r}, \mathbf{r}')$, which moreover are functions of the thermodynamic state, has been used. For a uniform fluid phase of average density ρ, these functionals reduce to ordinary functions of ρ which only depend on the modulus of the relative distance, i.e., $h(|\mathbf{r} - \mathbf{r}'|)$ and $c(|\mathbf{r} - \mathbf{r}'|)$. The Ornstein–Zernike equation then reads in the thermodynamic limit

$$h(|\mathbf{r} - \mathbf{r}'|) = c(|\mathbf{r} - \mathbf{r}'|) + \rho \int d\mathbf{r}'' c(|\mathbf{r} - \mathbf{r}''|)h(|\mathbf{r}'' - \mathbf{r}'|), \tag{7.59}$$

or

$$\tilde{h}(k) = \tilde{c}(k) + \rho\tilde{c}(k)\tilde{h}(k), \tag{7.60}$$

where $\tilde{h}(k)$ and $\tilde{c}(k)$ are the Fourier transforms of $h(r)$ and $c(r)$, e.g.,

$$\tilde{h}(k) = \int d\mathbf{r} e^{-i\mathbf{k}\cdot\mathbf{r}} h(r). \tag{7.61}$$

7.5 Density Functional Theory

Define the functional

$$A[\overline{\rho}_1] \equiv \mathcal{F}[\overline{\rho}_1] - \int_R d\mathbf{r}\, u(\mathbf{r})\overline{\rho}_1(\mathbf{r}), \tag{7.62}$$

which, after (7.48), yields $A[\rho_1] = \Omega[u]$ when $\overline{\rho}_1(r) = \rho_1(r)$. The first functional derivative of (7.62) at constant $u(\mathbf{r})$ is

$$\frac{\delta A[\overline{\rho}_1]}{\delta\overline{\rho}_1(\mathbf{r})} = k_B T \ln\left(\overline{\rho}_1(\mathbf{r})\Lambda^3\right) - k_B T \overline{c}_1(\mathbf{r}) - u(\mathbf{r}), \tag{7.63}$$

which, by virtue of (7.52), vanishes at the equilibrium density $\rho_1(\mathbf{r})$. The notation $\overline{c}_1(\mathbf{r})$ in (7.63) expresses that the functional is evaluated at the local density $\overline{\rho}_1(\mathbf{r})$. Further differentiation of (7.63) at constant $u(\mathbf{r})$ (with the same meaning for the notation $\overline{c}(\mathbf{r}, \mathbf{r}')$) yields

$$\beta\frac{\delta^2 A[\overline{\rho}_1]}{\delta\overline{\rho}_1(\mathbf{r})\delta\overline{\rho}_1(\mathbf{r}')} = \frac{\delta(\mathbf{r} - \mathbf{r}')}{\overline{\rho}_1(\mathbf{r})} - \overline{c}(\mathbf{r}, \mathbf{r}'). \tag{7.64}$$

At the equilibrium density the r.h.s. of (7.64) is positive:

$$\frac{\delta(\mathbf{r} - \mathbf{r}')}{\rho_1(\mathbf{r})} - c(\mathbf{r}, \mathbf{r}') = \beta\frac{\delta u(\mathbf{r})}{\delta\rho_1(\mathbf{r}')} > 0,$$

since

$$\frac{\delta\rho_1(\mathbf{r}')}{\delta u(\mathbf{r})} > 0, \tag{7.65}$$

as has been already shown in (6.82). Therefore, $A[\overline{\rho}_1]$ is a convex functional of $\overline{\rho}_1(\mathbf{r})$ and $\Omega[u]$ is the minimum value of $A[\overline{\rho}_1]$ at the equilibrium density $\rho_1(\mathbf{r})$, for a given $u(\mathbf{r})$. This variational principle is the cornerstone of density functional theory.

From (7.49), $A[\overline{\rho}_1]$ can be written as

$$A[\overline{\rho}_1] = k_B T \int\limits_R d\mathbf{r} \overline{\rho}_1(\mathbf{r}) \big(\ln(\overline{\rho}_1(\mathbf{r})\Lambda^3) - 1\big)$$

$$+ \mathcal{F}^{int}[\overline{\rho}_1] - \int\limits_R d\mathbf{r}\,\mu\,(\mathbf{r})\overline{\rho}_1(\mathbf{r}), \qquad (7.66)$$

and, hence, $\mathcal{F}^{int}[\overline{\rho}_1]$ plays a central role in density functional theory. It is clear that the complexity involved in $\mathcal{F}^{int}[\overline{\rho}_1]$ is the same as that of the partition function. Any practical implementation of density functional theory then requires some explicit approximation for the functional $\mathcal{F}^{int}[\overline{\rho}_1]$. Finally, note that the equilibrium local density $\rho_1(\mathbf{r})$ is the minimum of $A[\overline{\rho}_1]$, i.e.,

$$\left(\frac{\delta\mathcal{F}[\overline{\rho}_1]}{\delta\overline{\rho}_1(\mathbf{r})}\right)_{\overline{\rho}_1(\mathbf{r})=\rho_1(\mathbf{r})} - \mu + \phi(\mathbf{r}) = 0, \qquad (7.67)$$

which can also be interpreted as follows.

Consider a closed system of N interacting particles in a region R of volume V at temperature T (canonical ensemble) in a one-particle external potential $\phi(\mathbf{r})$. In this case, the Helmholtz free energy is a functional of the external potential $F[\phi]$ (see Sect. 5.5) and $\phi(\mathbf{r})$ induces a non-uniform local density of particles $\rho_1(\mathbf{r})$, which verifies following the normalization condition:

$$N = \int\limits_R d\mathbf{r}\rho_1(\mathbf{r}). \qquad (7.68)$$

Define the functional $\mathcal{F}[\overline{\rho}_1]$ as

$$F[\phi] = \mathcal{F}[\overline{\rho}_1] + \int\limits_R d\mathbf{r}\phi(\mathbf{r})\overline{\rho}_1(\mathbf{r}) \qquad (7.69)$$

and assume that the r.h.s. of (7.69) is minimized with the constraint

$$N = \int\limits_R d\mathbf{r}\overline{\rho}_1(\mathbf{r}).$$

As is well known, the result of this variational procedure can be obtained by removing the constraint using a Lagrange's multiplier μ and minimizing the functional

$$F[\phi] - \mu N = \mathcal{F}[\overline{\rho}_1] + \int_R d\mathbf{r}\, \phi(\mathbf{r})\overline{\rho}_1(\mathbf{r}) - \mu \int_R d\mathbf{r}\, \overline{\rho}_1(\mathbf{r}), \qquad (7.70)$$

leading to:

$$\left(\frac{\delta \mathcal{F}[\overline{\rho}_1]}{\delta \overline{\rho}_1(\mathbf{r})} \right)_{\rho_1(\mathbf{r})} - \mu + \phi(\mathbf{r}) = 0, \qquad (7.71)$$

which is (7.67). Observe that, in this case, once the solution $\rho_1(\mathbf{r})$ of (7.71) is found, Lagrange's multiplier must be eliminated by using (7.68).

7.6 Mean Field Theory

In the previous sections an approximation to the free energy of a system has been obtained through variational methods, perturbative methods or as a virial series. In this section a statistical approximation, called mean field theory, whose main hypothesis is that there are no correlations among the N identical particles of the system, is studied.

Consider for that purpose the joint probability density $\rho(\mathbf{r}^N)$ of finding the N particles at the points $\mathbf{r}^N$ of a closed region R of volume V at the temperature T, which is normalized in the following way:

$$\frac{1}{V^N} \int_R d\mathbf{r}^N \rho(\mathbf{r}^N) = 1. \qquad (7.72)$$

Note that $\rho(\mathbf{r}^N)$ is the marginal probability density of the canonical distribution, i.e.,

$$\rho(\mathbf{r}^N) = \frac{1}{Z^U} e^{-\beta U}, \qquad (7.73)$$

where U is the potential energy and Z^U the corresponding partition function.

If one assumes that the particles are not correlated, the probability density $\rho(\mathbf{r}^N)$ may be expressed as the product of N identical factors (see Appendix B), which are the individual probability densities, i.e.,

$$\rho(\mathbf{r}^N) = \prod_{j=1}^{N} \frac{\rho_1(\mathbf{r}_j)}{\rho}, \qquad (7.74)$$

where $\rho_1(\mathbf{r})$ is the local density of particles and $\rho = N/V$. From the normalization condition (7.72) it follows that

$$\frac{1}{V}\int_R d\mathbf{r}\rho_1(\mathbf{r}) = \rho. \tag{7.75}$$

In order to determine the intrinsic Helmholtz free energy functional, consider again Eq. (7.35). Since the local density of particles (7.75) is, in general, not uniform, $\mathscr{F}^{\mathrm{id}}[\rho_1]$ is the density functional (6.65), namely

$$\mathcal{F}^{\mathrm{id}}[\rho_1] = k_B T \int_R d\mathbf{r}\rho_1(\mathbf{r})\big(\ln(\rho_1(\mathbf{r})\Lambda^3) - 1\big). \tag{7.76}$$

Observe, further, that since the factorization (7.74) is independent of the interaction, the average value $\langle H^{\mathrm{int}}\rangle_\lambda$ is independent of λ (i.e., of the "charging parameter" which adds interaction to the system in (7.1) when $H^0 = H^{\mathrm{id}}$), i.e.,

$$\begin{aligned}
\int_0^1 d\lambda \langle H^{\mathrm{int}}\rangle_\lambda &= \int_0^1 d\lambda \frac{1}{V^N}\int_R d\mathbf{r}^N \rho(\mathbf{r}^N)\frac{1}{2}\sum_{i=1}^N\sum_{i\neq j=1}^N V(|\mathbf{r}_i - \mathbf{r}_j|)\\
&= \frac{1}{2}\sum_{i=1}^N\sum_{i\neq j=1}^N \frac{1}{N^2}\int_R d\mathbf{r}_i\rho_1(\mathbf{r}_i)\int_R d\mathbf{r}_j\rho_1(\mathbf{r}_j)V(|\mathbf{r}_i - \mathbf{r}_j|)\\
&= \frac{1}{2}\Big(1 - \frac{1}{N}\Big)\int_R d\mathbf{r}\rho_1(\mathbf{r})\int_R d\mathbf{r}'\rho_1(\mathbf{r}')V(|\mathbf{r} - \mathbf{r}'|) \tag{7.77}
\end{aligned}$$

In the thermodynamic limit, the intrinsic Helmholtz free energy functional $\mathscr{F}^{\mathrm{id}}[\rho_1]$ in mean field theory is, therefore,

$$\begin{aligned}
\mathcal{F}[\rho_1] = {}& k_B T \int d\mathbf{r}\rho_1(\mathbf{r})\big(\ln(\rho_1(\mathbf{r})\Lambda^3) - 1\big)\\
& + \frac{1}{2}\int d\mathbf{r}\rho_1(\mathbf{r})\int d\mathbf{r}'\rho_1(\mathbf{r}')V(|\mathbf{r} - \mathbf{r}'|). \tag{7.78}
\end{aligned}$$

For the sake of understanding why this theory is called mean field, consider the variational principle (7.71). After taking the thermodynamic limit, let $\phi(\mathbf{r}) \to 0$, where $\phi(\mathbf{r})$ is the external potential which produces the inhomogeneity. The variational principle then yields

$$k_B T \ln(\rho_1(\mathbf{r})\Lambda^3) + \int d\mathbf{r}'\rho_1(\mathbf{r}')V(|\mathbf{r} - \mathbf{r}'|) - \mu = 0, \tag{7.79}$$

i.e.,

$$\rho_1(\mathbf{r}) = \frac{e^{\beta\mu}}{\Lambda^3}e^{-\beta\phi_{\mathrm{mf}}(\mathbf{r})} \tag{7.80}$$

where the mean field potential $\phi_{mf}(\mathbf{r})$ has been defined as

$$\phi_{mf}(\mathbf{r}) = \int d\mathbf{r}' \rho_1(\mathbf{r}') V(|\mathbf{r} - \mathbf{r}'|). \tag{7.81}$$

Imposing the normalization condition to (7.80) one has

$$N = \int d\mathbf{r} \rho_1(\mathbf{r}) = \frac{e^{\beta\mu}}{\Lambda^3} \int d\mathbf{r} e^{-\beta\phi_{mf}(\mathbf{r})}, \tag{7.82}$$

and after elimination of $e^{\beta\mu}/\Lambda^3$ from (7.82) and substitution into (7.80) yields

$$\rho_1(\mathbf{r}) = N \frac{e^{-\beta\phi_{mf}(\mathbf{r})}}{\int d\mathbf{r}' e^{-\beta\phi_{mf}(\mathbf{r}')}}, \tag{7.83}$$

which is expression (5.41) for the local density of particles in an external potential $\phi_{mf}(\mathbf{r})$. Note that (7.81) and (7.83) are self-consistent equations in the local density $\rho_1(\mathbf{r})$ and the mean field potential $\phi_{mf}(\mathbf{r})$. This (fictitious) external potential is, according to (7.81), the average potential acting on a particle of the system due to the pair potential $V(|\mathbf{r} - \mathbf{r}'|)$.

Finally, note that in mean field theory $\mathscr{F}^{int}[\rho_1]$ reads

$$\mathcal{F}^{int}[\rho_1] = \frac{1}{2} \int d\mathbf{r} \rho_1(\mathbf{r}) \int d\mathbf{r}' \rho_1(\mathbf{r}') V(|\mathbf{r} - \mathbf{r}'|), \tag{7.84}$$

and from (7.50) it is found that the direct correlation function is

$$c(|\mathbf{r} - \mathbf{r}'|) = -\beta V(|\mathbf{r} - \mathbf{r}'|). \tag{7.85}$$

7.7 Numerical Simulations

A considerable part of the research in statistical physics is based on the so-called numerical simulation methods. The purpose of this section is to provide a few simple ideas of some known methods and their relation with the general principles of statistical physics. For a more detailed study, one should refer to some specialized texts which have been included in the Further Reading.

7.7.1 Molecular Dynamics

In this method, the classical trajectories of a system of N particles or molecules are determined, hence the name "Molecular Dynamics" or "MD." Let $C_t = (\mathbf{r}^N(t), \mathbf{p}^N(t))$ denote the mechanical state at time t of a system of N particles. If $a(C_t)$ is the value of a dynamical function at time t, a characteristic problem of MD consists in determining the time-averaged value of the dynamical function, a, namely

$$\bar{a} = \lim_{T \to \infty} \frac{1}{T} \int_0^T dt\, a(C_t). \tag{7.86}$$

As usual, to evaluate numerically (7.86) the time variable t, which is continuous, is replaced by a sequence of discrete values $t_1, \ldots, t_n, \ldots$ such that $t_{n+1} = t_n + \Delta t = t_1 + n\Delta t$. If $T = t_k = k\Delta t$, (7.86) may be approximated by

$$\bar{a} = \lim_{k \to \infty} \frac{1}{k} \sum_{n=1}^k a(C_{t_n}). \tag{7.87}$$

In practice, the sum in (7.87) is restricted to a large though finite number of terms, each of which follows from the previous one in a deterministic way upon integration of the equations of motion of the system. Thus, if Δt is small enough, the Taylor expansion of the position vector of the ith particle $\mathbf{r}_i(t_n \pm \Delta t)$ is written as:

$$\mathbf{r}_i(t_n \pm \Delta t) = \mathbf{r}_i(t_n) \pm \Delta t\, \mathbf{v}_i(t_n) + \frac{1}{2}(\Delta t)^2 \mathbf{a}_i(t_n) + O\big((\Delta t)^3\big), \tag{7.88}$$

where $\mathbf{v}_i(t_n)$ and $\mathbf{a}_i(t_n)$ are the velocity and acceleration of the particle, respectively, at time t_n. From (7.88) one has

$$\mathbf{r}_i(t_n + \Delta t) + \mathbf{r}_i(t_n - \Delta t) = 2\mathbf{r}_i(t_n) + (\Delta t)^2 \mathbf{a}_i(t_n) + O\big((\Delta t)^4\big), \tag{7.89}$$

and

$$\mathbf{r}_i(t_n + \Delta t) - \mathbf{r}_i(t_n - \Delta t) = 2\Delta t\, \mathbf{v}_i(t_n) + O\big((\Delta t)^3\big), \tag{7.90}$$

which may be written as

$$\mathbf{r}_i(t_n + \Delta t) = 2\mathbf{r}_i(t_n) - \mathbf{r}_i(t_n - \Delta t) + \frac{1}{m_i}(\Delta t)^2 \mathbf{F}_i(t_n) + O\big((\Delta t)^4\big), \tag{7.91}$$

and

$$\mathbf{v}_i(t_n) = \frac{1}{2\Delta t}[\mathbf{r}_i(t_n + \Delta t) - \mathbf{r}_i(t_n - \Delta t)] + O\big((\Delta t)^2\big). \tag{7.92}$$

In (7.91) $\mathbf{F}_i(t_n)$ is the force acting on particle i, of mass m_i, at time t_n since, according to Newton's equations of motion, $m_i a_i(t_n) = \mathbf{F}_i(t_n)$. This force, which is supposed to be known, is a function of the coordinates $\mathbf{r}^N(t_n)$. Equation (7.91) allows one to determine the position vector of the particles step by step, while from (7.92) one obtains the velocity of the particles step by step, although in a preceding interval Δt. Once the value of the dynamical function in each instant is known, one may compute (7.87) for a large number k of terms. This method of evaluation of (7.87) is known as the Verlet algorithm.

From a practical point of view, the algorithm presents some inconveniences. For instance, the interval Δt may neither be very small nor very large. Normally, Δt is optimized so that the total energy remains constant up to a given precision. Note that k cannot be excessively large, since this would substantially increase the computation time and, for the same reason, neither can N be very large, although only in the thermodynamic limit does one expect that the time average and the ensemble average over the microcanonical ensemble be equal (see Sect. 3.5). Since in this limit the intensive properties are independent of the geometry of the region in which the system is contained, periodic boundary condition is imposed on the system in the algorithm. It is well known that MD simulations lead to very good results, which is probably due to the fact that all these limiting processes behave properly and are rapidly convergent.

7.7.2 Monte Carlo Method

This is a probabilistic method, hence the name "Monte Carlo" (or "MC"), to evaluate multidimensional integrals of the form

$$\langle a \rangle = \frac{1}{V^N} \int_R d\mathbf{r}^N a(\mathbf{r}^N) \rho(\mathbf{r}^N), \tag{7.93}$$

which appear in the equilibrium statistical physics of systems of interacting particles. In (7.93), $a(\mathbf{r}^N)$ is a dynamical function that depends on the mechanical state $\mathbf{r}^N$ of the system, $\rho(\mathbf{r}^N)$ is a marginal probability density of a Gibbs ensemble (note that the integration over the momenta $\mathbf{p}^N$ may always be performed analytically so that the ensemble averages reduce to integrals of the type of (7.93)), and V is the volume of the region R wherein the system is contained.

If, as usual, the region R is discretized and divided into cells, the domain of integration in (7.93), R^N, is also discretized and the integral is replaced by a sum

$$\langle a \rangle = \sum_C a(C) p(C), \tag{7.94}$$

where C denotes a configuration $\mathbf{r}^N$, whose probability is $p(C)$. Note that the number of terms in the sum (7.94) is equal to the number of cells in R^N, which is finite. If one now considers an infinite sequence of configurations $\{C_1, C_2, \dots\}$ by the definition of probability, one has

$$p(C) = \lim_{k \to \infty} \frac{1}{k} \sum_{n=1}^{k} \delta_{C,C_n}, \tag{7.95}$$

namely $p(C)$ is the fraction of terms in the sequence for which $C_n = C$, provided the infinite sequence covers all possible configurations in a sufficiently regular way. Substitution of (7.95) into (7.94) leads to

$$\langle a \rangle = \sum_{C} a(C) \lim_{k \to \infty} \frac{1}{k} \sum_{n=1}^{k} \delta_{C,C_n},$$

$$= \lim_{k \to \infty} \frac{1}{k} \sum_{n=1}^{k} \sum_{C} a(C) \delta_{C,C_n}$$

$$= \lim_{k \to \infty} \frac{1}{k} \sum_{n=1}^{k} a(C_n) \tag{7.96}$$

and so $\langle a \rangle$ may be computed as the average of $a(C_n)$ over the infinite sequence.

Note that although (7.96) and (7.87) are analogous (the fact that in (7.87) the average is over the phase space $(\mathbf{r}^N, \mathbf{v}^N)$ while in (7.96) it is done only over $\mathbf{r}^N$ is irrelevant), there is a fundamental difference between them. Thus, while in (7.87) the sequence of mechanical states C_{t_n} is deterministic, because $C_{t_{n+1}}$ is obtained from C_{t_n} by solving Newton's equations of motion, in (7.96) the sequence of configurations C_n is probabilistic or stochastic so that, in principle, there is no rule to determine C_{n+1} from C_n. On the other hand, there are two aspects to emphasize about (7.94). The first one has to do with the fact that since in the systems analyzed in statistical physics N or V is large, so is the (finite) number of terms in (7.94), otherwise this sum could be evaluated directly. The second aspect is that, since in many systems $p(C)$ has a pronounced maximum, the contribution of the vast majority of terms of the sum to $\langle a \rangle$ is small, and so the choice in (7.96) of a completely random sequence C_n has no practical interest. For this reason, Metropolis et al. introduced a particular type of sampling ("importance sampling") which is detailed in the following. The method may be easily understood if the formal analogy between (7.87) and (7.96) is pursued further. Observe first that in MD all the mechanical states C_{t_n} that conserve the energy have the same probability and that as pointed out earlier $C_{t_{n+1}}$ is obtained from C_{t_n} by solving Newton's equations of motion. Since, as has been stated, in the MC method one has to assign a weight to each configuration C_n, it is necessary to obtain some "equations of motion" that allows the determination of C_{n+1} from C_n. In the Metropolis et al. algorithm, based on the theory of stochastic processes, the transition from C_n to C_{n+1} depends on the relative values of $p(C_n)$ and $p(C_{n+1})$.

Let $W(C', \wedge C, t + \Delta t)$ be the conditional probability that if at time t the system is in the configuration C', at time $t + \Delta t$ it will be found in the configuration C. If $p(C, t)$ denotes the probability of finding the system in configuration C at time t, then

$$p(C, t + \Delta t) = \sum_{C'} W(C', t|C, t + \Delta t) p(C', t). \tag{7.97}$$

If one subtracts from this equation the identity

$$p(C, t) = \sum_{C'} \delta_{C,C'} p(C', t), \tag{7.98}$$

and the result is divided by Δt, in the limit $\Delta t \to 0$ one finds

$$\frac{\partial p(C, t)}{\partial t} = \sum_{C'} T(C'|C) p(C', t), \tag{7.99}$$

where

$$T(C'|C) = \lim_{\Delta t \to 0} \frac{1}{\Delta t} [W(C', t|C, t + \Delta t) - \delta_{C,C'}], \tag{7.100}$$

and it has been assumed that $W(C', t/C, t + \Delta t)$ does not depend on t (stationary process). Equation (7.99), which determines the time evolution of the probability $p(C, t)$, is known as the master equation or central equation.

Note that from (7.100) it follows that

$$W(C', t|C, t + \Delta t) = \delta_{C,C'} + T(C'|C)\Delta t + O((\Delta t)^2), \tag{7.101}$$

if $C \neq C'$, $T(C'/C)$ is a conditional probability per unit of time, i.e., $T(C'/C) > 0$.
Since

$$\sum_{C} W(C', t|C, t + \Delta t) = 1, \tag{7.102}$$

from (7.101) and (7.102) one obtains

$$\sum_{C} T(C'|C) = 0, \tag{7.103}$$

i.e.,

$$T(C'|C') = -\sum_{C}{}^* T(C'|C) < 0, \tag{7.104}$$

where the asterisk indicates that the sum extends over all the configurations $C \neq C'$. If in the master equation the contributions $C \neq C'$ and $C = C'$ are separated, one obtains

$$\frac{\partial p(C,t)}{\partial t} = \sum_{C}{}^{*}T(C'|C)p(C',t) + T(C|C)p(C,t), \qquad (7.105)$$

which, with the aid of (7.104), may be written as

$$\frac{\partial p(C,t)}{\partial t} = \sum_{C'}{}^{*}[T(C'|C)p(C',t) - T(C|C')p(C,t)]. \qquad (7.106)$$

Note that this equation expresses that the probability of configuration C changes in time due to two terms of opposite sign. The first one is the flux of incoming probability from configurations $C' \neq C$, whereas the second is the flux of outgoing probability to configurations $C' \neq C$.

Assume that in its time evolution the system reaches a stationary or equilibrium state:

$$p_e(C) = \lim_{t \to \infty} p(C,t). \qquad (7.107)$$

In this case, from (7.106) it follows that

$$\sum_{C'}{}^{*}T(C'|C)p_e(C') = \sum_{C'}{}^{*}T(C|C')p_e(C), \qquad (7.108)$$

namely in the stationary state the incoming and outgoing fluxes of probability are equal.

One of the main objectives of the theory of stochastic processes is to derive the stationary solution of the master equation (7.108), $p_e(C)$, from the conditional probabilities per unit time $T(C|C')$ and $T(C'|C)$. Note, however, that since in an equilibrium system in order to evaluate the sum (7.94) one needs $p_e(C)$ (which is a marginal probability density of a Gibbs ensemble), Metropolis et al. addressed what may be referred to as the "inverse problem," namely to derive from $p_e(C)$ the conditional probabilities per unit time $T(C|C')$ which generate a sequence $p(C,t)$ that verifies (7.107). Observe that in the equilibrium statistical physics of interacting systems, one does not know the exact weight of each mechanical state (which would imply, for instance, that one might compute the partition function), but the relative weight between two of them is indeed known. For example, in the canonical ensemble the relative weight is $e^{-\beta \Delta E}$, where ΔE is the energy difference of the states. Since the solution to the inverse problem is not unique, Metropolis et al. assumed as a hypothesis that for any pair of configurations C and C' the detailed balance condition is verified:

$$T(C'|C)p_e(C') = T(C|C')p_e(C),$$ (7.109)

i.e.,

$$\frac{T(C'|C)}{T(C|C')} = \frac{p_e(C)}{p_e(C')}.$$ (7.110)

Note that a particular solution of (7.109) is

$$T(C|C') = \begin{cases} \frac{p_e(C')}{p_e(C)}, \; p_e(C) > p_e(C') \\ 1, \; p_e(C) \le p_e(C') \end{cases}$$ (7.111)

This is the Metropolis scheme by which the transition from a configuration C to a more probable one C' $(p_e(C') > p_e(C))$ is always accepted, whereas the transition to a less probable configuration $(p_e(C') < p_e(C))$ is only accepted with a probability equal to $p_e(C')/p_e(C)$. In this latter case a random number γ $(0 \le \gamma \le 1)$ is generated out of a uniform distribution, so that if $p_e(C')/p_e(C) > \gamma$ the new configuration is accepted, and if $p_e(C')/p_e(C) < \gamma$ it is rejected. Note that the probability $\gamma < p_e(C')/p_e(C)$ is then equal to $p_e(C)/p_e(C)$, as has been assumed in (7.111).

In summary, if one starts from an accepted configuration C_n, a new configuration C_{n+1} is build (which is usually done by varying slightly the position of one or more particles) and accepted or rejected with the aid of the Metropolis algorithm. If C_{n+1} is accepted a sequence ..., C_n, C_{n+1}, ... is constructed, but if C_{n+1}, is rejected it is replaced by C_n, and the sequence becomes instead, ..., C_n, C_n, ... so that C_n is counted twice. Upon repeating this process a final sequence of configurations (some of which may thus be equal), say $C_1, C_2, ..., C_n$ is obtained that for large enough n are generated with probability $p_e(C)$. From this sequence one evaluates (7.94) using (7.96). Observe, finally, that increasing n in a simulation is equivalent to increasing the time variable, which is real in MD but fictitious in the MC method.

Note that within the numerical simulation methods described in this section, in MD the energy is constant (microcanonical ensemble) while in the MC method the energy is variable and the relative weight of the states is that of the canonical ensemble. The generalization of the MC method to other ensembles and of the MD approach to out-of-equilibrium systems may be found in the textbooks included in the Further Reading.

Further Reading

1. D.A. McQuarrie, *Statistical Mechanics* (Harper & Row, New York, 1976). *Provides virial coefficients for some of the most current pair potentials.*
2. J.P. Hansen, I.R. McDonald, *Theory of Simple Liquids,* 2nd ed. (Academic Press, London, 1986). *Contains a detailed treatment of the density functional theory.*

3. M.P. Allen, D.J. Tildesley, *Computer Simulation of Liquids* (Clarendon Press, Oxford, 1987). *A classic introduction to the MD and MC simulation methods.*
4. G. Ciccotti, D. Frenkel, I.R. McDonald, *Simulation of Liquids and Solids* (North-Holland, Amsterdam, 1987). *A collection of historic landmark papers with comments.*

Chapter 8
Phases of Matter

Abstract The objective of statistical physics is to derive the macroscopic (thermo-dynamic) properties of matter from the laws of mechanics that govern the motion of its microscopic constituents. Note that in this context microscopic may have different meanings. Indeed, although it is common to consider that matter is made of atoms or molecules (something known as the atomic description), this is not the only possible description. For instance, one might consider from the outset that matter is made of electrons and nuclei (subatomic description) that interact through Coulomb forces. Clearly, the derivation of the thermodynamic properties of matter is much more involved when, instead of the atomic description, one uses the subatomic one. The reason is that, in this case, one should analyze successively how electrons and nuclei form atoms, how these atoms constitute molecules, and, finally, how the macro-scopic properties of matter may be derived from the molecular description. As a matter of fact, the subject of going from the subatomic description to the atomic one does not truly belong to the realm of statistical physics (rather it belongs to atomic and molecular physics) which usually takes as starting point the atomic and molecular interactions. There is yet a third description in the thermodynamic study of matter (supramolecular or mesoscopic description), which has sometimes been used to study systems having such a complex molecular architecture that it is very difficult to derive the intermolecular interactions from the atomic interactions. The properties of mesoscopic systems are nevertheless similar to those of atomic systems, but they differ in the relevant length scales: the angstrom in atomic systems (microscopic) and the micron in the supramolecular ones (mesoscopic). In the last few years, there has been a spectacular development of research in mesoscopic systems in what is called soft condensed matter (liquid crystals, colloidal dispersions, polymers, etc.), some examples of which are considered later in this chapter. Due to the existence of an interaction potential between the constituents of a non-ideal system, the latter may be found in different structures or phases whose relative stability depends on the ther-modynamic state, such as the pressure and the temperature. When these variables are changed, a phase transition may occur in which the structure of the system changes. These transitions will be studied in Chap. 9. In the present chapter a summary is provided of some common structures of matter.

© The Author(s), under exclusive license to Springer Nature Switzerland AG 2021   197
M. Baus and C. F. Tejero, *Equilibrium Statistical Physics*,
https://doi.org/10.1007/978-3-030-75432-7_8

8.1 Crystals

Consider first of all a material in the solid phase. A solid is usually formed by the union of crystalline elements with a certain spatial order. In what follows, the case is considered where the solid is formed by a single crystal. Although such monocrystalline solids exist in nature, their experimental preparation is not simple, because they usually contain defects, in particular on their surface. To simplify the study, it will be assumed that the monocrystal is perfect (it does not have any defects) and of infinite volume. A perfect crystal is an example of a phase with spatial order, which means that if one knows the average value of the position vector of a molecule, then the average values of the position vectors of all the other molecules are specified. If the molecules have spherical symmetry, their location in space is fixed by the position vectors of their centers of mass $\{\mathbf{r}_j\}$, which may be written as

$$\mathbf{r}_j = \mathbf{R}_j + \mathbf{u}_j, \tag{8.1}$$

where $\{\mathbf{R}_j\}$ are the position vectors of a Bravais lattice and $\{\mathbf{u}_j\}$ the displacement vectors (see Sect. 5.8). In a solid in equilibrium $\langle \mathbf{r}_j \rangle = \mathbf{R}_j$, where the average is taken over a Gibbs ensemble or, alternatively, $\langle \mathbf{u}_j \rangle = 0$. The set of vectors $\{\mathbf{R}_j\}$ defines the structure of a solid with spatial ordering, which for a crystal is a periodic lattice. There are also some solids in which the vectors $\{\mathbf{R}_j\}$ are the equilibrium positions of a quasiperiodic lattice formed by the superposition of periodic lattices whose lattice parameters are incommensurate. These solids are called quasicrystals and a simple example of a quasicrystal is studied in Appendix D.

8.1.1 Crystal Structure

The structure of a solid is characterized by the two-particle distribution function,

$$\rho_2(\mathbf{r}, \mathbf{r}') = \left\langle \sum_{j=1}^{N} \sum_{j \neq i=1}^{N} \delta(\mathbf{r} - \mathbf{r}_j)\delta(\mathbf{r}' - \mathbf{r}_i) \right\rangle, \tag{8.2}$$

which may be written as

$$\rho_2(\mathbf{r}, \mathbf{r}') = \rho_1(\mathbf{r})\rho_1(\mathbf{r}')g(\mathbf{r}, \mathbf{r}'), \tag{8.3}$$

where $g(\mathbf{r}, \mathbf{r}')$ is the pair correlation function and $\rho_1(\mathbf{r})$ is the local density of particles:

$$\rho_1(\mathbf{r}) = \left\langle \sum_{j=1}^{N} \delta(\mathbf{r} - \mathbf{r}_j) \right\rangle. \tag{8.4}$$

Since $g(\mathbf{r}, \mathbf{r}') \simeq 1$ when $|\mathbf{r} - \mathbf{r}'| > \sigma$ (where σ is the diameter of the molecule), a great majority of the structural properties of a crystal or a quasicrystal may thus be derived from $\rho_1(\mathbf{r})$.

Note that the local density of particles must be a function with very pronounced peaks in the vicinity of the equilibrium positions of the lattice $\{\mathbf{R}_j\}$ and must tend to zero when $\mathbf{r} \neq \mathbf{R}_j$. Since in a crystal all the vertices in the lattice are equivalent, $\rho_1(\mathbf{r})$ must have the following form:

$$\rho_1(\mathbf{r}) = \sum_{j=1}^{N} \phi(\mathbf{r} - \mathbf{R}_j), \tag{8.5}$$

where $\phi(\mathbf{r} - \mathbf{R}_j)$ is the local density around $\mathbf{R}_j$. In a quasicrystal the vertices are not equivalent, and so in the sum (8.5) the function $\phi(\mathbf{r} - \mathbf{R}_j)$ must be replaced by functions $\phi_j(\mathbf{r} - \mathbf{R}_j)$. As a first approximation, the function $\phi(\mathbf{r} - \mathbf{R}_j)$ of a crystal may be written as

$$\phi(\mathbf{r} - \mathbf{R}_j) = \left(\frac{\alpha}{\pi}\right)^{3/2} e^{-\alpha(\mathbf{r} - \mathbf{R}_j)^2}, \tag{8.6}$$

although it is known that $\phi(\mathbf{r} - \mathbf{R}_j)$ does not decrease as fast as a Gaussian nor has spherical symmetry (in (8.6), $\alpha(\mathbf{r} - \mathbf{R}_j)^2$ must be replaced by $(\mathbf{r} - \mathbf{R}_j) \cdot A \cdot (\mathbf{r} - \mathbf{R}_j)$, where A is a matrix that has all the symmetries of the Bravais lattice). Integrating over the whole crystal (which has been assumed to have an infinite volume, so that $N \to \infty$), one has

$$\frac{1}{V} \int d\mathbf{r} \rho_1(\mathbf{r}) = \rho, \tag{8.7}$$

where $\rho = N/V$ is the average density, which is an important parameter in the characterization of a crystal.

Since a crystal is obtained through repetition of a unit cell (Fig. 8.1), of n_1 particles and volume v_1, it follows that $N = \hat{N}n_1$ and $V = \hat{N}v_1$, where $\hat{N}$ is the number of unit cells and thus $\rho = n_1/v_1$. For instance, in a simple cubic (sc) lattice, whose lattice parameter is a, $v_1 = a^3$ and $n_1 = (1/8)8 = 1$, since each of the eight vertices belongs to eight unit cells. Therefore, $\rho = 1/a^3$, and so if one knows ρ, the lattice parameter is determined. In a body-centered cubic (bcc) lattice $v_1 = a^3$ and $n_1 = (1/8)8 + 1 = 2$, i.e., $\rho = 2/a^3$, and in a face-centered cubic (fcc) lattice $v_1 = a^3$ and $n_1 = (1/8)8 + (1/2)6 = 4$, i.e., $\rho = 4/a^3$ (note that in these lattices a is still the lattice parameter of the corresponding simple cubic lattice).

In a solid ρ has an upper limit, which is the density at which the atoms or molecules are in contact. When this happens, the distance between nearest-neighbor atoms is equal to the diameter of the atoms and one says that the solid reaches its maximum packing (note that if one forces two atoms to be at distances smaller than their diameter, the atoms ionize and the solid transforms into a different phase). If one

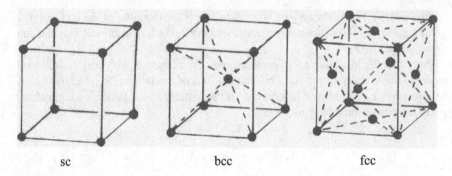

Fig. 8.1 Unit cell of the three cubic Bravais lattices: simple cubic (*sc*), body-centered cubic (*bcc*), and face-centered cubic (*fcc*)

Table 8.1 Some characteristics of the simple cubic lattice, the body-centered cubic lattice, and the face-centered cubic lattice

	n_1	r_{nn}/a	$\rho_{cp}\,\sigma^3$	η_{cp}
sc	1	1	1	0.524
bcc	2	$\sqrt{3}/2$	$3\sqrt{3}/4$	0.680
fcc	4	$\sqrt{2}/2$	$\sqrt{2}$	0.740

Number of particles in the unit cell, n_1, nearest-neighbor distance, r_{nn}, referred to lattice parameter of the simple cubic lattice, a, and density of close packing, ρ_{cp}, of hard spheres of diameter σ. Also it included the close packing fraction, $\eta_{cp} = \pi \rho_{cp}\sigma^3/6$, which is the fraction of the volume of space occupied by the spheres at their closest packing

assumes that the atoms are hard spheres of diameter a (which can, therefore, not interpenetrate), the density of maximum packing or "close packing" (cp) ρ_{cp} may be determined in a simple way. In fact, in a simple cubic lattice when $a = \sigma$ one has ρ_{cp} $\sigma^3 = 1$. Analogously, the distance between nearest neighbors (nn) r_{nn} in a bcc lattice is $r_{nn}^2 = 3(a/2)^2$, i.e., $r_{nn} = \sqrt{3a}/2$ and so if $r_{nn} = \sigma$ one has $\rho_{cp}\sigma^3 = 3\sqrt{3}/4$. In an fcc lattice $r_{nn}^2 = 2(a/2)^2$, i.e., $r_{nn} = \sqrt{2a}/2$ and so if $r_{nn} = \sigma$ one has $\rho_{cp}\sigma^3 = \sqrt{2}$. Since the latter is the highest value for the close packing density that one may obtain, the fcc lattice is called a closest packing lattice (Table 8.1).

8.1.2   Cell Theory

There exist different approximations for the determination of the thermodynamic properties of a crystal, such as its Helmholtz free energy. The simplest one is the cell theory of Lennard–Jones and Devonshire, which has also been applied to dense fluids but with less satisfactory results. An elementary way of describing a solid contained in a closed region R of volume V consists in assuming that each particle (atom or molecule) is localized in a region (Wigner–Seitz cell) R_1 of volume $v = V/N$ centered at its equilibrium position in the Bravais lattice. Notice that the region R_1 does not coincide with the unit cell of the Bravais lattice of the previous section, because in the Wigner–Seitz cell the particle is at the center of the cell. Since all the vertices of a perfect monocrystal are equivalent (which is only true in the thermodynamic limit, where the surface effects disappear) all the cells have the same shape. For instance, in a simple cubic lattice the Wigner–Seitz cell is a cube of side a, i.e., $v = a^3$. Since the vectors $\{\mathbf{R}_j\}$ are located at the centers of the cells $\{R_1\}$, the partition function of the interactions may be approximated in a mean field theory as

$$Z^{\text{int}}(\beta, V, N) = \frac{1}{V^N} \int_R d\mathbf{r}^N e^{-\beta H^{\text{int}}(\mathbf{r}^N)}$$

$$\simeq \frac{1}{v^N} \int_{R_1} d\mathbf{u}^N e^{-\beta \sum_{j=1}^N \varepsilon(\mathbf{u}_j)}; \tag{8.8}$$

with

$$H^{\text{int}}(\mathbf{r}^N) = \frac{1}{2} \sum_{i=1}^N \sum_{i \neq j=1}^N V(|\mathbf{r}_i - \mathbf{r}_j|) \tag{8.9}$$

where $V(|\mathbf{r}_i - \mathbf{r}_j|)$ is the pair potential interaction, $\varepsilon(\mathbf{u}_j)$ is the average potential acting on particle j due to the interaction with the particles in the neighboring cells (a potential which is independent of $\mathbf{R}_j$ since all cells are equivalent), $\mathbf{u}_j$ is the displacement vector and the last multiple integral in (8.8) extends over the region R_1, since each particle is kept inside one cell. Note that, in general, the relation between $\varepsilon(\mathbf{u}_j)$ and $V(|\mathbf{r}_i - \mathbf{r}_j|)$ is not evident. A case in which the approximation (8.8) turns out to be intuitively clear is when the pair potential is that of hard spheres, namely $V(r) = \infty$ ($r < \sigma$) and $V(r) = 0$ ($r > \sigma$) (see (8.42)). As is well known, this potential considers that the particles are spheres of diameter σ that cannot interpenetrate each other, and so the volume occupied by one sphere $v_0 = \pi\sigma^3/6$ cannot be overlapped by others. Therefore, this type of interaction is referred to as an excluded volume interaction. Note that although in a real system the repulsive part of the potential is not necessarily a hard-sphere one and that, in general, the potential also involves an attractive part, nowadays it is possible to produce systems of mesoscopic polystyrene spheres that interact through a hard-sphere potential (in fact, the discontinuity at $r \simeq \sigma$ is then less abrupt). Since these systems may form crystals when they are suspended in an inert medium, this is the main reason that justifies the study of crystalline solids using a hard-sphere interaction.

The major simplification involved in the approximation (8.8) is that the partition function may be factorized in the following form:

$$Z^{\text{int}}(\beta, V, N) = \left(\frac{v_f}{v}\right)^N \tag{8.10}$$

where

$$v_f = \int\limits_{R_1} d\mathbf{u}\, e^{-\beta\varepsilon(\mathbf{u})} \tag{8.11}$$

is, for the reason explained below, the free volume per particle. Indeed, if $V(r)$ is a hard-sphere potential, so will be $\varepsilon(\mathbf{u})$, and hence from (8.11) it follows that v_f is the volume of the region R_1 where $\varepsilon(\mathbf{u}) = 0$, i.e., it is the volume that a particle can occupy freely in a cell when it does not overlap with neighboring particles. In order to determine v_f one has to know how these neighboring particles are distributed around the particle being considered. Since $\varepsilon(\mathbf{u})$ is an average potential, one may assume as a first approximation that this potential has spherical symmetry, and so if r_1 is the average distance between the chosen particle and its neighbors, the free volume v_f is the volume a sphere of radius $r_1 - \sigma$, namely

$$v_f = \frac{4}{3}\pi (r_1 - \sigma)^3 \tag{8.12}$$

Since the exact expression of v_f depends on the form of the Wigner–Seitz cell, the inherent approximation involved in (8.12) is that this cell has approximately spherical symmetry (note that since the volume of the Wigner–Seitz cell and the one of a sphere only differ by a numerical factor, this difference is irrelevant in the study of the thermodynamic properties of the crystal).

Equation (8.12) may be expressed in terms of the average density of the crystal, since $\rho = 1/cr_1^2$, where c is a constant that depends on the geometry of the Wigner–Seitz cell and $\rho_{cp} = 1/c\sigma^3$, i.e.,

$$\frac{v_f}{v} = \left(\frac{r_1 - \sigma}{r_1}\right)^3 ;$$

or, alternatively,

$$\frac{v_f}{v} = \left(1 - \left(\frac{\rho}{\rho_{cp}}\right)^{1/3}\right)^3 , \tag{8.13}$$

so that $v_f \to v$ when $\rho \to 0$ and $v_f \to 0$ when $\rho \to \rho_{cp}$.

Therefore, in the cell theory, the Helmholtz free energy per particle of a crystal whose close packing density is ρ_{cp} as given in the thermodynamic limit by

$$f(T, \rho) = k_B T \left(\ln\left(\rho \Lambda^3\right) - 1\right) - 3k_B T \ln\left(1 - \left(\frac{\rho}{\rho_{cp}}\right)^{1/3}\right), \tag{8.14}$$

and hence the equation of state turns out to be given by

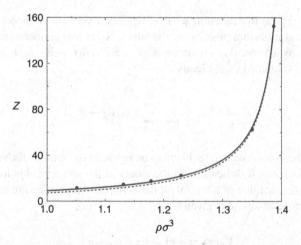

Fig. 8.2 Compressibility factor. $Z = \beta p/\rho$, in the cell theory (8.15) of a solid that crystallizes into a face-centered cubic structure ($\rho_{cp}\sigma^3 = \sqrt{2}$) as a function of $\rho\sigma^3$ (*continuous line*). The broken line is a simple fit ($Z \simeq 3\rho/(\rho_{cp} - \rho)$), proposed by Hall, to the high density values of Z obtained by Alder and Wainwright through numerical simulation (*dots*). *Source* B. J. Alder and T. E. Wainwright, J. Chem. Phys. 27, 1209 (1957), K. R. Hall, J. Chem. Phys. 57, 2252 (1972)

$$p = \frac{\rho k_B T}{1 - \left(\frac{\rho}{\rho_{cp}}\right)^{1/3}}. \tag{8.15}$$

Observe that these expressions are analytic functions of $\rho^{1/3}$ when $0 < \rho < \rho_{cp}$. Since the hard-sphere solid is thermodynamically unstable at low density, (8.14) and (8.15) are only relevant when the average density of the solid is greater than some limiting density (see Chap. 9). At high densities (8.15) is a very good approximation for the equation of state of a hard-sphere crystal, as clearly shown in Fig. 8.2.

8.1.3 van der Waals Theory

Consider now a crystal whose pair potential is the sum of a hard-sphere potential and an attractive potential. The Helmholtz free energy of the crystal in the van der Waals theory $F = F(T, V, N)$ is given by (7.25), where $F^0 = F^0(T, V, N)$ is the free energy of the cell theory (8.14). If the local density of particles of the reference system $\rho_1^0(\mathbf{r})$ is written as (8.5), from (7.25) one has

$$F = F^0 + \frac{1}{2}\sum_{i=1}^{N}\sum_{j=1}^{N}\int_R d\mathbf{r}\int_R d\mathbf{r}'\phi^0(\mathbf{r} - \mathbf{R}_i)\phi^0(\mathbf{r}' - \mathbf{R}_j)V_A(|\mathbf{r} - \mathbf{r}'|). \tag{8.16}$$

Since in a crystal the function $\phi^0(\mathbf{r} - \mathbf{R}_i)$ has a very pronounced peak in the vicinity of the equilibrium position of the lattice $\mathbf{R}_i$, at low temperatures it may be approximated by a Dirac delta function $\phi^0(\mathbf{r} - \mathbf{R}_i) \sim \delta(\mathbf{r} - \mathbf{R}_i)$ (classically this is true when $T = 0$) and so (8.16) reads

$$F = F^0 + \frac{1}{2} \sum_{i=1}^{N} \sum_{j=1}^{N} V_A\left(|\mathbf{R}_i - \mathbf{R}_j|\right). \tag{8.17}$$

Note that the double sum in (8.17) may be replaced by N times the sum over the equilibrium positions $\mathbf{R}_j$ when one of the atoms of the sum over i is located at the origin, due to the fact that in a crystal all the vertices of the lattice are equivalent.

If the attractive potential is given by

$$V_A(r) = -\varepsilon\left(\frac{\sigma}{r}\right)^n, \quad (r > \sigma), \tag{8.18}$$

where ε is a parameter with dimensions of energy and n is an integer, it follows that

$$F = F^0 - \frac{1}{2} N \varepsilon \frac{1}{x_1^n} \sum_{j=1}^{N} \left(\frac{x_1}{x_j}\right)^n, \tag{8.19}$$

where $x_j = |\mathbf{R}_j|/\sigma$. The Helmholtz free energy per particle of the crystal is then written as

$$\beta f(T, \rho) = \ln\left(\rho \Lambda^3\right) - 1 - 3 \ln\left(1 - \left(\frac{\rho}{\rho_{cp}}\right)^{1/3}\right) - \frac{1}{2t}\left(\frac{\rho}{\rho_{cp}}\right)^{n/3} M_n, \tag{8.20}$$

where one has used the fact that $x_1 = (\rho/\rho_{cp})^{1/3}$, introduced the dimensionless variable $t = k_B T/\varepsilon$, and defined the Madelung constant M_n as

$$M_n = \sum_{j=1}^{N} \left(\frac{x_1}{x_j}\right)^n, \tag{8.21}$$

some values of which (in the limit $N \to \infty$) have been included in Table 8.2 for different Bravais lattices and indices n.

8.1.4   Variational Theory

In some cases, the Helmholtz free energy of a crystal may be obtained, approximately, from the Gibbs–Bogoliubov inequality. Consider a solid whose Hamiltonian is given by:

Table 8.2 Madelung constant M_n of the simple cubic, body-centered cubic, and face-centered cubic lattices for different indices n

n	sc	bcc	fcc
4	16.53	22.64	25.34
6	8.40	12.25	14.45
8	6.95	10.36	12.80
10	6.43	9.56	12.31
12	6.20	9.11	12.13
14	6.10	8.82	12.06
16	6.05	8.61	12.03

Source N. W. Ashcroft and N. D. Mermin, *Solid State Physics*, Sawnders, Philadelphia

$$H = \sum_{j=1}^{N} \frac{\mathbf{p}_j^2}{2m} + \frac{1}{2} \sum_{i=1}^{N} \sum_{i \neq j=1}^{N} V\left(\left|\mathbf{r}_i - \mathbf{r}_j\right|\right). \tag{8.22}$$

In order to apply the variational method (7.16), one has to look for a simple reference system for which the upper bound of the Gibbs–Bogoliubov inequality may be evaluated analytically. If one takes as the reference system the Einstein solid (4.31) of Hamiltonian:

$$H^0 = \sum_{j=1}^{N} \left(\frac{\mathbf{p}_j^2}{2m} + \frac{1}{2} m \omega^2 (\mathbf{r}_j - \mathbf{R}_j)^2 \right), \tag{8.23}$$

the partition function Z^0 may be readily evaluated, since all the integrals are Gaussian, leading to

$$Z^0 = \frac{1}{\Lambda^{3N}} \left(\frac{\pi}{\alpha} \right)^{3N/2}, \tag{8.24}$$

where $\alpha = jimrn^2/2$ and Λ is the thermal de Broglie wavelength. The free energy of the Einstein solid is thus given by

$$F^0 = \frac{3}{2} N k_B T \ln\left(\frac{\alpha \Lambda^2}{\pi} \right). \tag{8.25}$$

Note further that, upon application of the equipartition theorem, one has

$$\left\langle \sum_{j=1}^{N} \frac{1}{2} m \omega^2 (\mathbf{r_j} - \mathbf{R}_j)^2 \right\rangle_0 = \frac{3}{2} N k_B T, \tag{8.26}$$

and hence

$$\langle H - H^0 \rangle_0 = \left\langle \frac{1}{2} \sum_{i=1}^{N} \sum_{i \neq j=1}^{N} V(|\mathbf{r}_i - \mathbf{r}_j|) \right\rangle_0 - \frac{3}{2} N k_B T$$

$$= \frac{1}{2} \int d\mathbf{r} \int d\mathbf{r}' \rho_2^0(\mathbf{r}, \mathbf{r}') V(|\mathbf{r} - \mathbf{r}'|) - \frac{3}{2} N k_B T, \qquad (8.27)$$

where $\rho_2^0(\mathbf{r}, \mathbf{r}')$ is the two-particle distribution function (8.2) of the reference system. Since the Einstein solid is an ideal system, $\rho_2^0(\mathbf{r}, \mathbf{r}') = \rho_1^0(\mathbf{r})\rho_1^0(\mathbf{r}')$, where $\rho_1^0(\mathbf{r})$ is the local density of particles, whose analytical expression may be derived from (8.4) and (8.23) with the result

$$\rho_1^0(\mathbf{r}) = \left(\frac{\alpha}{\pi} \right)^{3/2} \sum_{j=1}^{N} e^{-\alpha(\mathbf{r} - \mathbf{R}_j)^2}, \qquad (8.28)$$

which is a sum of normalized Gaussians of parameter α, in agreement with (8.6). Note that the higher the frequency of the oscillators or the lower the temperature, the Gaussians become narrower and the particles become more localized around the equilibrium positions of the atoms in the lattice, as follows from (8.28) since:

$$\left\langle (\mathbf{r}_j - \mathbf{R}_j)^2 \right\rangle_0 = \frac{3}{2\alpha} = \frac{3k_B T}{m\omega^2}. \qquad (8.29)$$

These fluctuations are usually connected to the nearest-neighbor distance in the lattice r_{nn} through the so-called Lindemann ratio L:

$$L = \sqrt{\frac{3}{2\alpha r_{nn}^2}}. \qquad (8.30)$$

Empirically it is known that a crystal melts when $L \gtrsim 0.15$.

The Gibbs–Bogoliubov inequality (7.16) reads in this case

$$F(\alpha) \leq \frac{3}{2} N k_B T \left(\ln \left(\frac{\alpha \Lambda^2}{\pi} \right) - 1 \right)$$

$$+ \frac{1}{2} \int d\mathbf{r} \int d\mathbf{r}' \rho_1^0(\mathbf{r}) \rho_1^0(\mathbf{r}') V(|\mathbf{r} - \mathbf{r}'|)$$

$$= \frac{3}{2} N k_B T \left(\ln \left(\frac{\alpha \Lambda^2}{\pi} \right) - 1 \right)$$

$$+ \frac{1}{2} \frac{1}{(2\pi)^3} \int d\mathbf{k} |\tilde{\rho}_1^0(\mathbf{k})|^2 \tilde{V}(k), \qquad (8.31)$$

where

$$\tilde{\rho}_1^0(\mathbf{k}) = \int d\mathbf{r} e^{-i\mathbf{k}\cdot\mathbf{r}} \rho_1^0(\mathbf{r}), \quad \tilde{V}(k) = \int d\mathbf{r} e^{-i\mathbf{k}\cdot\mathbf{r}} V(r), \qquad (8.32)$$

are the Fourier transforms of the local density of particles and of the pair potential, respectively. From (8.28) one has

$$\tilde{\rho}_1^0(\mathbf{k}) = e^{-k^2/4\alpha} \sum_{j=1}^{N} e^{-i\mathbf{k}\cdot\mathbf{R}_j}, \tag{8.33}$$

which in turn implies

$$\left|\tilde{\rho}_1^0(\mathbf{k})\right|^2 = e^{-k^2/2\alpha} \sum_{j=1}^{N}\sum_{i=1}^{N} e^{-i\mathbf{k}\cdot(\mathbf{R}_j-\mathbf{R}_i)}. \tag{8.34}$$

Since in a perfect crystal all the atoms are equivalent, the double sum in (8.34) may be replaced by N times the sum over the equilibrium positions $\mathbf{R}_j$, when one of the atoms involved in the sum over i is located at the origin, namely

$$\left|\tilde{\rho}_1^0(\mathbf{k})\right|^2 = N e^{-k^2/2\alpha} \sum_{j=1}^{N} e^{-i\mathbf{k}\cdot\mathbf{R}_j}, \tag{8.35}$$

and so (8.31) is written as

$$F(\alpha) \leq \frac{3}{2}Nk_BT\left(\ln\left(\frac{\alpha\Lambda^2}{\pi}\right) - 1\right)$$
$$+ N\frac{1}{2}\frac{1}{(2\pi)^3}\sum_{j=1}^{N}\int d\mathbf{k}\, e^{-k^2/2\alpha} e^{-i\mathbf{k}\cdot\mathbf{R}_j}\tilde{V}(k). \tag{8.36}$$

As an application, consider that $V(r)$ is the Yukawa potential of amplitude ε and range $1/K$:

$$V(r) = \varepsilon\frac{e^{-\kappa r}}{\kappa r} \tag{8.37}$$

in which case the integral in (8.36) may be determined analytically yielding ($f(\alpha) = F(\alpha)/N$):

$$\beta f(\alpha^*) \leq \frac{3}{2}\left[\ln\left(\frac{\alpha^*\kappa^2\Lambda^2}{\pi}\right) - 1\right]$$
$$+ \frac{1}{4t}e^{1/2\alpha^*}\sum_{j}\frac{1}{x_j}e^{-x_j}\text{erfc}\left(\frac{1-\alpha^*x_j}{\sqrt{2\alpha^*}}\right)$$
$$- \frac{1}{4t}e^{1/2\alpha^*}\sum_{j}\frac{1}{x_j}e^{x_j}\text{erfc}\left(\frac{1+\alpha^*x_j}{\sqrt{2\alpha^*}}\right), \tag{8.38}$$

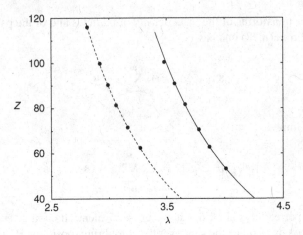

Fig. 8.3 Compressibility factor, $Z = \beta p/\rho$, obtained from the variational method of a crystal of point particles that interact through the Yukawa potential (8.37) as a function of the parameter λ ($\lambda^3 = k^3/\rho$) at the temperatures $t = 6.3 \times 10^{-4}$ (*continuous line*) and $t = 1.4 \times 10^{-3}$ (*broken line*). This system crystallizes into a bcc lattice and the dots are the values of MC simulations. *Source* E. J. Meijer and D. Frenkel, J. Chem. Phys. **94**, 2269 (1991)

where $x_j = kR_j$, $\alpha^* = \alpha/K^2$, $t = k_B T/\varepsilon$, erfc$(x) = 1 - $ erf(x), erf(x) being the error function, and the sum extends over all the shells of a Bravais lattice which are at a distance x_j of the origin of coordinates. Upon minimization of $f(\alpha)$ with respect to α for given values of the temperature, T, and of the density, ρ, one finds $\alpha^*(T, \rho)$, and so the Helmholtz free energy per particle of the solid is $f(T, \rho) = f(\alpha^*(T, \rho))$.

The interest of this result is that (8.37) models the effective pair potential interaction between charge-stabilized colloidal spheres suspended in a solvent which, in a first approximation, is the Derjaguin–Landau–Verwey–Overbeek (DLVO) potential (see Sect. 8.3.4). When the point particle limit of the DLVO potential is taken, a simple Yukawa pair potential is obtained. It is this limiting situation that many of the computer simulations have considered. This colloidal system may crystallize in a bcc or in a fcc lattice. Results for the compressibility factor $Z = \beta p/\rho$, where p is the pressure, of a crystal of point particles interacting through the Yukawa potential and which crystallizes in a bcc lattice are shown in Fig. 8.3.

8.2   Fluids

Under given conditions of pressure and temperature, a material may be found in a phase which is invariant under any translation or rotation (this is only rigorously true in the thermodynamic limit, where the surface effects disappear). In contrast with a crystal, in such a phase the particles are distributed randomly in space. When the interaction is of the excluded volume type, one says that the material is in a fluid phase. If the interaction also contains an attractive part, one similarly refers to a fluid as being the homogeneous and isotropic phase in which the effects of the repulsive part of the potential dominate over those of the attractive part (for more details see Sect. 8.2.4 and Chap. 9).

8.2.1 Dense Fluids

The virial expansion up to second order (see Sect. 7.3) is inadequate even for moderately dense fluids. Since it is not simple to determine the virial coefficients of order higher than two, one may adopt empirical criteria to sum the virial series. In this section two examples are considered.

Recall that the perturbative methods are based on the choice of a reference system which in the van der Waals theory (7.25) is that whose Hamiltonian is the sum of the kinetic energy and the repulsive part of the interaction potential. The free energy per particle, f^0, of the reference system in the second virial coefficient approximation (7.33) is thus given by:

$$\beta f^0 = \beta f^{id} + B_2 \rho, \tag{8.39}$$

where B_2 is the second virial coefficient of the repulsive part of the interaction potential. In the case of a repulsive square-well (SW) potential of amplitude ε and range σ, i.e.,

$$V(r) = \varepsilon \Theta(\sigma - r), \tag{8.40}$$

the second virial coefficient is given by

$$B_2 = \frac{2\pi}{3} \sigma^3 \left(1 - e^{-\beta \varepsilon}\right). \tag{8.41}$$

When $\varepsilon \to \infty$, the potential (8.40) is the hard-sphere potential:

$$V(r) = \begin{cases} \infty, & r < \sigma \\ 0, & r > \sigma \end{cases} \tag{8.42}$$

in which case

$$B_2 = 4v_0, \tag{8.43}$$

where $v_0 = \pi \sigma^3/6$ is the excluded volume around each particle due to the hard-sphere potential (the volume of a sphere of diameter σ). Since the hard-sphere potential prevents the spheres from overlapping, its effect is more important the denser the fluid is.

Consider two empirical approximations to sum the virial series of a hard-sphere fluid. In the first approximation, one starts from the virial expansion of the pressure at low densities $\beta\rho = \rho(1 + B_2\rho + \cdots)$, namely

$$\beta p \simeq \frac{\rho}{1 - B_2\rho}, \tag{8.44}$$

since $B_2\,\rho \ll 1$. Extrapolating empirically (8.44) to high densities, one has

$$\beta p = \frac{\rho}{1 - B_2\rho} = \rho \sum_{n=0}^{\infty} (B_2\rho)^n, \tag{8.45}$$

which in turn implies that $B_{n+1} = B_2^n$. With this approximation and since

$$\sum_{n=1}^{\infty} \frac{x^n}{n} = -\ln(1 - x), \, (x < 1),$$

the free energy per particle due to the repulsive part of the interaction potential, f_R^{int}, is given by

$$\beta f_R^{\text{int}} = \sum_{n=1}^{\infty} \frac{1}{n}(B_2\rho)^n = -\ln(1 - B_2\rho). \tag{8.46}$$

Observe that in this approximation due to van der Waals, the pressure tends to infinity when $\rho \to B_2^{-1}$, i.e., when the fraction of the volume of the Euclidean space occupied by the spheres (the packing fraction $\eta = \rho v_0$) is $n = 0.25$. This result is incorrect since, as was analyzed in Sect. 8.1.1, the maximum packing fraction in a system of hard spheres is that of a crystal whose structure is the one of a face-centered cubic lattice ($\eta_{\text{cp}} = 0.740$).

Another empirical approximation is due to Carnahan and Starling and is based on the values of the virial coefficients of a hard-sphere fluid. The analytical expression of the first three coefficients is known, $B_2 = 2\pi\sigma^3/2$,

$$\frac{B_3}{B_2^2} = \frac{5}{8}, \quad \frac{B_4}{B_2^3} = \frac{1}{2240\pi}\left(219\sqrt{2} + 4131\arccos\left(\frac{1}{\sqrt{3}}\right)\right) - \frac{89}{280},$$

and of the remainder, their numerical value up to B_{10}. This is an exception, since for most pair potentials analytical or numerical results are known only for B_2 and B_3.

The proposal made by Carnahan and Starling for the virial coefficients B_n of a hard-sphere fluid is

$$B_{n+1} = n(n + 3)\left(\frac{B_2}{4}\right)^n, \tag{8.47}$$

Table 8.3 Virial coefficients $b_n = B_n/(B_2)^{n-1}$ of the hard-sphere fluid obtained from the Carnahan–Starling (CS) equation and from analytical or numerical integration techniques (NI) for different values of n

n	b_n (CS)	b_n (NI)
3	5/8	5/8
4	0.28125	0.28695
5	0.109375	0.110252
6	0.039063	0.038882
7	0.013184	0.013024
8	0.004272	0.004183
9	0.001343	0.001309
10	0.000411	0.000404

Source N. Clisby and B. M. McCoy, J. Stat. Phys. **122**, 15 (2006)

which, as shown in Table 8.3, provides an excellent approximation to the exact values and to those obtained by numerical integration.

If one takes into account that

$$\sum_{n=1}^{\infty} x^n = \frac{x}{1-x}, \quad (x < 1),$$

and

$$\sum_{n=1}^{\infty} nx^n = x\frac{\partial}{\partial x}\sum_{n=1}^{\infty} x^n = \frac{x}{(1-x)^2}, \quad (x < 1),$$

the free energy f_R^{int} is given by

$$\beta f_R^{\text{int}} = \sum_{n=1}^{\infty} (n+3)\eta^n = \frac{\eta(4 - 3\eta)}{(1 - \eta)^2}, \tag{8.48}$$

where $\eta = \rho v_0$ is the packing fraction. The pressure p of the hard-sphere fluid is, therefore,

$$\frac{\beta p}{\rho} = \frac{1 + \eta + \eta^2 - \eta^3}{(1 - \eta)^3}, \tag{8.49}$$

which is known as the Carnahan–Starling equation of state.

Observe that (8.49) diverges at $\eta = 1$, which is an unphysical value for the packing fraction since the spheres cannot occupy the whole volume of the system. This divergence, however, is not important because the numerical simulations show that when $\eta \simeq 0.5$ the hard-sphere fluid crystallizes in a fcc lattice (see Chap. 9). Therefore, the hard-sphere fluid at high density is metastable with respect to the solid and the Carnahan–Starling equation ceases to be relevant. The compressibility factor of the

Fig. 8.4 Compressibility factor, $Z = \beta p/\rho$, of the Carnahan–Starling equation of state (*continuous line*) as a function of the packing fraction η. *The dashed lines* are the results obtained from the virial series when this series is approximated by $n = 2, 3, \ldots, 10$ terms. Note that, in these latter cases, Z increases with n

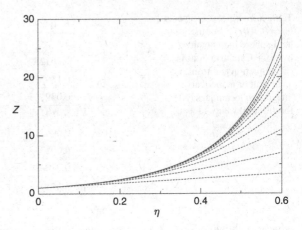

Carnahan–Starling and the van der Waals equations of state (8.46) and the one that is obtained by approximating the equation of state by $n = 2, 3, \ldots, 10$ terms of the virial expansion are displayed in Fig. 8.4.

8.2.2 Fluid Structure

As was analyzed in Sect. 6.8, the local density of particles $\rho_1(\mathbf{r})$ of a homogeneous phase is uniform (independent of $\mathbf{r}$) so from (8.7) it follows that

$$\rho_1(\mathbf{r}) = \rho, \tag{8.50}$$

where $\rho = N/V$ is the average density of particles. On the other hand, the two-particle distribution function (8.2) may be written as

$$\rho_2(\mathbf{r}, \mathbf{r}') = \rho^2 g(|\mathbf{r} - \mathbf{r}'|; T, \rho) = \rho^2[1 + h(|\mathbf{r} - \mathbf{r}'|; T, \rho)], \tag{8.51}$$

since the pair correlation function, $g(r; T, \rho)$, and the total correlation function, $h(r; T, \rho)$, depend on the modulus of the relative distance, of the density ρ, and of the temperature T (see Sect. 6.8).

In contrast with a solid, where the great majority of the structural information is contained in the local density of particles (8.5) and (8.6), the pair correlation function $g(r; T, \rho)$, or the total correlation function $h(r; T, \rho)$, contains all the information about the structure of a fluid. One may then obtain the thermodynamic properties of the fluid from the knowledge of these structural functions; for instance, the compressibility equation (6.101) shows that, if the total correlation function $h(r; T, \rho)$ is known, the isothermal compressibility coefficient can be determined and, after integration of the latter, also the equation of state.

The pair correlation function $g(r; \rho)$ of a hard-sphere fluid is shown in Fig. 8.5. This function has a similar behavior for fluids whose interaction contains a repulsive part, not necessarily of the hard-sphere type, and an attractive part. The main difference is that in the case of the hard-sphere fluid this function does not depend on temperature. Note that $g(r; \rho) = 0$ $(r < \sigma)$ since the probability of finding two particles at a distance smaller than the hard-sphere diameter is zero. On the other hand, as was already analyzed in Sect. 6.8, $g(r; \rho) \to 1$ $(r \gg \sigma)$. Moreover, the range of $g(r; \rho)$ (and, therefore, of $h(r; \rho)$) is of the order of several hard-sphere diameters. Hence, in the theory of fluids, the direct correlation function $c(r; T, \rho)$ is introduced because its range is smaller than that of $h(r; T, \rho)$. This function is defined through the Ornstein–Zernike equation (7.59):

$$h(r; T, \rho) = c(r; T, \rho) + \rho \int d\mathbf{r}' c(r'; T, \rho) h(|\mathbf{r} - \mathbf{r}'|; T, \rho), \qquad (8.52)$$

or, alternatively,

$$\tilde{h}(k; T, \rho) = \tilde{c}(k; T, \rho) + \rho \tilde{c}(k; T, \rho) \tilde{h}(k; T, \rho), \qquad (8.53)$$

which relates the Fourier transforms of both functions.

Note that from the compressibility (6.101) one has

$$\tilde{h}(0; T, \rho) = \int d\mathbf{r} h(r; T, \rho) = k_B T \chi_T(T, \rho) - \frac{1}{\rho}, \qquad (8.54)$$

which shows that the integral of the total correlation function diverges at the critical point because $\chi_T(T_c, \rho_c) = \infty$, i.e., $h(r; T_c, \rho_c)$ is long ranged. From (8.53) and (8.54) it follows, on the other hand, that

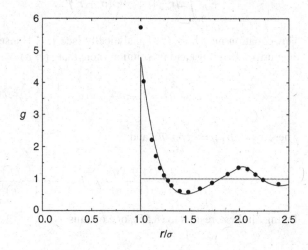

Fig. 8.5 Pair correlation function, $g(r; \rho)$, of a fluid of hard spheres of diameter σ as a function of the variable r/σ for $\eta = \pi \rho \sigma^3/6 = 0.49$. *The continuous line* is the solution of the Percus–Yevick equation and *the dots* are values obtained by MC simulations. *Source* J. P. Hansen and I. R. McDonald, *Theory of Simple fluids*, 2nd ed., Academic Press, London (1986)

$$\tilde{c}(0; \dot{T}, \rho) = \int d\mathbf{r} c(r; T, \rho) = \frac{1}{\rho} - \frac{1}{\rho^2 k_B T \chi_T(T, \rho)}, \tag{8.55}$$

and so at the critical point $\tilde{c}(0; T_c, \rho_c) = 1/\rho_c$, i.e., $c(r; T_c, \rho_c)$ is short ranged.

It is important to point out that the Ornstein–Zernike equation defines $c(r; T, \rho)$ so that, apart from the information obtained through (8.55), this function is as complicated as $h(r; T, \rho)$. In order to determine one or the other, an approximate "closure relation" is required. The best known is the Percus–Yevick closure given by

$$c(r; T, \rho) = \left(1 - e^{\beta V(r)}\right) g(r; T, \rho), \tag{8.56}$$

Note that for the hard-sphere potential (8.42), it follows from (8.56) that $c(r; \rho) = 0$ $(r > \sigma)$, where a is the diameter of the sphere, i.e., in this approximation $c(r; p)$ has the range of the potential. If, on the other hand, one defines the function $y(r; T, \rho)$ by the equation

$$y(r; T, \rho) = e^{\beta V(r)} g(r; T, \rho), \tag{8.57}$$

from (8.56) one has

$$c(r; T, \rho) = y(r; T, \rho) f(r; T), \tag{8.58}$$

where $f(r; T)$ is the Mayer function (7.42), and the Ornstein–Zernike equation transforms into a nonlinear integral equation for $y(r; T, \rho)$ which, in the case of hard spheres, reads

$$y(r; \rho) = 1 - \rho \int d\mathbf{r}' \, y(r'; \rho) \Theta(\sigma - r') y(|\mathbf{r} - \mathbf{r}'|; \rho) \Theta(|\mathbf{r} - \mathbf{r}'| - \sigma)$$
$$+ \rho \int d\mathbf{r}' \, y(r'; \rho) \Theta(\sigma - r') \tag{8.59}$$

This equation may be solved analytically (see J. P. Hansen and I. R. McDonald in the Further Reading) and its solution (note that $c(r; \rho) = -y(r; \rho)$ $(r < \sigma)$) is

$$c(x; \eta) = \left(-a - 6\eta b x - \frac{1}{2}\eta a x^3\right) \Theta(1 - x) \tag{8.60}$$

where $x = r/\sigma$, $\eta = \pi \rho \sigma^3 / 6$, and

$$a = \frac{(1 + 2\eta)^2}{(1 - \eta)^4}, \quad b = -\frac{\left(1 + \frac{1}{2}\eta\right)^2}{(1 - \eta)^4}. \tag{8.61}$$

Taking all these results to (8.55), one obtains

$$\frac{1}{\rho k_B T \chi_T(T, \rho)} = \frac{1 + 4\eta + 4\eta^2}{(1 - \eta)^4}, \tag{8.62}$$

whose integration with respect to ρ allows one to determine the equation of state:

$$\frac{\beta p}{\rho} = \frac{1 + \eta + \eta^2}{(1 - \eta)^3}, \tag{8.63}$$

which is known as the Percus–Yevick compressibility equation of state (PY-c).

Note that the equation of state of a fluid may be obtained in the canonical ensemble with the aid of the expression $p = k_B T \partial \ln Z(\beta, V, N)/\partial V$, i.e.,

$$p = k_B T \frac{\partial}{\partial V} \left[\ln \int_R d\mathbf{r}^N e^{-\beta H_N^{\text{int}}(\mathbf{r}^N)} \right], \tag{8.64}$$

where $H_N^{\text{int}}(\mathbf{r}^N)$ is the potential energy of interaction (8.9). Assume that the region R is a cube of side L ($V = L^3$) and consider the integral

$$A(L^3) \equiv \int_{L^3} d\mathbf{r}^N e^{-\beta H_N^{\text{int}}(\mathbf{r}^N)}. \tag{8.65}$$

One then has

$$A(\lambda^3 L^3) = \int_{\lambda^3 L^3} d\mathbf{r}^N e^{-\beta H_N^{\text{int}}(\mathbf{r}^N)} = \lambda^{3N} \int_{L^3} d\mathbf{x}^N e^{-\beta H_N^{\text{int}}(\lambda \mathbf{x}^N)}, \tag{8.66}$$

where the change of variable $\mathbf{r}_i = \lambda \mathbf{x}_i$ ($i = 1, 2, \ldots, N$) has been made and $\lambda \mathbf{x}^N \equiv \lambda \mathbf{x}_1;$ $\ldots, \lambda \mathbf{x}_N$. After derivation of the first and the third terms of (8.66) with respect to λ, particularizing the result for $\lambda = 1$ and dividing by $3V$, one finds

$$A'(V) = \rho A(V) - \frac{1}{6V} \beta \int_R d\mathbf{r}^N \sum_{i=1}^{N} \sum_{i \neq j=1}^{N} |\mathbf{r}_i - \mathbf{r}_j| V'(|\mathbf{r}_i - \mathbf{r}_j|) e^{-\beta H_N^{\text{int}}(\mathbf{r}^N)}, \tag{8.67}$$

where $\rho = N/V$ and the prime indicates a derivative with respect to the argument. From (8.64) and (8.67) it follows that

$$p = \rho k_B T - \frac{1}{6V} \int_R d\mathbf{r} \int_R d\mathbf{r}' |\mathbf{r} - \mathbf{r}'| V'(|\mathbf{r} - \mathbf{r}'|) \rho_2(\mathbf{r}, \mathbf{r}'), \tag{8.68}$$

where $\rho_2(\mathbf{r}, \mathbf{r}')$ is the two-particle distribution function. In the thermodynamic limit, from (8.51) and (8.57), one has

$$\frac{\beta p}{\rho} = 1 - \frac{1}{6}\beta\rho \int d\mathbf{r}\, r V'(r) g(r; T, \rho)$$

$$= 1 + \frac{1}{6}\rho \int d\mathbf{r}\, r y(r; T, \rho) f'(r; T), \qquad (8.69)$$

which is known as the virial equation of state.

Since for hard spheres $f'(r; T) = \delta(r - \sigma)$, from (8.69) the Percus–Yevick virial equation of state of a hard-sphere fluid (PY-v) is found:

$$\frac{\beta p}{\rho} = 1 - 4\eta c(1; \eta) = \frac{1 + 2\eta + 3\eta^2}{(1 - \eta)^2}. \qquad (8.70)$$

Note that although both the compressibility (8.54) and the virial equation of state (8.69) are exact, when an approximation is introduced to determine the structure of a fluid (for instance, the Percus–Yevick closure relation), the equations of state derived from each of them turn out to be different. This fact is known as the "thermodynamic inconsistency" of the virial and compressibility routes to the equation of state. The differences between (8.63), (8.70) and the Carnahan–Starling equation of state (8.49) (which is an excellent approximation to the simulation results) are shown in Fig. 8.6. Note that although the equations are practically identical at low densities ($\eta \ll 1$), at moderate densities ($\eta > 0.3$) the discrepancies are appreciable. This indicates that the Percus–Yevick closure relation is a good approximation only if the hard-sphere fluid is moderately dense.

Fig. 8.6 Compressibility factor, $Z = \beta p/\rho$, of a hard-sphere fluid as a function of the packing fraction η. *The continuous line* is the Carnahan–Starling equation of state, *the broken line* corresponds to the PY-c equation of state, and *the dotted line* corresponds to the PY-v equation of state. *The dots* indicate the results of numerical simulations. *Source* B. J. Alder and T. E. Wainwright, J. Chem. Phys. **27**, 1209 (1957)

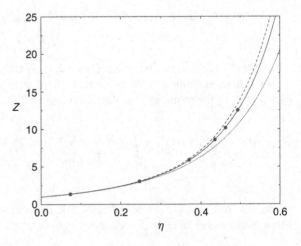

8.2.3 Fluids and Glasses

Up to this point, the equation of state of the hard-sphere fluid has been obtained from the sum of the virial series for the pressure, assuming that the virial coefficients obey a given law (Carnahan–Starling equation), or through the Percus–Yevick closure relation. Since, in general, this kind of information is not available for more complex systems, a third approximation is considered in this section.

A simple idea, due to van der Waals, is that, since the thermodynamic properties of a fluid of hard spheres of diameter a reduce to those of an ideal gas when $\sigma \to 0$, the Helmholtz free energy of a hard-sphere fluid may be approximated by the one of an ideal gas (5.36) in which the volume V is replaced by the "free volume" Vf, which is the volume accessible to the N spheres. If $v_0 = \pi\sigma^3/6$ is the volume of a hard sphere, the accessible volume is bounded by the region of V which is not occupied by the N spheres, namely $V_f \le V - Nv_0$. Therefore,

$$F(T, V, N) = F^{\text{id}}(T, V_f, N) = F^{\text{id}}(T, V, N) - Nk_BT \ln\left(\frac{V_f}{V}\right). \tag{8.71}$$

Since

$$\frac{V_f}{V} \le 1 - \rho v_0 + O(\rho^2), \tag{8.72}$$

the inequality (8.72) may be written as an equality:

$$\frac{V_f}{V} = 1 - \frac{\rho}{\rho_0}, \tag{8.73}$$

where ρ_0 is a constant satisfying $\rho < \rho_0 < \sqrt{2}/\sigma^3$, because V_f has to be positive. In the thermodynamic limit the Helmholtz free energy per particle is, therefore,

$$f(T, \rho) = k_BT\left[\ln(\rho\Lambda^3) - 1\right] - k_BT \ln\left(1 - \frac{\rho}{\rho_0}\right), \tag{8.74}$$

and so the pressure of the hard-sphere fluid in the van der Waals approximation is given by:

$$p = \frac{\rho k_BT}{1 - \frac{\rho}{\rho_0}}. \tag{8.75}$$

Note that the derivation of this equation of state is simple and that (8.75) has a certain parallelism with the equation of state of the cell theory (8.15). As has already been pointed out, $\rho_0\sigma^3 < \rho_{\text{cp}}\sigma^3 = \sqrt{2}$, which is the density of maximum packing of the most compact crystalline structure. The reason is that if $\rho_0 = \rho_{\text{cp}}$, the fluid and solid phases cannot exchange their stability, i.e., at any density the free energy of the fluid (8.74) would be smaller than that of the solid (8.14), and there would be no phase transition, contrary to what is observed in the simulations. As will be seen in the next chapter, an equilibrium disordered structure may only exist when $0 \leq \rho\sigma^3 \leq 0.943$, while in order for an equilibrium periodic structure of hard spheres to exist it is required that $1.041 \leq \rho\sigma^3 \leq \sqrt{2}$. It must be mentioned, however, that the accessible structure at high densities need not necessarily be a crystal. As a matter of fact, a method of great practical interest in order to avoid crystallization of a fluid consists in rapidly increasing the pressure in such a way that the particles (in this case the hard spheres) do not have time to reorganize and stay trapped in a disordered structure which, therefore, is not a phase in thermodynamic equilibrium (see Chap. 13). The density of maximum packing of this kind of structure, "random close packing" (rcp), is $\rho_{rcp}\sigma^3 \simeq 1.203$ (a result obtained from the simulations) although this value depends on how one constructs the packing. The structure obtained in this fashion in systems whose interaction is not necessarily that of hard spheres (note that in such a case the same result may also be obtained by abruptly decreasing the temperature) is called a glass phase (since ordinary window glasses are obtained using this method) which, as has already been mentioned, is not an equilibrium phase as the ones studied in this chapter. In a glass phase the time the system takes to reach equilibrium or relaxation time may be of the order of centuries. The most evident manifestation that the glass phase is an "intermediate phase" between the fluid and the crystal is that, in spite of being a disordered phase, its thermodynamic properties may be described using an equation of state which is similar to the one of cell theory, namely

$$p = \frac{\rho k_B T}{1 - \left(\frac{\rho}{\rho_{rcp}}\right)^{1/3}}, \qquad (8.76)$$

instead of an equation of state of equilibrium fluids such as (8.75).

The free energy of the solid, glass, and fluid phases of a hard-sphere system is represented in Fig. 8.7. The solid and glass phases are described using the cell theory, i.e., (8.15) and (8.76), while the fluid phase is described within the van der Waals approximation (8.75) with $\rho_0\sigma^3 \simeq 0.946$ (see Chap. 9). Note that the glass is stable with respect to the fluid in the interval $0.943 < \rho\sigma^3 < 1.203$, although it is metastable with respect to the solid in the whole density range.

8.2.4 Vapor and Liquid

In the previous sections the focus has been on a fluid with a repulsive interaction which, when abrupt enough, may be approximated by a hard-sphere potential. In real atomic fluids the interaction potential contains both a repulsive and an attractive part. The origin of the former are the repulsive forces between the electron clouds, while the latter arises due to the attractive forces of electric polarization (either permanent or induced) of these electronic clouds.

Fig. 8.7 Reduced free energy $F^* = \beta F \sigma^3 / V$ of the solid (*continuous line*), glass (*dotted line*) and fluid (*broken line*) phases of a system of hard spheres of diameter G as a function of $\rho\sigma^3$. For the sake of clarity, the figure has been divided into two parts:
a $0 \leq \rho\sigma^3 \leq 0.2$.
b $0.2 \leq \rho\sigma^3 \leq 1.4$

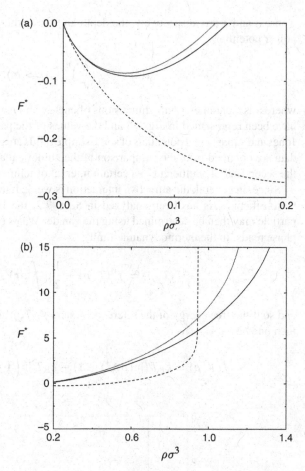

A simple example of this type of pair potential which adequately describes the thermodynamic properties of simple fluids is the Lennard–Jones (LJ) potential:

$$V(r) = 4\varepsilon \left[\left(\frac{\sigma}{r} \right)^{12} - \left(\frac{\sigma}{r} \right)^{6} \right], \tag{8.77}$$

which contains two parameters: σ, which is the distance at which $V(r) = 0$ and $-\varepsilon$, which is the value of the potential at the minimum $r_m = 2^{1/6}\sigma$. The term $1/r^{12}$ describes in a conventional way the repulsion, while the term $1/r^6$ is obtained from the induced dipole-induced dipole van der Waals interaction.

Two widely employed pair potential models in statistical physics are of the form $V(r) = V_R(r) + V_A(r)$, where $V_R(R)$ is the hard-sphere potential (8.42), σ is the diameter of the sphere and $V_A(r)$ is either the square-well potential,

$$V_A(r) = -\varepsilon[\Theta(\sigma + \delta - r) - \Theta(\sigma - r)], \tag{8.78}$$

where ε and δ are, respectively, the depth and the width of the well, or the inverse power potential,

$$V_A(r) = -\varepsilon\left(\frac{\sigma}{r}\right)^n, \quad (r > \sigma), \tag{8.79}$$

where ε is a parameter with dimensions of energy and n an integer. These potentials have been represented in Fig. 8.8 and the values of the parameters for the Lennard–Jones and square-well potentials of some simple fluids are gathered in Table 8.4. These data are obtained from the comparison of the numerical and experimental values of the second virial coefficient in a certain interval of temperature.

Since, in general, the attractive interaction is weak, it may be treated using perturbation theory. As has been analyzed in Sect. 7.2, the Helmholtz free energy per particle may then be determined using the van der Waals (7.25), which for a uniform phase reads, in the thermodynamic limit,

$$f(T, \rho) = f^0(T, \rho) + \frac{1}{2}\rho \int d\mathbf{r} V_A(r), \tag{8.80}$$

and so if the free energy of the reference system $f^0(T, \rho)$ is approximated by (8.74), then one has

$$f(T, \rho) = k_B T\left(\ln\left(\rho\Lambda^3\right) - 1\right) - k_B T \ln\left(1 - \frac{\rho}{\rho_0}\right) - a\rho, \tag{8.81}$$

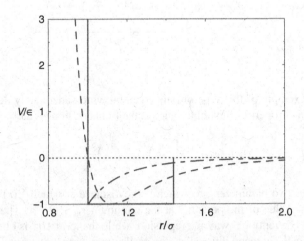

Fig. 8.8 Lennard–Jones potential (*broken line*), square-well potential (*continuous line*), and inverse power potential with $n = 6$ (*dash-dotted line*). The hard-sphere diameter in the last two potentials has been taken to be equal to the variable σ of the Lennard–Jones potential and the width of the square-well $\delta = 79/55$ has been chosen in such a way that the area of the well is equal to the integral of the negative part of the Lennard–Jones potential (represented are $V(r)/\varepsilon$ vs. r/σ)

Table 8.4 Parameters of the Lennard–Jones (LJ) and square-well (SW) potentials obtained from the comparison of numerical and experimental values of the second virial coefficient in a certain interval of temperature

		δ/σ	σ	ε/k_B
Ar	LJ		3.504	117.7
Ar	SW	1.70	3.067	93.3
Kr	LJ		3.827	164.0
Kr	SW	1.68	3.278	136.5
Xe	LJ		4.099	222.3
Xe	SW	1.64	3.593	198.5

The parameters σ and ε/k_B are expressed in angstroms and Kelvin, respectively

Source A. E. Sherwood and J. M. Prausnitz, J. Chem. Phys. **41**, 429 (1964)

where a is a positive constant (since $V_A(r) < 0$) which measures the cohesion between the particles induced by the attractive part of the potential:

$$a = -\frac{1}{2} \int d\mathbf{r} V_A(r). \qquad (8.82)$$

A remarkable fact about (8.81) is that the relative contribution to the Helmholtz free energy coming from the attractive and repulsive parts depends on temperature. According to this, once the density ρ is fixed, there exists a value of the temperature, T_0, for which both contributions compensate, i.e.,

$$k_B T_0 = -\frac{a\rho}{\ln\left(1 - \frac{\rho}{\rho_0}\right)}. \qquad (8.83)$$

When $T = T_0$ the free energy is that of an ideal system, even though there is interaction between the particles. If $T \gg T_0$ the deviation with respect to the ideal behavior is due mainly to the repulsive part of the potential, while if $T \ll T_0$ it is the attractive part of the potential that contributes most to the non-ideal behavior. As will be analyzed in the next chapter, at low density the free energy (8.81) describes at high temperature the fluid phase of the previous sections and at low temperature the gas or vapor phase, which is the natural continuation of the fluid phase. At low temperature and high density, on the contrary, (8.81) describes a new disordered phase in which the attraction is essential and which is called the liquid phase. As has been indicated, in the vapor phase the repulsion dominates over the attraction whereas in the liquid phase the opposite happens. Since both phases are described by the same free energy, it is possible to go from one phase to the other in a continuous way. That is the reason why the title of the thesis of van der Waals is "on the continuity of the liquid and gas phases." In that thesis he derived the equation of state,

$$p = \frac{\rho k_B T}{1 - b\rho} - a\rho^2, \tag{8.84}$$

which is called the van der Waals equation of state, to study the liquid–vapor transition. This equation may be derived using (8.81), with $\rho_0 = 1/b$, and the thermodynamic relation $p = \rho^2 \partial f / \partial \rho$.

8.3 Mixtures

Although the systems analyzed so far are formed by particles of a single species, it is well known that many systems in nature are a mixture of various components. According to the fundamental equation of thermodynamics, the internal energy E of a multicomponent system is a function of the entropy S, of the volume V (assuming that this is the only external parameter), and of the numbers of particles of the different species $\{N_\alpha\}$, namely $E = E(S, V, \{N_\alpha\})$. All the thermodynamic information contained in this equation may be taken over to the Helmholtz free energy $F = F(T, V, \{N_\alpha\})$ through a Legendre transformation (see Appendix A).

8.3.1 Binary Mixtures

In the description of a mixture, we may use as intensive variables the partial densities $\rho_\alpha = N_\alpha / V$ that verify the condition $\sum_\alpha \rho_\alpha = \rho = N/V$, where ρ is the total density and $N = \sum_\alpha N_\alpha$, or the number fractions $x_\alpha = N_\alpha / N$, that verify $\sum_\alpha x_\alpha = 1$. Therefore, in the thermodynamic description of a mixture one must clearly specify the variables that are used. The case of a disordered phase, which is later particularized to the binary mixture case, is analyzed in this subsection.

The Helmholtz free energy (5.36) of an ideal system in the thermodynamic limit,

$$N k_B T \left(\ln\left(\frac{N\Lambda^3}{V}\right) - 1 \right),$$

may be generalized to the case of a mixture as

$$\begin{aligned}
F^{\text{id}}(T, V, \{N_\alpha\}) &= k_B T \sum_\alpha N_\alpha \left(\ln\left(\frac{N_\alpha \Lambda_\alpha^3}{V}\right) - 1 \right) \\
&= k_B T V \sum_\alpha \rho_\alpha \left(\ln\left(\rho_\alpha \Lambda_\alpha^3\right) - 1 \right) \\
&= k_B T N \sum_\alpha x_\alpha \left(\ln x_\alpha + \ln\left(\rho \Lambda_\alpha^3\right) - 1 \right), \tag{8.85}
\end{aligned}$$

where $\Lambda_\alpha = h/\sqrt{2\pi m_\alpha k_B T}$ is the thermal de Broglie wavelength of species α.

Similarly, the virial expansion up to second order, (7.33) and (7.45), can be generalized for a mixture as

$$F^{\text{int}}(T, V, \{N_\alpha\}) = k_B T V \sum_\alpha \sum_\gamma \rho_\alpha \rho_\gamma b_{\alpha\gamma}, \tag{8.86}$$

i.e.,

$$B_2 = \sum_\alpha \sum_\gamma x_\alpha x_\gamma b_{\alpha\gamma}, \tag{8.87}$$

with

$$b_{\alpha\gamma} = -\frac{1}{2} \int d\mathbf{r}\left(e^{-\beta V^{\alpha\gamma}(r)} - 1\right), \tag{8.88}$$

where $V^{\alpha\gamma}(r)$ is the total interaction potential between particles of species α and γ.

An interesting problem consists in determining whether a mixture is thermodynamically stable or metastable with respect to the system of the separate species, i.e., whether the free energy of the mixture is smaller or greater than the sum of the free energies of each of the components. In order to simplify the equations, consider a binary mixture in the second virial coefficient approximation. The Helmholtz free energy of the mixture is according to (8.85) and (8.86), given by

$$F(T, V, N_1, N_2) = k_B T V \sum_{\alpha=1}^{2} \rho_\alpha \left(\ln\left(\rho_\alpha \Lambda_\alpha^3\right) - 1 + \sum_{\gamma=1}^{2} \rho_\gamma b_{\alpha\gamma}\right), \tag{8.89}$$

so that the free energy per particle $f(T, \rho, x) = F(T, V, N_1, N_2)/N$, where $N = N_1 + N_2$, $\rho = N/V$, and $x = x_1$ is the number fraction of one of the components ($x_2 = 1 - x$), turns out to be

$$\begin{aligned}
f(T, \rho, x) = {}& k_B T x \left(\ln x + \ln'\left(\rho \Lambda_1^3\right) - 1\right) \\
& + k_B T (1-x)\left(\ln(1-x) + \ln\left(\rho \Lambda_2^3\right) - 1\right) \\
& + k_B T \rho\left(x^2 b_{11} + 2x(1-x)b_{12} + (1-x)^2 b_{22}\right),
\end{aligned} \tag{8.90}$$

where one has used the fact that $b_{12} = b_{21}$.

The free energy per particle of component 1, if considered separately at the same density and temperature as those of the mixture is, according to (8.90), $f(T, \rho, 1)$, while that of component 2 is $f(T, \rho, 0)$. If the "free energy of mixing" is defined by the expression

$$F^{\text{mix}}(T, V, N_1, N_2) = N f(T, \rho, x) - N_1 f(T, \rho, 1) - N_2 f(T, \rho, 0),$$

then the free energy of mixing per particle $f^{mix}(T, \rho, x)$ is given by

$$f^{mix}(T, \rho, x) = f(T, \rho, x) - xf(T, \rho, 1) - (1 - x)f(T, \rho, 0). \qquad (8.91)$$

From (8.90) it follows that

$$\begin{aligned} f^{mix}(T, \rho, x) = & k_B T\{x \ln x + (1 - x)\ln(1 - x)\} \\ & + k_B T\rho x(1 - x)(2b_{12} - b_{11} - b_{22}). \end{aligned} \qquad (8.92)$$

The first term in the r.h.s. of (8.92) is the ideal contribution to the "entropy of mixing," while the second one is the contribution of the "energy of mixing" in the second virial coefficient approximation. Note that the interaction term does not depend separately on the variables b_{ij}, but only on the coefficient,

$$b = -\frac{1}{2}(\{b_{11} - b_{12}\} + \{b_{22} - b_{12}\}), \qquad (8.93)$$

i.e., on the asymmetries between the different species and, therefore, such a mixture is called regular. If $f^{mix}(T, \rho, x)$ is negative for certain values of T, ρ, and x, the mixture is the thermodynamically stable phase. It may occur that when modifying one of these variables $f^{mix}(T, \rho, x)$ becomes positive and the mixture then will be metastable with respect to the pure phases. One says then that in the mixture a demixing phase transition takes place (cf. Chap. 9).

8.3.2   Colloidal Suspensions

Colloidal suspensions are multicomponent systems composed of mesoscopic colloidal particles dispersed in a molecular solvent which often contains other smaller particles as well, i.e., polymers, ions of a dissociated salt, etc. The statistical treatment of such mixtures is generally complicated because the widely separated length and time scales between the colloids and the remaining constituents inhibits a treatment of all the components on an equal footing. Indeed, a typical linear dimension of a colloidal particle is of the order of μm, whereas a typical atomic radius is of the order of nm. This large difference in size has important physical consequences. First, characteristic timescales for the motion of the colloidal particles are much slower and one typically observes the system to evolve at macroscopic timescales in contrast with the rapid motion of the small particles. From the experimental point of view, the size of a colloidal particle, which is of the same order of magnitude as the wavelength of visible light, allows to analyze the structure of a colloidal fluid by light-scattering experiments, instead of the traditional neutron-scattering experiments which have to be performed in the case of atomic fluids. Also, colloidal crystals exhibit a rigidity against mechanical deformations which is many orders of magnitude smaller than their atomic counterparts, i.e., they are "soft" solids which can be easily "shear-melted." Moreover, as indicated above, a melted colloidal crystal will take a macroscopic time to recrystallize, opening up the possibility of a detailed study of the crystallization kinetics.

For equilibrium systems, when the colloids are monodisperse, i.e., all particles have the same size, a colloidal suspension can often be viewed as a one-component system of "supramolecules" described by an "effective" Hamiltonian. This effective Hamiltonian results from tracing out in the partition function the degrees of freedom of the remaining, microscopic, constituents, and, hence, it depends on the thermodynamic state of the mixture. In order to illustrate this concept, let us consider a binary mixture of N_1 large particles of mass M_1 and coordinates $\mathbf{R}^{N_1}$, and N_2 small particles of mass m_2 and coordinates $\mathbf{r}^{N_2}$, in a region R of volume V and being at temperature T. The full two-component potential energy is

$$U\left(\mathbf{R}^{N_1}, \mathbf{r}^{N_2}\right) = U_{11}\left(\mathbf{R}^{N_1}\right) + U_{12}\left(\mathbf{R}^{N_1}, \mathbf{r}^{N_2}\right) + U_{22}\left(\mathbf{r}^{N_2}\right)$$
$$\equiv U_{11} + U_{12} + U_{22} \tag{8.94}$$

where U_{11} (U_{22}) contains the interactions between the large (or small) particles, while U_{12} describes the interactions between the large and the small particles. If the goal is to eliminate the small particles from the picture, the partition function of the mixture can be written, by first performing the elementary integration over the momenta of both components, as

$$Z(\beta, V, N_1, N_2) = \frac{1}{\Lambda_1^{3N_1} N_1!} \int_R d\mathbf{R}^{N_1} e^{-\beta[U_{11}(\mathbf{R}^{N_1}) + \Delta F(\beta, V, N_1, N_2; \mathbf{R}^{N_1})]}, \tag{8.95}$$

where $\Delta F\left(\beta, V, N_1, N_2; \mathbf{R}^{N_1}\right)$ has been defined by

$$e^{-\beta \Delta F(\beta, V, N_1, N_2; \mathbf{R}^{N_1})} \equiv \frac{1}{\Lambda_2^{3N_2} N_2!} \int_R d\mathbf{r}^{N_2} e^{-\beta[U_{12}(\mathbf{R}^{N_1}, \mathbf{r}^{N_2}) + U_{22}(\mathbf{r}^{N_2})]}, \tag{8.96}$$

with Λ_1 (Λ_2) denoting the thermal de Broglie wavelength of the large (or small) particles. Note that this formal transformation of the partition function of the mixture allows to identify the r.h.s. of (8.95) as the partition function of a one-component system of large particles described by an effective potential energy:

$$U_{11}^{\text{eff}}\left(\beta, V, N_1, N_2; \mathbf{R}^{N_1}\right) = U_{11}\left(\mathbf{R}^{N_1}\right) + \Delta F\left(\beta, V, N_1, N_2; \mathbf{R}^{N_1}\right)$$
$$\equiv U_{11} + \Delta F. \tag{8.97}$$

The effective potential energy (8.97) consists of the direct interactions U_{11} and a free energy term ΔF which indirectly includes the effects of the small particles. An important difference with respect to atomic systems, where the interactions are determined by the electronic structure, is that in colloidal systems the effective interactions depend on the thermodynamic state, here β, V, N_1, and N_2. A change, say, in the temperature or in the composition of the mixture modifies the effective inter-actions which can thus be controlled externally. As a matter of fact, the complexity of the mixture, with different size scales, is conserved in the exact effective potential energy (8.97) because (8.95) and (8.96) are merely formal transformations of the partition function of the mixture. Moreover, effective colloidal interactions are not necessarily pairwise additive, even if the underlying interactions in (8.94) are strictly pairwise. However, in order to take advantage of the well-known methods developed in statistical physics for atomic systems, one has always to resort to approximations in order to express ΔF in terms of effective pair potential interactions which depend on the intensive thermodynamic variables, here $V_{\text{eff}}(|\mathbf{R} - \mathbf{R}'|; \beta, \rho_1, \rho_2)$, where ρ_1 and ρ_2 are the number densities of the components. Two well-known examples of effective pair potentials will be introduced below. In general, problems with the pair potential picture are to be expected only at very high concentrations of colloids.

Further, it has to be emphasized that, even in the pair potential approximation, ΔF contains a term which does not depend on the coordinates of the large particles, i.e.,

$$\Delta F = F_0(\beta, V, N_1, N_2)$$

$$+ \frac{1}{2} \sum_{i=1}^{N} \sum_{i \neq j=1}^{N} V_{\text{eff}}(|\mathbf{R}_i - \mathbf{R}_j|; \beta, \rho_1, \rho_2). \tag{8.98}$$

The term $F_0(\beta, V, N_1, N_2)$ must be extensive in the thermodynamic limit and, therefore, can be written as $V f_0(\beta, \rho_1, \rho_2)$ where $f_0(\beta, \rho_1, \rho_2)$ is some function of its arguments. For this reason, $V f_0(\beta, \rho_1, \rho_2)$ is known as the volume term. The first-principles determination of $V_{\text{eff}}(|\mathbf{R} - \mathbf{R}'|; \beta, \rho_1, \rho_2)$ and $f_0(\beta, \rho_1, \rho_2)$ is generally a formidable task which can be solved only approximately in most cases. The lack of dependence of the volume term on the coordinates of the large particles moreover implies that the average of any dynamical function (e.g., the pair correlation function) in unaffected by this term, i.e., the volume term plays no role in the structure of the one-component effective system. Nevertheless, the volume term affects the phase diagram and the thermodynamic properties, because of its dependence on the intensive variables β, ρ_1, and ρ_2. Finally, observe that the foregoing arguments can be easily transposed to other multicomponent systems and also to different Gibbs ensembles.

8.3.3 Asakura–Oosawa Potential

The addition of a non-adsorbing polymer to a suspension of colloidal particles dispersed in a solvent can cause an effective attraction between the particles by the depletion mechanism. This phenomenon can be interpreted in terms of a volume restriction whereby the exclusion of polymer particles between two neighboring colloidal particles produces a net attraction between them (see Fig. 8.9).

Consider that the suspension is in equilibrium with a reservoir containing solely the solvent and the polymer molecules. If p and p_R denote the pressure of the system and of the reservoir, the osmotic pressure difference π is given by $\pi = p - p_R$. Let us assume for simplicity that the colloidal particles and the polymer molecules can be considered as hard spheres of diameters σ and σ' ($\sigma > \sigma'$), respectively. On an isolated colloidal particle the polymer suspension exerts an isotropic osmotic pressure π. But if two colloidal particles approach each other so that the center-to-center separation r is smaller than $\sigma + \sigma'$, polymer molecules will be excluded from a well-defined region between the particles (the depletion region). The resulting effect is an unbalanced osmotic pressure driving the particles together. Integration of this osmotic pressure over the portion of available surface area of the two particles gives rise to the following effective pair potential between two colloidal hard spheres:

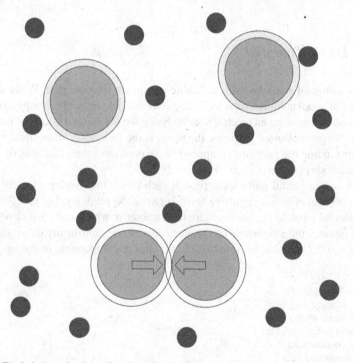

Fig. 8.9 Depletion mechanism. Two large hard spheres of diameter σ in a sea of small spheres of diameter σ'. The spherical shell with thickness $\sigma'/2$ surrounding each large sphere is called the depletion layer. When the depletion layers do not overlap (upper case), there is an isotropic osmotic pressure acting on each large sphere. When the depletion layers overlap (lower case), there is an unbalanced osmotic pressure driving the large spheres together

$$V_{\text{eff}}(r; \pi) = \begin{cases} \infty, & r < \sigma \\ V_{\text{AO}}(r; \pi), & \sigma < r < \sigma + \sigma' \\ 0, & r > \sigma + \sigma' \end{cases} \qquad (8.99)$$

In (8.99) $V_{\text{AO}}(r; \Pi) = -\pi\Omega(r; \sigma + \sigma')$ is the Asakura–Oosawa depletion potential, with $\Omega(r; \sigma + \sigma')$ denoting the volume of the overlapping depletion zones, i.e.,

$$\Omega(r; x) = \frac{\pi x^3}{6}\left(1 - \frac{3r}{2x} + \frac{r^3}{2x^3}\right), \quad (x = \sigma + \sigma'). \qquad (8.100)$$

The effective pair potential (8.99) consists, therefore, of a hard-sphere repulsion and an attractive contribution. The range of the attractive part can be monitored by modifying σ' (the radius of gyration of the polymer, see Sect. 8.5.1) while the depth is controlled by π (i.e., by the polymer concentration).

8.3.4 DLVO Potential

Consider colloidal particles with ionizable groups on their surface. When they are dispersed in a polar liquid such as water, some of the ionizable groups dissociate and the colloids acquire an electric charge. Since the discharged counterions remain near the charged colloidal particles, the result is the formation of an electric double layer surrounding the particles composed of the counterions and the ions of any salt added to the suspension (Fig. 8.10).

When two colloidal particles approach each other, the overlap of these double layers causes an effective repulsive force between the particles. Let us assume that each colloidal particle is a hard sphere of diameter σ which carries a charge $-Ze$, where e denotes the magnitude of the electronic charge, uniformly distributed over their surface. Let us also assume that the remaining components in the suspension

Fig. 8.10 Two colloidal particles of charge $-Ze$ in a sea of microions of charge $+ e$. The overlap of the electric double layers causes an effective repulsive force between the colloidal particles

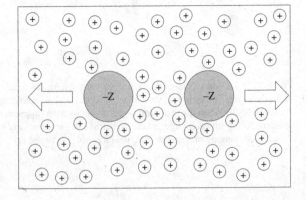

are point particle microions (counterions of charge $+e$ and fully dissociated pairs of monovalent salt ions of charge $\pm e$) and that the solvent is a continuum of dielectric constant ε. The point particle microions screen the Coulomb repulsion between the colloids in such a way that the resulting effective pair potential between two charged hard spheres is

$$V_{\text{eff}}(r; \kappa) = \begin{cases} \infty, & r < \sigma \\ V_{\text{DLVO}}(r; \kappa), & r > \sigma \end{cases} \tag{8.101}$$

where $V_{\text{DLVO}}(r; \kappa)$ is the Derjaguin–Landau–Verwey–Overbeek (DLVO) potential:

$$V_{\text{DLVO}}(r; \kappa) = \frac{Z^2 e^2}{\varepsilon} \left(\frac{e^{\kappa \sigma/2}}{1 + \kappa \sigma/2} \right)^2 \frac{e^{-\kappa r}}{r}. \tag{8.102}$$

In (8.102) κ is the Debye screening parameter:

$$\kappa^2 = \frac{4\pi e^2}{\varepsilon k_B T} (Z\rho + 2\rho_s) \tag{8.103}$$

with ρ (ρ_s) denoting the number density of the colloids (salt). The effective pair potential (8.101) consists, therefore, of a hard-sphere potential and a repulsive state-dependent contribution. The range and depth of the DLVO potential can be controlled externally by modifying the intensive thermodynamic variables ρ, ρ_s, and T.

8.4 Liquid Crystals

Many systems in nature are formed by non-spherical molecules whose interaction cannot be described by a central pair potential. Sometimes the anisotropy of the molecules plays absolutely no role and so such systems may be treated with the methods that have been described in Chap. 7. In other cases, such as a system with molecules very elongated in one direction, the molecular anisotropy manifests itself at a macroscopic scale in the form of phases with anisotropic properties. These phases, which are specific of very oblate or prolate molecules, are called mesophases since they have a degree of order which is intermediate between the one of the (disordered) fluid phase and that of the (ordered) crystalline phase. Two of these mesophases are the nematic phase and the smectic phase. In the former, which is invariant under all translations, the molecules have a degree of orientational order, since they tend to align along a common axis called the nematic director. In the latter, which is invariant under certain translations, the molecules are situated on parallel planes and they may have moreover some degree of orientational order within the planes and, sometimes, also between consecutive planes. Since these mesophases have both properties of the liquid phase (such as fluidity) and of the crystalline phase (such as birefringence), they are also called liquid crystals.

In this section only the simplest mesophase is considered, namely a nematic liquid crystal. Note that in order to describe an elongated molecule, on top of the position vector of the center of mass $\mathbf{r}$, one needs a unitary vector $\mathbf{u}$ in the direction of the elongation. For instance, if the molecule is cylindrical $\mathbf{u}$ has the direction of the axis of the cylinder and if it is ellipsoidal, the direction of the principal axis of the ellipsoid. In what follows it will be assumed that the molecule has symmetry of rotation around u (like, e.g., an ellipsoid of revolution) and that the directions $\mathbf{u}$ and $-\mathbf{u}$ represent the same state of orientation. In a system of N identical molecules, since the centers of mass are distributed uniformly in space, the nematic phase may be considered as a mixture of molecules with different orientations or unit vectors, where each species of the mixture corresponds to one orientation. Let $\rho_1(\mathbf{u})$ be the average density of molecules with orientation u, which is normalized in the following way:

$$\int d\mathbf{u}\rho_1(\mathbf{u}) = \rho, \tag{8.104}$$

with

$$\int d\mathbf{u} \cdots = \frac{1}{4\pi} \int_0^{2\pi} d\phi \int_0^{\pi} d\theta \sin\theta \ldots, \tag{8.105}$$

where (θ, ϕ) are the polar angles of $\mathbf{u}$ and $\rho = N/V$. If the mixture is described with the second virial coefficient approximation (8.89), the variational Helmholtz free energy functional of the nematic phase may then be written as

$$\frac{\beta\mathcal{F}(T, V, N; [\rho_1])}{V} = \int d\mathbf{u}\rho_1(\mathbf{u})\big[\ln\big(\rho_1(\mathbf{u})\Lambda^3\big) - 1\big]$$
$$+ \int d\mathbf{u}\rho_1(\mathbf{u}) \int d\mathbf{u}'\rho_1(\mathbf{u}')b(\mathbf{u}, \mathbf{u}'), \tag{8.106}$$

where the sum over species has been replaced by an integral over orientations, since u is a continuous variable, and Λ also includes now the orientational degrees of freedom. In (8.106) $b(\mathbf{u}, \mathbf{u}')$ is, after (8.88),

$$b(\mathbf{u}, \mathbf{u}') = -\frac{1}{2} \int d\mathbf{r}\big(e^{-\beta V(\mathbf{r};\mathbf{u},\mathbf{u}')} - 1\big), \tag{8.107}$$

where $V(\mathbf{r}; \mathbf{u}, \mathbf{u}')$ is the interaction potential between two molecules with orientations $\mathbf{u}$ and $\mathbf{u}'$, and $\mathbf{r}$ is the relative position vector of the centers of mass.

Note that $\mathcal{F}\big(T, V, N, [\overline{\rho}_1]\big)$ is a functional of the density $\overline{\rho}_1(\mathbf{u})$ whose equilibrium value, $\rho_1(\mathbf{u})$, is, according to (7.71), given by

$$\ln\left(\rho_1(\mathbf{u})\Lambda^3\right) + 2\int d\mathbf{u}'\rho_1(\mathbf{u}')b(\mathbf{u}, \mathbf{u}') = \beta\mu, \tag{8.108}$$

which is a nonlinear integral equation for $\rho_1(\mathbf{u})$, where the constant $\beta\mu$ should be chosen such as to comply with the normalization condition (8.104). In order to solve (8.108) one needs the expression for $b(\mathbf{u}, \mathbf{u}')$. In the following two limiting cases are considered in which $b(\mathbf{u}, \mathbf{u}')$ may be obtained in an approximate manner, which in turn allows one to determine the Helmholtz free energy.

8.4.1 Maier–Saupe Theory

In the case of a weak interaction potential, i.e., $\beta V(\mathbf{r}; \mathbf{u}, \mathbf{u}') \ll 1$, (8.107) may be approximated by

$$b(\mathbf{u}, \mathbf{u}') \simeq \frac{1}{2}\beta \int d\mathbf{r}V(\mathbf{r}; \mathbf{u}, \mathbf{u}'). \tag{8.109}$$

If, on the other hand, $V(\mathbf{r}; \mathbf{u}, \mathbf{u}')$ depends on $r = |\mathbf{r}|$ and on the scalar products of $\mathbf{r}$, $\mathbf{u}$, and $\mathbf{u}'$, one may write

$$b(\mathbf{u} \cdot \mathbf{u}') = \sum_{l=0}^{\infty} a_l P_l(\mathbf{u} \cdot \mathbf{u}') \tag{8.110}$$

where $P_l(\mathbf{u} \cdot \mathbf{u}')$ is the Legendre polynomial of order l and argument $\mathbf{u} \cdot \mathbf{u}'$. The multipolar expansion (8.110) takes into account the anisotropy of the electronic clouds in non-spherical molecules. When, as in a nematic, $\mathbf{u}$ and $-\mathbf{u}$ denote the same orientational state, the series (8.110) only contains the even terms. Therefore,

$$\int d\mathbf{u}'\rho_1(\mathbf{u}')b(\mathbf{u} \cdot \mathbf{u}') = \sum_{l=0}^{\infty} a_{2l} \int d\mathbf{u}'\rho_1(\mathbf{u}') P_{2l}(\mathbf{u} \cdot \mathbf{u}'), \tag{8.111}$$

where

$$a_{2l} \equiv \frac{1}{2}\int_{-1}^{1} dx\, P_{2l}(x)b(x), \quad x = \mathbf{u} \cdot \mathbf{u}'. \tag{8.112}$$

If, finally, one assumes that in the multipolar expansion (8.111) only the first two terms, $P_0(x) = 1$ and $P_2(x) = (3x^2 - 1)/2$, are relevant, then the result is the Maier–Saupe approximation,

$$\int d\mathbf{u}'\rho_1(\mathbf{u}')b(\mathbf{u} \cdot \mathbf{u}') \simeq \rho a_0 + a_2 \int d\mathbf{u}'\rho_1(\mathbf{u}') P_2(\mathbf{u} \cdot \mathbf{u}') \tag{8.113}$$

where use has been made of the normalization condition (8.104). As shown later on, the dipolar approximation (8.113) is sufficient to obtain a nematic phase.

When the nematic has cylindrical symmetry, $\rho_1(\mathbf{u})$ only depends on the angle formed by $\mathbf{u}$ and the nematic director $\mathbf{n}$ ($\mathbf{n}^2 = 1$), and so

$$\rho_1(\mathbf{u}) = \rho h(x), \quad x = \mathbf{u} \cdot \mathbf{n} = \cos \theta, \tag{8.114}$$

where it has been assumed that $\mathbf{n}$ has the direction of the polar z-axis and $h(x)$ is the angular distribution of the molecules at equilibrium. From the sum rule,

$$P_l(\mathbf{u} \cdot \mathbf{u}') = \frac{4\pi}{2l+1} \sum_{m=-l}^{l} Y_l^m(\theta, \phi) Y_l^{-m}(\theta', \phi'), \tag{8.115}$$

where $Y_l^m(\theta, \phi)$ are the spherical harmonics, the integral

$$I \equiv \int d\mathbf{u}' \rho(\mathbf{u}') P_l(\mathbf{u} \cdot \mathbf{u}')$$

in (8.113) reads

$$I = \frac{\rho}{2l+1} \sum_{m=-l}^{l} Y_l^m(\theta, \phi) \int_0^{2\pi} d\phi'$$

$$\times \int_0^{\pi} d\theta' \sin \theta' h(\cos \theta') Y_l^{-m}(\theta', \phi'). \tag{8.116}$$

Note that since $Y_l^m(\theta, \phi) \simeq e^{im\phi}$, the only nonzero contribution to (8.116) corresponds to $m = 0$, and so taking into account that

$$Y_l^0(\theta', \phi') = \sqrt{\frac{2l+1}{4\pi}} P_l(\cos \theta'), \tag{8.117}$$

yields

$$I = \frac{1}{2} \rho P_l(x) \int_{-1}^{1} dx' h(x') P_l(x'). \tag{8.118}$$

Using (8.118), then (8.108) may be cast in the following form:

$$\ln h(x) + \rho a_2 P_2(x) \int_{-1}^{1} dx' h(x') P_2(x') = C, \tag{8.119}$$

where C is a constant and hence from the normalization condition,

$$\frac{1}{2}\int_{-1}^{1} dx h(x) = 1,$$ (8.120)

it follows that the angular distribution at equilibrium is given by

$$h(x) = \frac{e^{q_2 P_2(x)}}{\int_0^1 dx' e^{q_2 P_2(x')}},$$ (8.121)

where use has been made of the fact that $P_2(x) = P_2(-x)$ and, in turn, $h(x) = h(-x)$. In (8.121)

$$q_2 = -\rho a_2 \int_{-1}^{1} dx' h(x') P_2(x'),$$ (8.122)

is an order parameter (see Chap. 9) that minimizes the free energy and which is obtained from (8.121) and (8.122), or, alternatively, from the following equation

$$2\varepsilon_2 = \frac{q_2 \int_0^1 dx e^{q_2 P_2(x)}}{\int_0^1 dx P_2(x) e^{q_2 P_2(x)}}$$ (8.123)

where $\varepsilon_2 = -\rho a_2$. From (8.123) it follows that

$$\lim_{q_2 \to 0+} \varepsilon_2 = \frac{1}{2} \lim_{q_2 \to 0+} \frac{q_2 \int_0^1 dx e^{q_2 P_2(x)}}{\int_0^1 dx P_2(x) e^{q_2 P_2(x)}} = \frac{1}{2 \int_0^1 dx P_2^2(x)} = \frac{5}{2},$$ (8.124)

and there exists a solution $q_2 \neq 0$ of (8.123) when $\varepsilon_2 \geq 2.244$ (see Fig. 9.9), which implies that the dipolar interaction (which is the one that induces the orientation) has to be strong enough for the existence of a thermodynamically stable nematic phase. Comparing the angular distribution (8.121) with (7.83), one concludes that the Maier–Saupe theory is a mean field theory in which the orientational order of each molecule is produced by the mean field potential $\phi(x) = -k_B T q_2 P_2(x)$. This potential is attractive (it has a minimum) when the molecules orient in the direction of the director and repulsive (it has a maximum) when they orient perpendicularly to the director (Figs. 8.11 and 8.12).

Fig. 8.11 Angular distribution $h = h(\cos\theta)$ in the Maier–Saupe theory as a function of θ for $q_2 = 1$ (*dotted line*), $q_2 = 2$ (*broken line*), and $q_2 = 3$ (*continuous line*)

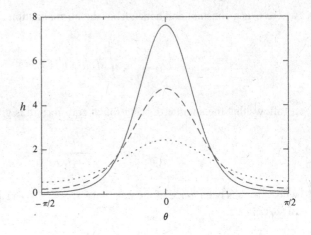

Fig. 8.12 Mean field potential $\beta\phi = \beta\phi\,(\cos\theta)$ in the Maier–Saupe theory as a function of θ for $q_2 = 1$ (*dotted line*), $q_2 = 2$ (*broken line*), and $q_2 = 3$ (*continuous line*)

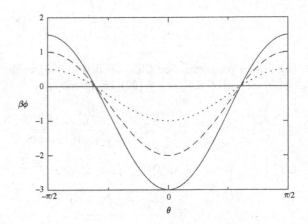

8.4.2 Onsager Theory

Assume now that, on the contrary, the interaction potential is that of a hard body, namely $V(\mathbf{r}; \mathbf{u}, \mathbf{u}') = \infty$ if the molecules overlap and $V(\mathbf{r}; \mathbf{u}, \mathbf{u}') = 0$ if they do not. In this case, from (8.107) one has

$$b(\mathbf{u} \cdot \mathbf{u}') = \frac{1}{2} E(\mathbf{u} \cdot \mathbf{u}') \qquad (8.125)$$

where $E(\mathbf{u} \cdot \mathbf{u}')$ is the excluded volume of two molecules with orientations $\mathbf{u}$ and $\mathbf{u}'$, whose analytical expression may only be determined exactly for some molecules of simple geometry.

Thus, in the case of spherocylinders (Fig. 8.13), i.e., cylinders of length L and diameter of the base D closed at their extremes by two hemispheres of diameter D one has, according to Onsager,

Fig. 8.13 Spherocylinder, i.e., a cylinder of length L and diameter of the base D closed at its extremes by two hemispheres of diameter D

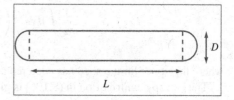

$$E(\mathbf{u} \cdot \mathbf{u}') = 8v_0 + 2L^2 D\sqrt{1 - (\mathbf{u} \cdot \mathbf{u}')^2}, \tag{8.126}$$

where $v_0 = \pi D^2 L/4 + \pi D^3/6$ is the volume of the spherocylinder. In his original theory, Onsager assumed that the spherocylinders were very much elongated ($L \gg D$), i.e.,

$$E(\mathbf{u} \cdot \mathbf{u}') = 2L^2 D\sqrt{1 - (\mathbf{u} \cdot \mathbf{u}')^2} + O\left(\frac{v_0}{L^2 D}\right). \tag{8.127}$$

It may be shown that in the limit of hard rods, $D/L \to 0$, if the packing fraction $\eta = \rho v_0 \simeq \pi \rho D^2 L/4 \to 0$ in such a way that $\eta L/D$ remains finite, the second virial coefficient approximation (8.106) is exact.

In order to determine the variational Helmholtz free energy functional (8.106), one may consider the following approximation. As one would expect that in the nematic phase, $h(\theta) = h(\cos \theta)$ has a very pronounced maximum when the molecules are oriented parallel ($\theta \simeq 0$ or $\theta \simeq \pi$) and $h(\theta) = h(\theta - \pi)$, the function $h(\theta)$, with $\rho h(\theta) = \rho_1(\mathbf{u})$, may be written as

$$h(\theta) = \frac{1}{N(\alpha)} \begin{cases} e^{-\alpha\theta^2/2}, & 0 \le \theta \le \pi/2 \\ e^{-\alpha(\pi-\theta)^2/2}, & \pi/2 \le \theta \le \pi \end{cases} \tag{8.128}$$

where $N(\alpha)$ is a constant that normalizes (8.120), namely

$$N(\alpha) = \int_0^{\pi/2} d\theta \sin\theta e^{-\alpha\theta^2/2}. \tag{8.129}$$

When $\alpha \gg 1$, the maximum of $h(\theta)$ is very pronounced and the Helmholtz free energy may be evaluated asymptotically. Thus, with the change of variable $y = \alpha\theta^2/2$, the normalization constant may be approximated by

$$N(\alpha) = \frac{1}{\alpha} \int_0^{\alpha\pi^2/8} dy e^{-y} \frac{\sin\sqrt{\frac{2y}{\alpha}}}{\sqrt{\frac{2y}{\alpha}}}$$

$$\simeq \frac{1}{\alpha} \int_0^\infty dy e^{-y}\left(1 - \frac{y^2}{3\alpha^2}\cdots\right) \simeq \frac{1}{\alpha}, \tag{8.130}$$

where the upper limit $\alpha\pi^2/8$ has been replaced by infinity.

The entropy contribution in (8.106) is given by

$$\int d\mathbf{u}\rho_1(\mathbf{u})\left(\ln\left(\rho_1(\mathbf{u})\Lambda^3\right) - 1\right) = \rho\left(\ln\left(\rho\Lambda^3\right) - 1\right)$$

$$+ \frac{1}{2}\rho\int_0^\pi d\theta \sin\theta h(\theta)\ln h(\theta), \tag{8.131}$$

with

$$\frac{1}{2}\int_0^\pi d\theta \sin\theta h(\theta)\ln h(\theta) = \int_0^{\pi/2} d\theta \sin\theta h(\theta)\ln h(\theta)$$

$$= \int_0^{\pi/2} d\theta \sin\theta h(\theta)\left(-\ln N(\alpha) - \frac{1}{2}\alpha\theta^2\right)$$

$$= -\ln N(\alpha) + \frac{1}{N(\alpha)}\alpha\frac{\partial}{\partial\alpha}N(\alpha)$$

$$\simeq \ln\alpha - 1, \tag{8.132}$$

while the contribution of the second virial coefficient is

$$\frac{1}{2}\int d\mathbf{u}\rho_1(\mathbf{u})\int d\mathbf{u}'\rho_1(\mathbf{u}')E(\mathbf{u}\cdot\mathbf{u}') = \rho^2\left[4v_0 + L^2 DI(\alpha)\right], \tag{8.133}$$

with

$$I(\alpha) = \int_0^{\pi/2} d\theta \sin\theta \int_0^{\pi/2} d\theta' \sin\theta' h(\theta)h(\theta')\sqrt{1 - \cos^2(\theta - \theta')}. \tag{8.134}$$

Since $\alpha \gg 1$ in (8.134) it is only necessary to consider $\theta \simeq 0$ and $\theta' \simeq 0$, i.e., $\sin\theta \simeq \theta$, $\sin\theta' \simeq \theta'$ and $\cos(\theta - \theta') \simeq \cos\theta\cos\theta'$, and hence

$$\sqrt{1 - \cos^2\theta\cos^2\theta'} \simeq \sqrt{1 - (1 - \theta^2)(1 - \theta'^2)} \simeq \sqrt{\theta^2 + \theta'^2}$$

With the changes of variable $y = \alpha\theta^2/2$ and $y' = \alpha\theta'^2/2$, (8.134) may be approximated by

$$I(\alpha) \simeq \sqrt{\frac{2}{\alpha}} \int_0^\infty dy e^{-y} \int_0^\infty dy' e^{-y'} \sqrt{y + y'}.$$

Since

$$\int_0^\infty dy' e^{-y'} y' \sqrt{y + y'} = e^y \int_y^\infty du e^{-u} \sqrt{u} = \sqrt{y} + \frac{\sqrt{\pi}}{2} \text{erfc}(\sqrt{y}) e^y,$$

one has

$$\int_0^\infty dy e^{-y} \left(\sqrt{y} + \frac{\sqrt{\pi}}{2} \text{erfc}(\sqrt{y}) e^y \right) = \frac{3\sqrt{\pi}}{4},$$

i.e.,

$$I(\alpha) \simeq \frac{3}{2} \sqrt{\frac{\pi}{2\alpha}}. \tag{8.135}$$

The Helmholtz free energy (note that upon parameterizing $h(\theta)$ according to (8.128), the functional (8.106) becomes a function of α) then reads

$$\frac{\beta F(V, N; \alpha)}{V} = \rho \left(\ln(\rho \Lambda^3) - 1 \right) + 4 v_0 \rho^2 + \rho (\ln \alpha - 1)$$
$$+ \frac{3}{2} L^2 D \rho^2 \sqrt{\frac{\pi}{2\alpha}}, \tag{8.136}$$

which has a minimum at

$$\alpha = \frac{\pi}{2} \left(\frac{3}{4} L^2 D \rho \right)^2. \tag{8.137}$$

For $a \rho v_0 = 5$, with $a v_0 = \pi L^2 D / 4$, the minimum is located at $\alpha = 35.81$, which justifies the asymptotic development ($\alpha \gg 1$) of this section.

Substituting (8.137) into (8.136) one finds:

$$\frac{\beta F(V, N)}{V} = \rho \ln(\rho \Lambda^3) + 4 v_0 \rho^2 + \rho \ln \left(\frac{\pi}{2} \left(\frac{3}{4} L^2 D \rho \right)^2 \right), \tag{8.138}$$

which is the Helmholtz free energy of a nematic phase of spherocylinders with a high degree of orientation in the Onsager theory.

8.5 Polymers

Except for the section dedicated to liquid crystals, throughout this chapter only systems of particles (atoms, molecules, colloids, ...) with spherical symmetry have been considered. In nature there exist very long molecules which, because they are flexible, cannot be studied as hard bodies of fixed geometry but rather as deformable tubes or strings. Such giant molecules are called polymers when their atomic structure may be represented as a periodic repetition of a certain group of atoms referred to as the monomer. For instance, polyethylene is a molecule with a structure CH_3- $(CH_2)_n-CH_3$ which contains many ethylene groups ($CH_2=CH_2$), with a large degree of polymerization, n. The typical value of n lies in the range $10^2 < n < 10^5$, although in some polymers of biological interest, such as DNA ($n \simeq 10^7$), this number may be even larger ($n \simeq 10^9$).

8.5.1 Radius of Gyration

In this section, the simplest case of a polymer solution, constituted by the mixture of N_A polymers of the form $A_1 - A_2 - \cdots - A_n$ ($n \gg 1$) and a solvent of N_B (small and spherical) molecules of type B, is considered. For simplicity, it will be assumed that the monomers A_k ($k = 1, ..., n$) are identical spherical molecules of type A separated by a fixed distance b that represents the chemical bond (A–A) between two consecutive monomers. It will be further assumed that the structure of the polymer is a linear flexible chain with a fixed positional order along the chain, although the bond between A_{k-1} and A_k can rotate around the bond between A_k and A_{k+1}. In a real solution, the different polymers have, in general, different degrees of polymerization and thus the solution is referred to as polydisperse. In what follows it will be admitted that the solution is monodisperse (i.e., n is fixed) and very dilute ($N_A \ll N_B$), so that one can neglect the polymer–polymer interactions. The properties of the solution may then be determined from the behavior of a single polymer. Note that since $n \gg 1$, one may apply the methods of statistical physics even to a single chain in the solution. Let $\mathbf{r}_1, ..., \mathbf{r}_n$ be the position vectors of the n monomers of a polymer (Fig. 8.14), which may be characterized by its mass and its length. If each monomer has a mass $\bar{m}_A$ the mass of the polymer is nm_A. Concerning its length, two types of measures may be considered. Let $\mathbf{b}_k = \mathbf{r}_{k+1} - \mathbf{r}_k$ be the vector in the direction of the kth bond, whose modulus is constant, $|\mathbf{b}_k| = b$, although its orientation may change. Consider the dynamical function

$$\mathbf{R} = \sum_{k=1}^{n-1} \mathbf{b}_k = \mathbf{r}_n - \mathbf{r}_1, \tag{8.139}$$

which is a vector whose length measures the end-to-end distance of the polymer. In an isotropic solution, the probabilities of $\mathbf{R}$ and of $-\mathbf{R}$ are equal, and so $\langle \mathbf{R} \rangle = 0$. In

Fig. 8.14 Schematic representation of a linear polymer chain of $(n-1)$ segments

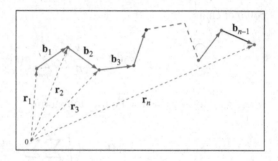

spite of that one may associate to (8.139) a measure R_0 related to the fluctuation of $\mathbf{R}$, that is $R_0^2 = \langle \mathbf{R}^2 \rangle$. If, as a first approximation, one assumes that the bonds are not correlated $\langle \mathbf{b}_k \cdot \mathbf{b}_l \rangle = \delta_{kl} \langle \mathbf{b}_k \rangle \cdot \langle \mathbf{b}_l \rangle = \delta_{kl} b^2$, one has

$$R_0^2 = \sum_{k=1}^{n-1} \langle \mathbf{b}_k^2 \rangle = (n-1)b^2, \tag{8.140}$$

namely

$$R_0 = \sqrt{n}\, b, \tag{8.141}$$

because $n - 1 \simeq n$ for $n \gg 1$, which in the language of polymer physics is expressed by saying that R_0 grows ("scales") as $\sqrt{n}$. Note that since $n \gg 1$, in polymer physics one frequently analyzes how the variables depend on the degree of polymerization n. The scaling law (8.141) is a first example.

From (8.139) to (8.141) it follows that this measure of the length of a polymer is that of a random walk of $n - 1$ steps of the same length, b. In reality, this walk is not completely random for two main reasons. The first one is that in a polymer there exists interaction between consecutive bonds, $\langle \mathbf{b}_k \cdot \mathbf{b}_{k+1} \rangle \neq 0$, and thus the term $\langle \mathbf{b}_k \cdot \mathbf{b}_l \rangle$ is not zero, but a decreasing function of $|k - l|$. It may be shown, however, that this effect does not modify the scaling law $R_0 \sim \sqrt{n}$, which may then be expressed as $R_0 = \sqrt{n}\, b_e$, where b_e is an effective bond length $b_e \neq b$. The second and most important reason is that the random walk has to avoid the superposition of bonds since they cannot cross, i.e., around each bond there is a volume excluded to any other bond. When the two effects are combined, the scaling law that one obtains is

$$R_0 = n^\nu b_e, \tag{8.142}$$

where the Flory scaling exponent $\nu \simeq 0.588$ (a value close to 3/5) is not the classical exponent, 1/2, of (8.141). It must be pointed out that the numerical value of this exponent has not been obtained from (8.142), since R_0 cannot be measured directly, but from a similar equation involving a different measure of the size of the polymer, called its radius of gyration R_g. Consider the correlation function of an isolated polymer chain

$$g_0(\mathbf{r}) = \left\langle \frac{1}{n} \sum_{i=1}^{n} \sum_{j=1}^{n} \delta(\mathbf{r} - \mathbf{r}_{ij}) \right\rangle, \tag{8.143}$$

where $\mathbf{r}_{ij} = \mathbf{r}_i - \mathbf{r}_j$, whose Fourier transform,

$$\tilde{g}_0(\mathbf{k}) = \left\langle \frac{1}{n} \sum_{i=1}^{n} \sum_{j=1}^{n} e^{-i\mathbf{k}\cdot\mathbf{r}_{ij}} \right\rangle, \tag{8.144}$$

may be measured in X-ray scattering experiments. In an isotropic solution $\tilde{g}_0(\mathbf{k}) = \tilde{g}_0(k)$, where $k = |\mathbf{k}|$, so that the expansion of (8.144) when $k \to 0$ reads

$$\tilde{g}_0(k) = \left\langle \frac{1}{n} \sum_{i=1}^{n} \sum_{j=1}^{n} \left(1 - \frac{1}{2}(\mathbf{k} \cdot \mathbf{r}_{ij})^2 + \cdots \right) \right\rangle$$
$$= \tilde{g}(0)\left(1 - \frac{1}{3}k^2 R_g^2 + \cdots \right). \tag{8.145}$$

where it has been assumed that $\mathbf{k}$ has the direction of the z-axis. Equation (8.145) is known as the Guinier formula. In (8.145) R_g is the radius of gyration:

$$R_g^2 = \frac{3}{2}\left\langle \frac{1}{n^2} \sum_{i=1}^{n} \sum_{j=1}^{n} z_{ij}^2 \right\rangle = \frac{1}{2}\left\langle \frac{1}{n^2} \sum_{i=1}^{n} \sum_{j=1}^{n} r_{ij}^2 \right\rangle, \tag{8.146}$$

(note that to obtain the last equality use has been made of the isotropy of the solution) which may also be written as

$$R_g^2 = \frac{1}{n}\left\langle \sum_{i=1}^{n} (\mathbf{r}_i - \mathbf{r}_0)^2 \right\rangle, \tag{8.147}$$

where $\mathbf{r}_0$ is the center of mass vector of the polymer, i.e.,

$$\mathbf{r}_0 = \frac{1}{n} \sum_{j=1}^{n} \mathbf{r}_j. \tag{8.148}$$

From the experimental determination of $\tilde{g}_0(k)$, one may obtain in the limit $k \to 0$ the value of the radius of gyration R_g of an isolated polymer which is known to obey the scaling law:

$$R_g = n^\nu b_e, \tag{8.149}$$

where $v \simeq 0.558$ is the scaling exponent. Note that since, according to (8.147), R_g^2 is the average value of the sum of the quadratic deviations of the position vectors of the monomers with respect to $\mathbf{r}_0$, the volume of the sphere, $4\pi R_g^3/3$, is in a certain way a measure of the spatial extension of the polymer.

8.5.2 Flory–Huggins Theory

The thermodynamic properties of a dilute polymer solution can be determined, in first approximation, from the regular solution theory (see Sect. 8.3.1) or Flory–Huggins theory. Note that if the solution contains N_A polymers and, therefore, nN_A monomers, the former are indistinguishable, although the n monomers of each polymer are distinguishable because they are ordered along the chain. Hence, a solution of N_A polymers and N_B solvent molecules (which are assumed to be identical and thus indistinguishable) contains $N = N_A + N_B$ "particles" but $N_T = nN_A + N_B$ molecules. The thermodynamic properties of the solution are, therefore, extensive with respect to N_T instead of N (the Helmholtz free energy per molecule, F/N_T, is finite in the thermodynamic limit). Since $N_T - N = (n-1)N_A \gg 1$ this difference is important. The free energy of the solution is, according to (8.89), given by

$$F(T, V, N_A, N_B) = k_B T V \rho_A \left(\ln\left(\rho_A \Lambda_A^3\right) - 1\right)$$
$$+ k_B T V \rho_B \left(\ln\left(\rho_B \Lambda_B^3\right) - 1\right)$$
$$- V\left(\rho_A^2 a_{AA} + \rho_B^2 a_{BB} + 2\rho_A \rho_B a_{AB}\right), \qquad (8.150)$$

where it has been assumed that the interaction potentials $V^{\alpha\gamma}(r)$ are weak and attractive $(\beta V^{\alpha\gamma}(r) \equiv \beta V_A^{\alpha\gamma}(r) \ll 1)$, namely

$$b_{\alpha\gamma} = -\frac{1}{2} \int d\mathbf{r} \left(e^{-\beta V_A^{\alpha\gamma}(r)} - 1\right) \simeq \frac{1}{2}\beta \int d\mathbf{r} V_A^{\alpha\gamma}(r) \equiv -\beta a_{\alpha\gamma},$$

where

$$a_{\alpha\gamma} = -\frac{1}{2} \int d\mathbf{r} V_A^{\alpha\gamma}(r) \qquad (8.151)$$

are positive constants that measure the cohesion energy due to the attractive interactions $(V_A^{\alpha\gamma}(r) < 0)$ between the different species. Note that although in polymers there must also exist a repulsive interaction, this is not considered explicitly in this theory because the excluded volume effects are much more complicated in this case than for simple fluids (see below).

In (8.150) $\rho_A = N_A/V$, $\rho_B = N_B/V$ and Λ_A and Λ_B are, respectively, the thermal de Broglie wavelengths of the polymers (assumed to be spherical and of mass nm_A, where m_A is the mass of the monomer) and of the solvent molecules (which have also been assumed to be spherical and of mass m_B). Recall that the properties of the mixture are independent of Λ_A and Λ_B. Using the notation

$$f = \frac{F(T, V, N_A, N_B)}{N_T}, \quad x \equiv x_A = n\frac{N_A}{N_T}, \quad x_B = 1 - x_A, \quad \rho = \frac{N_T}{V},$$

where x_A is the number fraction of monomers (not that of polymers) in the solution, $\rho a_{AA} = n^2 \varepsilon_{AA}/2$, $\rho a_{AB} = n\varepsilon_{AB}/2$, and $\rho a_{BB} = \varepsilon_{BB}/2$, where the parameters $\varepsilon_{\alpha\gamma}$ are molecule–molecule interaction energies (positive since the interaction is attractive), then (8.150) may be cast in the form

$$\begin{aligned}
\beta f(T, \rho, x) = {} & \frac{x}{n}\left(\ln(\rho\Lambda_A^3) + \ln\left(\frac{x}{n}\right) - 1\right) \\
& + (1 - x)\left(\ln(\rho\Lambda_B^3) + \ln(1 - x) - 1\right) \\
& - \frac{1}{2}\beta\left(x^2\varepsilon_{AA} + (1 - x)^2\varepsilon_{BB} + 2x(1 - x)\varepsilon_{AB}\right),
\end{aligned} \tag{8.152}$$

and so the free energy of mixing per molecule is given by

$$\beta f^{\mathrm{mix}}(T, x) = \frac{1}{n}x \ln x + (1 - x)\ln(1 - x) + \chi x(1 - x), \tag{8.153}$$

where χ is the Flory–Huggins mixing parameter:

$$\chi = \frac{1}{2}\beta(\varepsilon_{AA} + \varepsilon_{BB} - 2\varepsilon_{AB}). \tag{8.154}$$

Note that χ is independent of the density and inversely proportional to the temperature. In order to stress the importance of the $1/n$ term in (8.153), consider the equation of state (the osmotic pressure) of the solution. The osmotic pressure n is defined as the difference in pressure between the polymer solution and the pure solvent when both phases, separated by a membrane which is only permeable to the solvent, are in equilibrium. The condition for equilibrium of the solvent on both sides of the membrane is

$$\mu_B(T, p + \pi, x) = \mu_B(T, p, 0), \tag{8.155}$$

where $\mu_B(T, p, x)$ is the chemical potential of the solvent in the polymer solution, namely

$$\mu_B(T, p, x) = \frac{\partial G(T, p, N_A, N_B)}{\partial N_B}, \tag{8.156}$$

where $G(T, p, N_A, N_B)$ is the Gibbs free energy of the solution

$$G(T, p, N_A, N_B) = F(T, V, N_A, N_B) + pV, \tag{8.157}$$

with $F(T, V, N_A, N_B)$ given by (8.150). If instead of N_A and N_E one chooses as independent variables N_T and x, namely $F(T, V, N_T, x)$, from (8.156), (8.157), and (8.152) it follows that

$$\mu_B(T, p, x) = \left(\frac{\partial F(T, V, N_T, x)}{\partial N_T}\right)_{T,\rho,x} \left(\frac{\partial N_T}{\partial N_B}\right)_{N_A}$$
$$+ \left(\frac{\partial F(T, V, N, x)}{\partial x}\right)_{T,\rho,N_T} \left(\frac{\partial x}{\partial N_B}\right)_{N_A}$$
$$+ \frac{p}{\rho}\left(\frac{\partial N_T}{\partial N_B}\right)_{N_A}. \qquad (8.158)$$

Since

$$\left(\frac{\partial N_T}{\partial N_B}\right)_{N_A} = 1, \quad \left(\frac{\partial x}{\partial N_B}\right)_{N_A} = -\frac{x}{N_T},$$

and

$$\left(\frac{\partial F(T, V, N_T, x)}{\partial N_T}\right)_{T,\rho,x} = f(T, \rho, x),$$

one can write

$$\mu_B(T, p, x) = f(T, \rho, x) - x\frac{\partial f(T, \rho, x)}{\partial x} + \frac{p}{\rho}$$
$$= f^{\text{mix}}(T, x) - x\frac{\partial f^{\text{mix}}(T, x)}{\partial x}$$
$$+ f(T, \rho, 0) + \frac{p}{\rho}, \qquad (8.159)$$

i.e.,

$$\pi = \rho\left(x\frac{\partial f^{\text{mix}}(T, x)}{\partial x} - f^{\text{mix}}(T, x)\right), \qquad (8.160)$$

so that from (8.153) one finally obtains

$$\pi = \rho k_B T\left(\frac{1}{n}x - x - \chi x^2 - \ln(1 - x)\right). \qquad (8.161)$$

When $x \ll 1$, performing a Taylor expansion of the term inside the brackets of (8.161) leads to

$$\pi = \rho k_B T \left(\frac{1}{n}x + \left(\frac{1}{2} - \chi \right)x^2 + \frac{1}{3}x^3 + \cdots \right), \tag{8.162}$$

i.e., to dominant order

$$\pi \simeq \rho k_B T \frac{1}{n}x = \frac{N_A}{V}k_B T, \tag{8.163}$$

which indicates that the solution behaves as an ideal gas, a result known as the van't Hoff law.

In the case of a polymer solution ($n \gg 1$), this law is modified due to the smallness of $1/n$, yielding

$$\pi = \rho k_B T \left(\left(\frac{1}{2} - \chi \right)x^2 + \frac{1}{3}x^3 + \cdots \right), \tag{8.164}$$

a result that may be easily tested experimentally since it does not depend on the degree of polymerization. The osmotic pressure π of solutions of poly(α-methylstyrene) of different molecular weights in toluene as a function of the concentration c is presented in Fig. 8.15. Note that when the molecular weight ($M \sim n$) increases all the curves tend to a universal curve independent of M. The same result follows from (8.162) since when $n \gg 1$ the osmotic pressure is given by (8.164) which is independent of n.

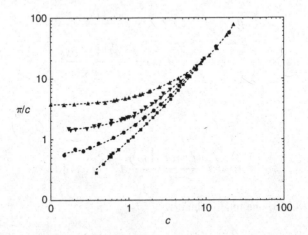

Fig. 8.15 Osmotic pressure π of solutions of poly(α-methylstyrene) of different molecular weights M (in units of the molecular weight of the monomer) in toluene at 25 °C as a function of the concentration c. The molecular weight M $\times$ 10^{-4} is 7.08 (triangles), 20 (*inverted triangles*), 50.6 (*circles*), and 182 (*squares*). The units are π/c (cm) and c (g cm^{-3}). The curves are only drawn to guide the eye. *Source* I. Noda, N. Kato, T. Kitano and M. Nagasawa, Macromolecules, **14**, 668 (1981)

Fig. 8.16 Compressibility factor $Z = \beta\pi/\rho$ of a polymer solution in the Flory–Huggins theory as a function of the number fraction of monomers in the solution x for $x = 1$ (*dash-dotted line*), $\chi = 1/2$ (*broken line*), and $\chi = 1/4$ (*continuous line*). Note that when $x = 1$ the osmotic pressure is negative

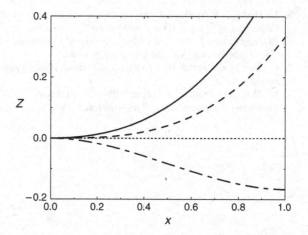

Note, further, that at the θ-temperature for which $2\chi = 1$, the quadratic term in (8.164) vanishes and the osmotic pressure is given by

$$\pi \simeq \frac{1}{3}\rho k_B \theta x^3 \quad (T = \theta), \tag{8.165}$$

whenever $n \gg 1$. From (8.164) and (8.165) it follows that the polymer solution behaves in a qualitatively different manner than a solution of small molecules, which corresponds to (8.162) with $n = 1$. One says that the polymer is in a good solvent if $\chi < 1/2$, in a bad solvent if $\chi > 1/2$, and in a θ-solvent if $\chi = 1/2$. Note that the Flory–Huggins theory is qualitatively correct only in a good solvent or in a θ-solvent, since in a bad solvent the osmotic pressure may become negative (Fig. 8.16). The reason is that in a good solvent the polymer prefers to form contacts with the solvent molecules and "stretches," while in a bad solvent it tends to avoid such contacts and hence adopts the form of a "coil." In the latter case, the excluded volume effects, which are not accounted for in the Flory–Huggins theory, will dominate.

Further Reading

1. J.O. Hirschfelder, C.F. Curtiss, R.B. Bird, *Molecular Theory of Gases and Liquids* (Wiley, New York, 1954). *Contains a detailed presentation of the cell theory.*
2. T.M. Reed, K.E. Gubbins, *Applied Statistical Mechanics* (McGraw-Hill Kogakusha Ltd, Tokyo, 1973). *Discusses the relation between the pair potential and the polarization of the electronic clouds surrounding the atoms.*
3. N.W. Ashcroft, N.D. Mermin, *Solid State Physics* (Saunders, Philadelphia, 1976). *Provides a good introduction to the different Bravais lattices.*
4. L.V. Woodcock, Ann. N. Y. Acad. Sci. **371**, 274 (1981). *Discusses the glass phase of hard-sphere systems.*

5. J.P. Hansen, I.R. McDonald, *Theory of Simple Fluids*, 2nd edn. (Academic Press, London, 1986). *Contains a derivation of the solution of the PY-equations.*
6. P.G. de Gennes, J. Prost, *The Physics of Liquid Crystals* (Clarendon Press, Oxford, 1993). *A classic introduction to all the known mesophases.*
7. C. Janot, *Quasicrystals* (Clarendon Press, Oxford, 1994). *A first introduction to the nonperiodic crystal structures.*
8. M. Doi, *Introduction to Polymer Physics* (Clarendon Press, Oxford, 1996). *Contains an elementary derivation of the Flory–Huggins theory.*

Chapter 9
Phase Transitions

Abstract In the previous chapter, some common structures of molecular systems were studied. The number of structures of a system increases as the symmetry of the molecules decreases. Under certain conditions of pressure and temperature, some of these structures may form thermodynamically stable phases (their free energies satisfy the stability conditions analyzed in Chap. 2). When two or more phases are simultaneously stable, the equilibrium phase is the one with the lowest free energy, while the others are metastable phases. When the thermodynamic conditions are modified, the free energies of the different phases change accordingly and in some instances the equilibrium phase becomes metastable with respect to another phase (formerly metastable) which is now the new equilibrium phase. This change of stability is called a phase transition, whose study is the subject of this chapter.

9.1 Structural Transitions

When the phase transition takes place between two different structures, the transition is called a structural phase transition as, for example, the order–disorder transitions between a completely disordered phase and a partially ordered phase. Thus, in the transition between the fluid phase and the crystalline (solid) phase, also known as the melting–freezing transition, the fluid is the disordered phase and the crystal is the ordered phase. The order is due to the absence of some translation and rotation symmetry elements. Another example is the transition between the isotropic and the nematic phases of a fluid, since the latter has a broken symmetry of rotation. Not all the structural transitions are, however, order–disorder transitions. Thus, when a nematic fluid with orientational order crystallizes into a completely ordered crystal, both phases are ordered. Also, in the transition between a bcc crystal and a fcc crystal, both phases are completely ordered. In conclusion, in a structural phase transition, the two phases have a different degree of order due to the different elements of symmetry that are broken. In what follows two examples of structural transitions are considered.

9.1.1 Fluid–Solid Transition

This is a universal phase transition since, no matter how complex its atoms or molecules may be, a system always exists in a fluid and in a solid phase. In the former, the centers of mass of the molecules and their orientations are distributed randomly in space, and the phase is, therefore, uniform and isotropic. In the latter, the molecules are located on a periodic lattice and the structure is that of a crystal or solid. The location of the fluid–solid transition (F–S) in the phase diagram depends of course on the molecular interaction potential. A system of spherical molecules interacting through the hard-sphere potential (8.42) which may be described in a simple way using the cell theory (Sect. 8.1.2) and the free volume theory (Sect. 8.2.4) will be considered in this subsection. The Helmholtz free energy per particle of the fluid phase is then given by (8.74), i.e.,

$$\beta f_F(T, \rho) = \ln(\rho\Lambda^3) - 1 - \ln\left(1 - \frac{\rho}{\rho_0}\right), \tag{9.1}$$

and that of the solid phase is given by (8.14), i.e.,

$$\beta f_S(T, \rho) = \ln(\rho\Lambda^3) - 1 - 3\ln\left(1 - \left(\frac{\rho}{\rho_{cp}}\right)^{1/3}\right). \tag{9.2}$$

Note that, according to (9.2), from all the possible crystalline structures, the equilibrium phase at high density is the most compact one (for instance, a fee phase) for which $\rho_{cp}\sigma^3 = \sqrt{2}$, where σ is the diameter of the spheres.

The existence of a fluid–solid transition in a hard-sphere system was suggested by Kirkwood and Monroe in 1941 and verified in 1957 by MD simulations. In the case of colloidal particles (which behave approximately as hard spheres), this transition has moreover been observed experimentally (Fig. 9.1). All these results show that $\rho_F\sigma^3 = 0.943$ and $\rho_S\sigma^3 = 1.041$, where ρ_F and ρ_S are the densities of the fluid and solid phases at coexistence. Note that, since it concerns an excluded volume interaction, the transition is independent of temperature since the interaction partition function does not depend on T. This happens because the potential is infinitely repulsive when the spheres overlap and zero when there is no contact between them (see (8.42)). Since, on the other hand, the interaction energy is zero, the thermodynamic relation $F = E - TS$ reduces to $F = 3Nk_BT/2 - TS = T(3Nk_B/2 - S)$, and the state of minimum free energy coincides with the one of maximum entropy T appearing merely as a scale factor. This is why it is said that such a transition is entropy-driven.

Note that from (9.1) and (9.2), it follows that the Helmholtz free energy of the hard-sphere system may be written as

$$\beta f(T, \rho) = \ln(\rho\Lambda^3) - 1 - \ln\left(1 - \frac{\rho}{\rho_0}\right)\Theta(\rho^* - \rho)$$

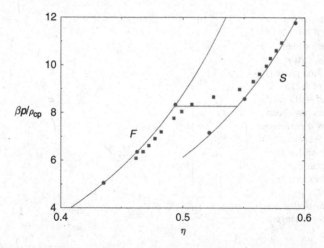

Fig. 9.1 Reduced pressure $\beta p/\rho_{cp}$ of a system of hard spheres as a function of the packing fraction η. The *horizontal line* indicates the pressure at the fluid–solid coexistence obtained by MD. The *circles* are the results of numerical simulations, and the *squares* are the experimental values of sedimenting polystyrene spheres with a diameter of 720 nm in a saline solution obtained from X-ray densitometry. *Source* A. P. Gast and W. B. Russel, Physics Today **51**, 24 (1998); B. J. Alder and T. E. Wainwright, J. Chem. Phys. **27**, 1209 (1957)

$$- 3\ln\left(1 - \left(\frac{\rho}{\rho_{cp}}\right)^{1/3}\right)\Theta(\rho - \rho^*), \tag{9.3}$$

where ρ^* is the density at which the change in stability occurs, namely

$$1 - \frac{\rho^*}{\rho_0} \doteq \left(1 - \left(\frac{\rho^*}{\rho_{cp}}\right)^{1/3}\right)^3. \tag{9.4}$$

By defining the variables

$$x_F = \frac{\rho_F}{\rho_0}, \quad x_S = \left(\frac{\rho_S}{\rho_{cp}}\right)^{1/3}, \tag{9.5}$$

from (9.1) and (9.2) it follows that the conditions of mechanical equilibrium and chemical equilibrium are, respectively, given by (see Chap. 2)

$$\frac{\rho_0}{\rho_{cp}}\left(\frac{x_F}{1 - x_F}\right) = \frac{x_S^3}{1 - x_S}, \tag{9.6}$$

and

Fig. 9.2 Reduced free energy $F^* \equiv \beta F \sigma^3 / V$ of the hard-sphere system as a function of $\rho \sigma^3$. At $\rho^* \sigma^3 \simeq 0.944$, the fluid phase (F) and the solid phase (S) exchange their stability (the stable branches have been represented by *a continuous line* and the metastable branches by *a broken line*). The dots indicate the densities at coexistence ($\rho_F \sigma^3 \simeq 0.907$, $\rho_S \sigma^3 \simeq 1.195$), and *the dotted line* is Maxwell's double tangent

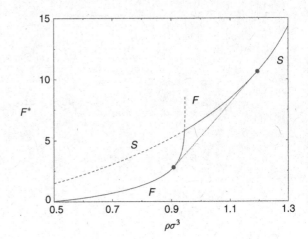

$$\ln\left(\frac{\rho_0}{\rho_{\mathrm{cp}}}\right) + \ln\left(\frac{x_F}{1 - x_F}\right) + \frac{1}{1 - x_F} = 3\ln\left(\frac{x_S}{1 - x_S}\right) + \frac{1}{1 - x_S}. \qquad (9.7)$$

Note that the density ρ^* at which the exchange of stability takes place and the densities at coexistence, ρ_F and ρ_S, depend on ρ_0, which is the highest value of the density for which a fluid may exist and which, as analyzed in Sect. 8.2.3, is unknown. It is clear that $\rho_0 \sigma^3 < \rho_{\mathrm{cp}} \sigma^3 = \sqrt{2}$. On the other hand, $\rho_0 \sigma^3 > 3/2\pi$, since the lowest order of the virial expansion of the pressure is $B_2 \rho = 2\pi \rho \sigma^3 / 3 < 1$. Therefore, as a simple approximation, one may assume that $\rho_0 \sigma^3$ is the arithmetic mean of these values, namely

$$\rho_0 \sigma^3 = \frac{1}{2}\left(\frac{3}{2\pi} + \sqrt{2}\right) \simeq 0.946. \qquad (9.8)$$

Substituting (9.8) into (9.4), (9.6), and (9.7) it follows that $\rho^* \sigma^3 \simeq 0.944$, $\rho_F \sigma^3 \simeq 0.907$, and $\rho_S \sigma^3 \simeq 1.195$, respectively. The densities at coexistence are, therefore, in good agreement with the results obtained both experimentally and by the simulations. The fluid–solid transition of hard spheres analyzed in this section is shown in Fig. 9.2.

9.1.2   Isotropic–Nematic Transition

When the molecules are not spherical, the fluid–solid transition of the previous subsection takes place in successive stages. Since the intermediate phases have, in general, properties that are common to both the fluid and the crystal phases, they are called liquid crystal phases. The transition between a uniform isotropic fluid which is completely disordered (whose properties are indicated by a subscript I), and a uniform fluid with nematic orientational order (whose properties are indicated by a subscript N) is studied in this subsection. If the molecules are hard spherocylinders, the transition may be studied using the Onsager theory of Sect. 8.4.2. Within the second virial coefficient approximation, the Helmholtz free energy per particle of the isotropic phase is given by

$$\beta f_I(T, \rho) = \ln(\rho \Lambda^3) - 1 + \frac{1}{2}\rho \int d\mathbf{u} \int d\mathbf{u}' E(\mathbf{u} \cdot \mathbf{u}'), \qquad (9.9)$$

which follows from (8.106) and (8.126) when $\rho(\mathbf{u}) = \rho$ (note that in this case the functional reduces to a function of the density). Using (8.127) and

$$\int d\mathbf{u} \int d\mathbf{u}' \sqrt{1 - (\mathbf{u} \cdot \mathbf{u}')^2} = \frac{1}{2} \int_0^\pi d\theta \sin^2 \theta = \frac{\pi}{4}, \qquad (9.10)$$

since in the scalar product $\mathbf{u} \cdot \mathbf{u}'$ one of the vectors may be arbitrarily chosen to be in the direction of the polar axis, one has

$$\beta f_I(T, \rho) = \ln\left(\frac{\Lambda^3}{a v_0}\right) - 1 + \ln(a\rho v_0) + 4\rho v_0 + a\rho v_0, \qquad (9.11)$$

where v_0 is the volume of the spherocylinder, of length L and diameter D, and the constant $a \equiv \pi L^2 D / 4 v_0$ has been introduced.

On the other hand, according to (8.138), the Helmholtz free energy per particle of the nematic phase reads

$$\beta f_N(T, \rho) = \ln\left(\frac{\Lambda^3}{a v_0}\right) + 3\ln(a\rho v_0) + 4\rho v_0 + \ln\left(\frac{9}{2\pi}\right). \qquad (9.12)$$

Therefore, the density ρ^* at which the exchange of stability occurs is given by

$$a\rho^* v_0 - 2\ln(a\rho^* v_0) = 1 + \ln\left(\frac{9}{2\pi}\right), \qquad (9.13)$$

which may be solved numerically, yielding $a\rho^* v_0 \simeq 4.256$. Note that although the Onsager theory has been derived from a low-density virial expansion, it may be shown that in the hard-rod limit $L/D \to \infty$ ($a \to \infty$) if $\rho v_0 \to 0$ and $x = a\rho v_0 < \infty$, the theory (9.11) is exact, since the remaining terms of the virial expansion vanish identically in this limit.

In the Onsager limit

$$a \to \infty, \quad \rho v_0 \to 0, \quad x = a\rho v_0 < \infty, \qquad (9.14)$$

the conditions of mechanical equilibrium and chemical equilibrium between the two phases, which follow from (9.11) and (9.12), are

$$x_I(1 + x_I) = 3x_N, \tag{9.15}$$

and

$$\ln x_I + 2x_I = 3 \ln x_N + 3 + \ln\left(\frac{9}{2\pi}\right), \tag{9.16}$$

where x_I and x_N are the values of x for the two phases, I and N, at coexistence. These equations may be solved numerically yielding $x_I = 3.604$ and $x_N = 5.532$, which represents a relative change of density of 38.4% at the transition. According to (8.137) $\alpha = \pi x_N^2/2 \simeq 48$, which justifies the asymptotic expansion $\alpha \gg 1$. The exact results of the Onsager theory are, on the other hand, $x_I = 3.290$ and $x_N = 4.189$, which shows the validity of the approximation made in Sect. 8.4.2. The isotropic–nematic transition of this section in the Onsager limit is shown in Fig. 9.3.

Note finally that the packing fractions at coexistence of the isotropic phase, $\eta_I = x_I/a$, and of the nematic phase, $n_N = x_N/a$, decrease with L/D as

$$\eta_I = \frac{3.604}{L/D}, \quad \eta_N = \frac{5.532}{L/D},$$

which is the qualitative behavior found experimentally as shown in Fig. 9.4.

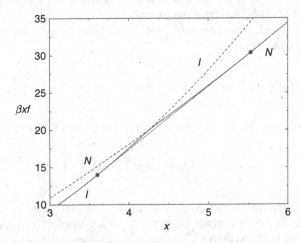

Fig. 9.3 Reduced free energy βxf of a system of hard rods as a function of $x = a\rho v_0$ in the Onsager limit. At $x^* \simeq 4.256$ the isotropic (I) and nematic (N) phases exchange their stability (the stable branches have been represented with a continuous line and the metastable branches by *a broken line*). The *dots* indicate the densities at coexistence ($x_I \simeq 4.256, x_N \simeq 5.532$), and *the dotted line* is Maxwell's double tangent

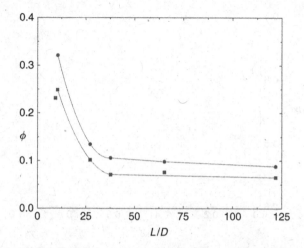

Fig. 9.4 Volume fraction ϕ at coexistence of the isotropic and nematic phases of a suspension of poly(bencyl-L-glutamate) in dioxane as a function of the estimated value of L/D. These molecules behave approximately as long rods and so $\phi \sim D/L$. The *circles* correspond to the nematic phase and the *squares* to the isotropic phase. The lines have been drawn to guide the eye. *Source* P. G. de Gennes and J. Prost, *The Physics of Liquid Crystals*, Clarendon Press, Oxford (1993)

9.2 Isostructural Transitions

The phase transitions that take place between two (ordered or disordered) phases with the same symmetries are called isostructural transitions. In this section two examples are considered.

9.2.1 Liquid–Vapor Transition

The best-known isostructural transition is probably the liquid–vapor transition in which both the dilute phase (the vapor, V) and the dense phase (the liquid, L) are uniform and isotropic. Since the two phases are disordered phases and they only differ in density, both are described by the same Helmholtz free energy. Thus, in the van der Waals approximation, the free energy per particle is given by (8.81)

$$\beta f(T, \rho) = \ln(\rho \Lambda^3) - 1 - \ln\left(1 - \frac{\rho}{\rho_0}\right) - \beta a \rho. \qquad (9.17)$$

Since $a > 0$, the free energy $\rho f(T, \rho)$ is not always a convex function of the density, and at low temperatures (Fig. 9.5), it is formed by two convex branches separated by a concave branch along which the system is unstable. In this case, from the method of the double tangent construction of Maxwell analyzed in Sect. 2.8.3, it follows

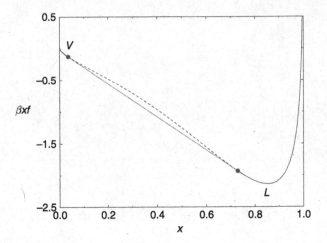

Fig. 9.5 Reduced free energy $\beta x f$ of the disordered phases (vapor, V, and liquid, L) in the van der Waals approximation as a function of $x = \rho/\rho_0$ for $t = k_B T/a\rho_0 = 0.2$. The convex branches (*continuous lines*) are separated by a concave branch (*broken line*) where the system is unstable. *The dotted line* is Maxwell's double tangent, and the points indicate the coexistence densities at that temperature ($x_V = 0.036$ y $x_L = 0.729$)

that there are two solutions $p_V(T)$ and $p_L(T)$ having the same pressure and chemical potential.

Introducing the variables

$$x_V = \frac{\rho_V(T)}{\rho_0}, \quad x_L = \frac{\rho_L(T)}{\rho_0}, \quad t = \frac{k_B T}{a\rho_0}, \tag{9.18}$$

from (9.17) it follows that the conditions for mechanical equilibrium and chemical equilibrium are $p(T, x_V) = p(T, x_L)$ and $\mu(T, x_V) = \mu(T, x_L)$, where

$$\beta p(T, x) = \rho_0 \left(\frac{x}{1-x} - \frac{1}{t} x^2 \right), \tag{9.19}$$

and

$$\beta \mu(T, x) = \ln\left(\rho_0 \Lambda^3\right) - 1 + \ln\left(\frac{x}{1-x}\right) + \frac{1}{1-x} - \frac{2}{t} x. \tag{9.20}$$

Note that since the pressure in each phase has to be positive, one has

$$x_V(1 - x_V) < t, \quad x_L(1 - x_L) < t, \tag{9.21}$$

and since $x_V < x_L$, from (9.21) it follows that when $t \to 0$, then $x_V \sim O(t) \to 0$ and $x_L \sim 1 - O(t) \to 1$. As the temperature is increased, the width of the transition $x_L - x_V$ decreases until, for a certain critical value $t_c = 8/27$, $x_L = x_V = x_c = 1/3$, as

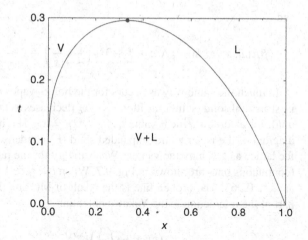

Fig. 9.6 Binodals in the (x, t) phase diagram of the liquid–vapor (L–V) transition in the van der Waals theory, with $x = \rho/\rho_0$ and $t = k_B T/a\rho_0$. When $t = t_c = 8/27$, the binodals coincide at $x_c = 1/3$

shown in Fig. 9.6. The behavior of the liquid–vapor transition in the vicinity of this "critical" value of the temperature is analyzed in Sect. 9.6.2.

9.2.2   Solid–Solid Transition

An isostructural transition may also take place between two ordered phases. Consider, for instance, the Helmholtz free energy per particle of a crystal in the van der Waals theory (see Sect. 8.1.3)

$$\beta f(T, \rho) = \ln(\rho\Lambda^3) - 1 - 3\ln\left(1 - \left(\frac{\rho}{\rho_{cp}}\right)^{1/3}\right) - \frac{1}{2t}\left(\frac{\rho}{\rho_{cp}}\right)^{n/3} M_n. \quad (9.22)$$

At low temperatures, the free energy (9.22) is not necessarily an increasing function of the density and so van der Waals loops similar to those of the liquid–vapor transition appear. If $\rho_E(T)$ and $\rho_C(T)$ are the coexistence densities of the expanded solid (E) (the one with the smaller density) and of the condensed solid (C) (the one with the greater density) and one introduces the variables

$$x_E = \left(\frac{\rho_E(T)}{\rho_{cp}}\right)^{1/3}, \quad x_C = \left(\frac{\rho_C(T)}{\rho_{cp}}\right)^{1/3}, \quad (9.23)$$

the conditions of mechanical equilibrium and chemical equilibrium are $p(T, x_E) = p(T, x_C)$ and $\mu(T, x_E) = \mu(T, x_C)$, where

$$\beta p(T, x) = \rho_{cp} x^3 \left(\frac{1}{1 - x} - \frac{1}{6t} n x^n M_n\right), \quad (9.24)$$

and

$$\beta\mu(T,x) = \ln\left(\rho_{cp}\Lambda^3\right) - 1 + 3\ln\left(\frac{x}{1-x}\right) + \frac{1}{1-x} - \frac{2}{3t}nx^n M_n. \tag{9.25}$$

In much the same way as occurs for the liquid–vapor transition, the width of the isostructural solid–solid transition $x_C - x_E$ decreases as the temperature is increased until, for a certain critical value t_c, $x_C = x_E = x_c$. The binodals in the (x, t) phase diagram of the isostructural expanded solid (E)-condensed solid (C) transition of a fcc lattice as given by the van der Waals theory for the potential (8.18) with $n = 50$ (continuous line) are shown in Fig. 9.7. When $t = t_c \simeq 1.026$, the binodals coincide at $x_c \simeq 0.963$. The broken line is the result of MC simulations when the attractive part of the interaction is the Yukawa potential:

$$V_A(r) = -\varepsilon\left(\frac{\sigma}{r}\right)e^{\kappa\sigma(1-r/\sigma)}, \quad (r > \sigma)$$

with $\kappa\sigma = 67$, where $1/K$ is a length that measures the range of the attractive part. In this case, the binodals coincide at $x_c \simeq 0.937$, $t_c \simeq 0.715$.

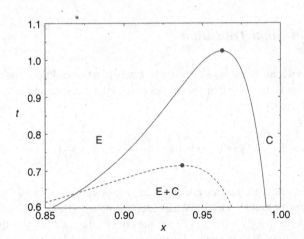

Fig. 9.7 Binodals in the (x, t) phase diagram of the isostructural expanded solid-condensed solid (E-C) transition of a face-centered cubic lattice. *The continuous line* corresponds to the van der Waals theory for an inverse power potential with an index $n = 50$, and *the broken line* is the result of MC simulations for an attractive Yukawa potential with (see text) $\kappa\sigma = 67$. *Source* P. G. Bolhuis, M. H. J. Hagen and D. Frenkel, Phys. Rev. E **50**, 4880 (1994)

9.3 Symmetry Breaking and Order Parameters

As was analyzed in Chap. 3, some equilibrium states have less symmetry than the Hamiltonian of the system (which, in general, is invariant under translations and rotations) and so they are referred to as states of broken symmetry. The different phases of matter may, therefore, be classified in terms of their broken symmetries. Thus, from a completely symmetric disordered phase (uniform and isotropic), one obtains ordered phases (non-uniform and/or anisotropic), for each element of symmetry that is eliminated. The sequence fluid, nematic, smectic, and crystal, is an example of the above. This method of classification may also be applied to the transition between two phases, and hence, if the phases differ in certain symmetry elements, one says that the transition is one of symmetry breaking. According to this criterion, all the structural transitions, such as the order–disorder (liquid–solid) and those of order 1–order 2 between two crystalline structures (bcc–fcc), are symmetry-breaking transitions. The isostructural transitions, however, are not since both phases have the same structure, with the same degree of order (fcc–fcc) or disorder (liquid–vapor), and so it is said that these transitions conserve the symmetry.

The symmetry-breaking transitions may, in general, be described by means of an "order parameter," which is a variable that quantifies the degree of order in the phases, and so it is usually defined in such a way that it is zero for the phase with the greater symmetry. In some cases, such as in the bcc–fcc transition, there is no phase with a greater symmetry, and so no order parameter exists (since this parameter may only be defined when the symmetry group of one phase is a subgroup of the symmetry group of the other phase). Two representative examples of phase transitions that may be described with an order parameter are analyzed in this section.

Consider a fluid with uniaxial anisotropy and assume, for greater generality, that $b(\mathbf{u} \cdot \mathbf{u}')$ in (8.106) is written as

$$b_l(\mathbf{u} \cdot \mathbf{u}') = a_0 + a_l P_l(\mathbf{u} \cdot \mathbf{u}') \tag{9.26}$$

where l ($l \neq 0$) may be even or odd. For instance, $l = 2$ is the Maier–Saupe fluid and $l = 1$ is the Heisenberg fluid, where $P_1(\mathbf{u} \cdot \mathbf{u}') = \mathbf{u} \cdot \mathbf{u}'$ is the so-called Heisenberg exchange interaction. The first case has already been analyzed in (8.4.1), and the second one provides a simple model to study the paramagnetic–ferromagnetic transition. If the average density of molecules with orientation u is written as $\rho_1(\mathbf{u}) = \rho\, h(x)$, with $x = \cos\theta$, where $h(x)$ is the angular distribution function, the variational Helmholtz free energy per particle (8.106) reads (see (8.118))

$$\beta f_l(T, \rho; [h]) = \frac{1}{2} \int\limits_{-1}^{1} dx\, h(x)\left(\ln(\rho\Lambda^3) + \ln h(x) - 1\right) + \rho a_0$$

$$+ \frac{1}{4}\rho a_l \left(\int_{-1}^{1} dx \, h(x) P_l(x) \right)^2, \tag{9.27}$$

whose equilibrium solution $h_l(x)$ (notice the change of notation with respect to Sect. 8.4.1) is obtained by minimizing the functional, namely

$$\ln h_l(x) + \rho a_l P_l(x) \int_{-1}^{1} dx' h_l(x') P_l(x') = C, \tag{9.28}$$

where C is a constant. Since $\ln h_l(x) \sim P_l(x)$, if $h(x)$ is parameterized as

$$h(x) = \frac{1}{N_l(q)} e^{q P_l(x)}, \tag{9.29}$$

with

$$q \equiv -\rho a_l \int_{-1}^{1} dx' h(x') P_l(x'), \tag{9.30}$$

where $N_l(q)$ in (9.29) is obtained from the normalization condition (8.120):

$$N_l(q) = \frac{1}{2} \int_{-1}^{1} dx \, e^{q P_l(x)}, \tag{9.31}$$

the functional (9.27) becomes a function of the parameter q:

$$\beta f_l(T, \rho; q) = \ln(\rho \Lambda^3) - 1 + \rho a_0 + \frac{1}{4\varepsilon_l} q^2 - \ln N_l(q), \tag{9.32}$$

where

$$\varepsilon_l \equiv \varepsilon_l(T, \rho) = -\rho a_l. \tag{9.33}$$

Upon minimizing (9.32) with respect to q, the equilibrium value of the parameter q, q_l, is obtained from the equation

$$q = 2\varepsilon_l \frac{\partial \ln N_l(q)}{\partial q}. \tag{9.34}$$

Note that in the disordered phase $(h_l(x) = 1)$ $q_l = 0$, i.e., q_l is an order parameter and

$$\beta f_I(T, \rho; q_l = 0) = \ln(\rho\Lambda^3) - 1 + \rho a_0 \equiv f_I(T, \rho)$$

is the free energy of the isotropic phase (l). In the phase of lower symmetry with orientational order, $q_l \neq 0$ and the solution of (9.34) $q_l(\varepsilon_l) = q_l(T, \rho)$ is a function of the density and of the temperature. The variational free energy difference between the phase with orientational order and the isotropic phase,

$$\Delta f_I(T, \rho; q) = f_I(T, \rho; q) - f_I(T, \rho), \tag{9.35}$$

evaluated at equilibrium,

$$\Delta f_I(T, \rho) \equiv \Delta f_I(T, \rho; q_l), \tag{9.36}$$

allows one to determine the relative stability of the phases.

Note that, as a consequence of the underlying symmetries, it is possible to derive some qualitative properties of the transition. Since $P_l(x) = (-1)^l P_l(-x)$, it follows that $N_l(q) = N_l((-1)^l q)$, and so if l is odd $N_l(q)$ and $\Delta f_I(T, \rho; q)$ are even functions of q. On the other hand, if q_l is a solution of (9.34), so is $-q_l$, from which it follows that $h_l(-x) \neq h_l(x)$ (in fact, $h_l(-x) = 1/h_l(x)$) and, therefore, x and $-x$ denote different physical states (with different probability). When l is even, one has $N_l(-q) \neq N_l(q)$, and so the series expansion of $\Delta f_I(T, \rho; q)$ in powers of q contains odd and even terms. Since, on the other hand, $h_l(-x) = h_l(x)$, x and $-x$ are the same physical state and if q_l is a solution of (9.34), $-q_l$ is not, in general, a solution of this equation.

The symmetry properties $N_l(q) = N_l(-q)$ (when l is odd) and $N_l(-q) \neq N_l(q)$ (when l is even) are also shared by the variational Helmholtz free energy, which allows one to distinguish, according to Landau, two types of symmetry-breaking transitions.

9.4 Landau Theory

In order to obtain the series expansion of (9.35) with respect to the parameter q, note that (9.31) may be written as

$$N_l(q) = 1 + \sum_{n=1}^{\infty} c_n(l) q^n, \quad c_n(l) \equiv \frac{1}{2} \frac{1}{n!} \int_{-1}^{1} dx [P_l(x)]^n. \tag{9.37}$$

Since

$$P_1(x) = x, \quad P_2(x) = \frac{1}{2}(3x^2 - 1),$$

Table 9.1 Some coefficients in the expansion (9.36) for $l = 1$ and $l = 2$

	$l = 1$	$l = 2$
$c_2(l)$	1/6	1/10
$c_3(l)$	0	1/105
$c_4(l)$	1/120	1/280

$$P_3(x) = \frac{1}{2}(5x^3 - 3x), \quad P_4(x) = \frac{1}{8}(35x^4 - 30x^2 + 3),$$

one has

$$c_1(l) = 0 \ (l \neq 0), \quad c_2(l) = \frac{1}{2(2l + 1)},$$

$$c_3(2l + 1) = 0, \quad c_3(2l) > 0,$$

and

$$c_2^2(l) > 2c_4(l), (l = 1, 2),$$

as may be verified using the numerical values gathered in Table 9.1.

In this way, the series expansion of $\Delta f_l(T, \rho; q)$ is given by

$$\beta \Delta f_l(T, \rho; q) = \frac{1}{4\varepsilon_l}q^2 - \sum_{n=2}^{\infty} c_n(l)q^n + \frac{1}{2}\left(\sum_{n=2}^{\infty} c_n(l)q^n\right)^2 + \cdots, \tag{9.38}$$

or, alternatively,

$$\beta \Delta f_l(T, \rho; q) = q^2\big(a_2(l) + a_3(l)q + a_4(l)q^2 + \cdots\big). \tag{9.39}$$

The coefficients $a_i(l)$ $(i = 2, 3, 4)$ in (9.39) are given by

$$a_2(l) = \frac{1}{4\varepsilon_l} - c_2(l), \tag{9.40}$$

i.e., $a_2(l) \geq 0$ when $\varepsilon_l \leq l + 1/2$ and $a_2(l) < 0$ when $\varepsilon_l > l + 1/2$,

$$a_3(l) = -c_3(l), \tag{9.41}$$

i.e., $a_3(l) = 0$ if l is odd and $a_3(l) < 0$ if l is even, and

$$a_4(l) = \frac{1}{2}c_2^2(l) - c_4(l) > 0. \tag{9.42}$$

The values of the order parameter which make (9.39) to be an extremum are obtained from the equation

$$\left(\frac{\partial \beta \Delta f_l(T, \rho; q)}{\partial q}\right)_{q=q_l} = 2a_2(l)q_l + 3a_3(l)q_l^2 + 4a_4(l)q_l^3 + \cdots = 0, \quad (9.43)$$

and the extrema are a minimum when

$$\left(\frac{\partial^2 \beta \Delta f_l(T, \rho; q)}{\partial q^2}\right)_{q=q_l} = 2a_2(l) + 6a_3(l)q_l + 12a_4(l)q_l^2 + \cdots > 0. \quad (9.44)$$

From (9.43) and (9.44) it immediately follows that there are two branches of solutions. The first one,

$$q_l = 0, \quad a_2(l) > 0, \quad (9.45)$$

corresponds to the disordered phase and the second one,

$$a_2(l) + \frac{3}{2}a_3(l)q_l + 2a_4(l)q_l^2 = 0, \quad \frac{3}{2}a_3(l)q_l + 4a_4(l)q_l^2 > 0, \quad (9.46)$$

to the ordered phase. Note that the inequality in (9.46) is obtained by substituting the condition of extremum into (9.44). Observe now how the symmetry of the ordered phase affects the transition.

9.4.1 Continuous Transitions

When $l = 2n + 1$ ($n = 0, 1, 2, \ldots$), (9.46) has two solutions,

$$q_{2n+1}^{\pm} = \pm\sqrt{-\frac{a_2(2n + 1)}{2a_4(2n + 1)}}, \quad a_2(2n + 1) < 0, \quad (9.47)$$

which correspond to the same ordered phase since the free energy is an even function of q_{2n+1}. Since the domain of stability of the ordered phase ($q_{2n+1} \neq 0$, $a_2(2n + 1)$ < 0, $\varepsilon_{2n+1} > 2n + 3/2$) is complementary to that of the disordered phase ($q_{2n+1} = 0$, $a_2(2n + 1) > 0$, $\varepsilon_{2n+1} < 2n + 3/2$), the coexistence of phases is only possible when the following equality is verified:

$$\varepsilon_{2n+1} = 2n + \frac{3}{2}, \quad (9.48)$$

which is the equation of a straight line (see (8.112) and (9.33)) in the (ρ, T) plane, since $\varepsilon_l \sim \rho\beta$, i.e., (9.48) is the coexistence line of the order–disorder transition. When

$\varepsilon_{2n+1} \to 2n + 3/2$ from the ordered phase, $q^{\pm}_{2n+1} \to 0$, until the order parameter becomes zero at coexistence and the two phases become identical. Since the order parameter goes to zero continuously in the transition, these symmetry-breaking phase transitions are called, generically, continuous transitions.

9.4.2 Discontinuous Transitions

When $l = 2n$ ($n = 1, 2, \ldots$), (9.46) has two solutions,

$$q^{\pm}_{2n} = -\frac{3a_3(2n)}{8a_4(2n)}\left(1 \pm \sqrt{1 - \frac{32}{9}\frac{a_2(2n)a_4(2n)}{a_3^2(2n)}}\right), \tag{9.49}$$

which verify the following condition, obtained after elimination of $a_3(2n)$ in (9.43) and subsequent substitution in (9.44):

$$\left(q^{\pm}_{2n}\right)^2 > \frac{a_2(2n)}{2a_4(2n)}. \tag{9.50}$$

If one defines

$$D \equiv \frac{32}{9}\frac{a_2(2n)a_4(2n)}{a_3^2(2n)}, \tag{9.51}$$

then (9.50) may be cast in the form

$$1 - D \pm \sqrt{1 - D} > 0, \tag{9.52}$$

from which the following cases follow: (a) When $D > 1$, the solutions $q^{\pm}_{2n}$ are imaginary, which is not possible since $h_{2n}(x)$ has to be real, (b) if $0 \leq D < 1$, the inequality is only satisfied by $q^{\pm}_{2n}$, and (c) when $D < 0$, both solutions satisfy the inequality but $q^{-}_{2n} < 0$, which in turn would imply that this ordered phase would be less probable than the disordered phase. Therefore, the only physical solution for $D \leq 1$ is the root q^{+}_{2n} of (9.49), and so the ordered and disordered phases may coexist in the finite domain:

$$\frac{2}{4n + 1} \leq \frac{1}{\varepsilon_{2n}} \leq \frac{2}{4n + 1} + \frac{9}{8}\frac{a_3^2(2n)}{a_4(2n)}. \tag{9.53}$$

Since the order parameter is always different from zero,

$$q^{+}_{2n} \geq \frac{-3a_3(2n)}{8a_4(2n)}, \tag{9.54}$$

this parameter changes discontinuously at the transition. Examples of this kind of symmetry-breaking transitions have already been considered in Sect. 9.1, and such transitions are called, generically, discontinuous transitions.

9.5 Bifurcation Theory

The results of the Landau theory are based on the series expansion (9.39) of the variational free energy $\beta \Delta f_l(T, \rho; q)$ in powers of the order parameter. Note that although in the vicinity of a continuous transition ($q_l \to 0$), the series $\beta \Delta f_l(T, \rho; q)$ converges, far from the transition or for a discontinuous transition ($q_l \neq 0$) the series might not. The interest of the Landau theory is that, even if the series may not be a convergent one, the qualitative results that follow from it are correct. To see this, consider again (9.34) whose number of solutions, $q_l = q_l(\varepsilon_l)$, determines the number of phases. Since (9.34) always has as solution the trivial solution $q_l = 0$ (the disordered phase), each solution $q_l \neq 0$ corresponds to an ordered phase. One says then that these phases are obtained by bifurcation from the trivial solution and hence (9.34) is called the bifurcation equation. The solutions to this equation may be represented in the (ε_l, q_l) plane in what is known as a bifurcation diagram, which is analogous to the phase diagrams of Sect. 2.8.

As a first example, let $l = 1$, in which case (9.31) and (9.34) read, respectively, as

$$N_1(q) = \frac{1}{q} \sinh q,$$ (9.55)

and

$$\frac{1}{2\varepsilon_1} = \frac{1}{q_1}\left(\coth q_1 - \frac{1}{q_1}\right) \equiv \frac{1}{q_1}L(q_1),$$ (9.56)

where $L(q_1)$ is the Langevin function. The solutions of (9.56) may be determined numerically, and the results are shown in the bifurcation diagram displayed in Fig. 9.8.

Observe that when $\varepsilon_1 < 3/2$, the stable phase is the disordered phase ($q_1 = 0$), and if $\varepsilon_1 > 3/2$, the stable phase is the ordered phase ($q_1 \neq 0$), which is degenerate since $\pm q_1$ denotes the same phase. When $\varepsilon_1 = 3/2$, the order parameter goes to zero, and the two phases coexist and are identical. Therefore, $\varepsilon_1 = 3/2$ is the bifurcation point at which the new solution $q_1 \neq 0$ appears. Notice that in this case the bifurcation and coexistence points coincide. Due to the form of this diagram the bifurcation is called a fork bifurcation.

As a second example consider now $l = 2$, in which case (9.31) reads

$$N_2(q) = \sqrt{\frac{2}{3q}} e^q D\left(\sqrt{\frac{3q}{2}}\right),$$ (9.57)

Fig. 9.8 Bifurcation diagram of the Heisenberg fluid ($l = 1$) in the (ε_1, q_1) plane. The curves B and L are the solutions of the bifurcation equation (9.56) and of the Landau theory, respectively. In the interval $0 \leq \varepsilon_1 \leq 3/2$ both curves coincide

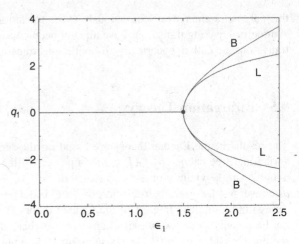

where $D(x)$ is the Dawson function:

$$D(x) = e^{-x^2} \int_0^x dy \, e^{y^2}. \tag{9.58}$$

On the other hand, (9.34) reduces in this case to

$$\frac{1}{\varepsilon_2} = \frac{1}{q_2} \left(\frac{\sqrt{\frac{3}{2q_2}}}{D\left(\sqrt{\frac{3q_2}{2}}\right)} - \frac{1}{q_2} - 1 \right), \tag{9.59}$$

whose numerical solution leads to the bifurcation diagram presented in Fig. 9.9, which shows that if $\varepsilon_2 < 2.24$, the stable phase is the disordered phase ($q_2 = 0$), if $\varepsilon_2 > 2.5$, the stable phase is the ordered phase ($q_2 \neq 0$), and in the interval $2.24 < \varepsilon_2 < 2.5$, both phases are stable. The bifurcation point is, therefore, $\varepsilon_2 = 2.24$, and the corresponding bifurcation diagram is known as a Hopf bifurcation. The coexistence point is obtained from the conditions of mechanical equilibrium and chemical equilibrium of the phases. Note that, in this case, the bifurcation point and the point of coexistence do not coincide.

The qualitative results obtained from the bifurcation diagrams are similar to the ones of the Landau theory, although there exist quantitative differences. For instance, in the Landau theory the bifurcation point when $l = 2$ is found at $\varepsilon_2 = 70/33 \simeq 2.12$ (instead of at $\varepsilon_2 \simeq 2.24$) and the order parameter changes from zero to $q_2 = 5/2$ (instead of $q_2 \simeq 1.45$). Therefore, the Landau theory provides a simple and qualitatively correct description of the symmetry-breaking phase transitions.

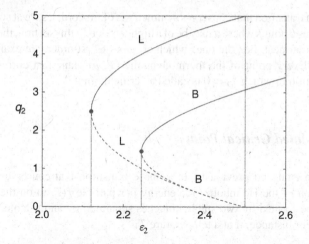

Fig. 9.9 Bifurcation diagram of the Maier–Saupe fluid ($l = 2$) in the (ε_2, q_2) plane. The *curves B and L* are the solutions of the bifurcation equation (9.59) and of the Landau theory, respectively. *The broken curves* indicate the non-physical solutions, and *the continuous curves* are the physical solutions. In the intervals $2.24 < \varepsilon_2 < 2.5$ (B) and $2.12 < \varepsilon_2 < 2.5$ (L) the two phases are stable and at some point within these intervals coexistence takes place

9.6 Critical Points

As was indicated in Chap. 8, the different phases of matter may be classified by their broken symmetries; for instance, the disordered phase (which has the same symmetries as the Hamiltonian), the intermediate phases or mesophases (in which one or more symmetries are lacking), and the completely ordered phase (in which all the symmetries are broken). In a similar fashion, a transition between two phases may be classified as structural (with symmetry breaking) or isostructural (without symmetry breaking), depending on whether or not the phases differ in their elements of symmetry.

From an experimental point of view it is common to classify phase transitions as transitions with a latent heat ($l_Q \neq 0$) or without a latent heat ($l_Q = 0$), since it is easier to measure l_Q than to determine a change of symmetry. Recall that according to the Clausius–Clapeyron equation (2.75), when $l_Q \neq 0$ the first derivatives of the Gibbs free energy ($s = -\partial\mu/\partial T$, $v = \partial\mu/\partial p$) are discontinuous ($s_1 \neq s_2$, $v_1 \neq v_2$). For that reason Ehrenfest introduced a classification that is determined by the continuity of the thermodynamic potential. According to Ehrenfest, a phase transition is of order n if the first derivative of the thermodynamic potential that is not continuous is an n-th derivative. In this way, if $l_Q \neq 0$ the transition is of first order and if $l_Q = 0$ it is, at least, of second order. However, it should be pointed out that the Ehrenfest thermodynamic classification is not always equivalent to the classification based on symmetry breaking. Hence, a symmetry-breaking transition is of first order when it is discontinuous, and it is at least of second order when it is continuous. On the other hand, a transition that conserves symmetry is of first order, save at some isolated thermodynamic points where it may be of a higher order. In this section, the transitions without a latent heat, i.e., the ones which are not of first order, are examined more thoroughly. Every point of the thermodynamic (T, p) plane that corresponds to a phase transition in which $l_Q = 0$ is called a "critical point."

9.6.1 Isolated Critical Points

As has been analyzed previously, in a phase transition that conserves symmetry, the projection of the Helmholtz free energy per particle $f(T, v)$ on the isothermal plane may be formed by two stable branches separated by an unstable branch (see Fig. 2.11). For instance, if at a temperature T,

$$\frac{\partial^2 f(T, v)}{\partial v^2} \begin{cases} > 0, \ 0 < v < v_2' \\ < 0, \ v_2' < v < v' \\ > 0, \ v_1' < v \end{cases} \tag{9.60}$$

where $v_1' = v_1'(T)$ and $v_2' = v_2'(T)$, then a phase with $v < v_2'$ may coexist with a phase with $v > v_1'$. Since the variation of the specific volume at the transition, $\Delta v = v_1 - v_2$, is $\Delta v > v_1' - v_2' > 0$ (see Fig. 2.11), the transition is discontinuous. It may happen, however, that upon changing the temperature, the difference $v_1' - v_2'$ decreases and vanishes for a certain value $T = T_c$. Since, according to (9.60), the derivative $\partial^2 f(T, v)/\partial v^2$ changes sign at v_1' and v_2' at that temperature, one then has that $\partial^2 f(T, v)/\partial v^2 = 0$ at $T = T_c$. Note further that, according to the construction of Maxwell's double tangent (see Sect. 2.8.3), when $T = T_c$ one has $\Delta v = 0$ ($v_1 = v_2 = v_c$, where v_c is the critical volume), and so at the critical temperature T_c the discontinuous transition becomes continuous. In order to analyze the stability of this special thermodynamic state, consider the convexity condition of $f(T, v)$ in the interval $v_2 < v < v_1$ (see Appendix A and Fig. 9.10):

$$f(T, \lambda v_1 + (1 - \lambda)v_2) < \lambda f(T, v_1) + (1 - \lambda)f(T, v_2), \quad (0 < \lambda < 1), \tag{9.61}$$

which expresses the fact that the function $f(T, v)$ has as an upper bound, the straight line joining the points $(v_2, f(T, v_2))$ and $(v_1, f(T, v_1))$.

If $v_2 < v''$, $v' < v_1$ and $v' = v'' \pm \varepsilon$ ($\varepsilon \to 0$), the convexity condition in the interval between v' and v'' is written as

$$f(T, v'' \pm \lambda \varepsilon) < \lambda f(T, v'' \pm \varepsilon) + (1 - \lambda)f(T, v''). \tag{9.62}$$

Expanding in Taylor series both sides of the inequality, one has ($v = v''$)

Fig. 9.10 Graphical representation of a convex function $f(v)$ in an interval $v_2 < v < v_1$. Note that f (*continuous line*) has as an upper bound the straight line (*broken line*) that joins both points

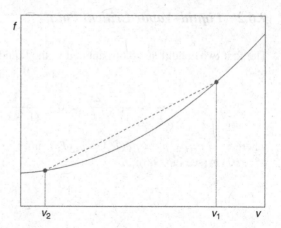

$$\sum_{n=1}^{\infty} \frac{1}{n!}(\pm\lambda\varepsilon)^n \frac{\partial^n f(T,v)}{\partial v^n} < \lambda \sum_{n=1}^{\infty} \frac{1}{n!}(\pm\varepsilon)^n \frac{\partial^n f(T,v)}{\partial v^n}, \tag{9.63}$$

or, alternatively,

$$0 < \sum_{n=2}^{\infty} \frac{1}{n!}\left(\lambda - \lambda^n\right)(\pm\varepsilon)^n \frac{\partial^n f(T,v)}{\partial v^n}. \tag{9.64}$$

Since $\lambda > \lambda^n$ ($n \geq 2$) and $(\pm\varepsilon)^n$ may be positive or negative, the inequality is satisfied when the first nonzero derivative is even and positive. If this occurs for $n = 2$, one obtains the usual convexity condition

$$\frac{\partial^2 f(T,v)}{\partial v^2} > 0. \tag{9.65}$$

Since at the critical point $(\partial^2 f(T,v)/\partial v^2)_c = 0$ (where the subscript c indicates an expression evaluated at $T = T_c$ and $v = v_c$), at this point the inequality (9.64) is satisfied only when the third derivative vanishes and the fourth one is positive. Thus, the Helmholtz free energy is convex at the critical point when

$$\left(\frac{\partial^2 f(T,v)}{\partial v^2}\right)_c = 0, \quad \left(\frac{\partial^3 f(T,v)}{\partial v^3}\right)_c = 0, \quad \left(\frac{\partial^4 f(T,v)}{\partial v^4}\right)_c > 0. \tag{9.66}$$

The solutions to the first two equations (9.66), i.e., $\{T_c, v_c\}$ are called critical states and in them the discontinuous transition becomes continuous. Generally, this takes place at an isolated point (T_c, p_c) of the coexistence curve $p = p(T)$, called the critical point (see Fig. 2.15).

9.6.2   Liquid–Vapor Critical Point

The first two conditions (9.66) applied to the liquid–vapor transition of Sect. 9.2.1
read

$$\frac{1}{(1-x_c)^2} - \frac{2x_c}{t_c} = 0, \qquad \frac{1}{(1-x_c)^3} - \frac{1}{t_c} = 0, \qquad (9.67)$$

where $x_c = \rho_c/\rho_0$, $\rho_c = \rho_V(T_c) = \rho_L(T_c)$, and $t_c = k_B T_c/a\rho_0$. The unique solution
of (9.67) is (see Fig. 9.6):

$$x_c = \frac{1}{3}, \quad t_c = \frac{8}{27}, \qquad (9.68)$$

and so from the equation of state,

$$p = \rho k_B T \left(\frac{1}{1-x} - \frac{1}{t}x \right), \qquad (9.69)$$

one has that the compressibility factor, $\beta p/\rho$, at the critical point is independent of
the model (i.e., of ρ_0 and a):

$$\frac{\beta_c p_c}{\rho_c} = \frac{3}{8} = 0.375; \qquad (9.70)$$

where p_c is the critical pressure. This behavior agrees qualitatively with the experimental results, although for the majority of simple fluids $\beta_c p_c/\rho_c \simeq 0.294$. Therefore, if (8.84) is written in terms of the reduced variables $\overline{p} = p/p_c$, $\overline{\rho} = \rho/\rho_c$, and
$\overline{T} = T/T_c$, one has

$$\overline{p} = \frac{8\overline{\rho}\overline{T}}{3-\overline{\rho}} - 3\overline{\rho}^2, \qquad (9.71)$$

which is independent of the characteristics of the fluid, i.e., of the particular values
of p_c, ρ_c, and T_c. For that reason, if two fluids have the same reduced variables, it is
said that they find themselves in "corresponding states."

If in the vicinity of the critical point ($\overline{p} = 1$, $\overline{\rho} = 1$, and $\overline{T} = 1$), one writes
$\overline{p} = 1 - \delta\overline{p}$, $\overline{\rho} = 1 - \delta\overline{\rho}$, and $\overline{T} = 1 - \delta\overline{T}$, and this leads to

$$\frac{\Delta s}{3k_B/2} = \frac{\Delta v}{4v_c} = \begin{cases} 0, & T \geq T_c \\ \sim (\delta\overline{T})^{1/2}, & T = T_c - 0 \end{cases}, \qquad (9.72)$$

i.e., the discontinuity in Δv and Δs tends to zero when $T \to T_c - 0$. Experimental
and numerical results show that $\Delta v \sim (\delta\overline{T})^\beta$ where $\beta \simeq 0.35$, which differs from

the value $\beta = 1/2$ of (9.72), and so the van der Waals theory does not provide an adequate description of this "critical exponent" (see Chap. 10).

Note, further, that although at the critical point the liquid–vapor transition is continuous, some thermodynamic variables are discontinuous at the transition, as for instance the isothermal compressibility coefficient χ_T,

$$\chi_T = \begin{cases} \left(-6p_c\delta\overline{T}\right)^{-1}, & T = T_c + 0 \\ \left(12p_c\delta\overline{T}\right)^{-1}, & T = T_c - 0 \end{cases} \tag{9.73}$$

which diverges when $T \simeq T_c$ with an amplitude that is discontinuous, while the constant pressure specific heat c_p

$$\frac{c_p}{3k_B/2} = \begin{cases} 1 + 2(-9\delta\overline{T})^{-2/3}, & T = T_c + 0 \\ 1 + (3\delta\overline{T})^{-1}, & T = T_c - 0 \end{cases} \tag{9.74}$$

diverges with an amplitude and a critical exponent, which are both discontinuous. In both cases, the "critical exponents" of the van der Waals theory also differ from those experimentally determined. It thus follows that, in spite of its simplicity, the van der Waals theory provides a qualitatively but not quantitatively correct description of the liquid–vapor transition, including its critical point.

9.6.3 Solid–Solid Critical Point

The first two conditions of (9.66), applied to the solid–solid transition of Sect. 9.2.2, read

$$\frac{3 - 2x_c}{(1 - x_c)^2} - \frac{n(n + 3)}{6t_c} x_c^n M_n = 0, \tag{9.75}$$

and

$$\frac{2(2 - x_c)}{(1 - x_c)^3} - \frac{n^2(n + 3)}{6t_c} x_c^{n-1} M_n = 0, \tag{9.76}$$

where $x_c^3 = \rho_c/\rho_{cp}$, $\rho_c = \rho_C(T_c) = \rho_E(T_c)$, and $t_c = k_B T_c/\varepsilon$. The unique solution to (9.75) and (9.76) is given by

$$x_c = \frac{1}{4(n + 1)}\left[5n + 4 - \sqrt{n^2 + 16(n + 1)}\right], \tag{9.77}$$

$$t_c = \frac{n(n + 3)}{6} M_n \frac{x_c^n(1 - x_c)^2}{3 - 2x_c}. \tag{9.78}$$

Table 9.2 Critical values x_c and t_c of the isostructural solid–solid transition of a crystal whose structure is the face-centered cubic lattice for different indices n (see Fig. 9.7)

n	x_c	t_c
50	0.963	1.026
100	0.981	1.052
150	0.987	1.062
200	0.990	1.067

The compressibility factor, $\beta p / \rho$, at the critical point is

$$\frac{\beta_c p_c}{\rho_c} = \frac{4(n+1)^2}{n+3} \frac{\sqrt{n^2 + 16(n+1)} - n - 4}{\left(\sqrt{n^2 + 16(n+1)} - n\right)^2}, \tag{9.79}$$

which is a function of the index n. Therefore, there is no "corresponding states" law as in the case of fluids. The values of x_c and t_c of a fcc lattice for different indices n are shown in Table 9.2.

The isostructural solid–solid transition has been recently obtained using simulation methods (see Fig. 9.7). This transition is thermodynamically stable in systems in which the range of the attractive part of the interaction potential is small compared with the range of the repulsive part. For instance, in the case of a square-well attractive potential of width δ, i.e., $V_A(r) = -\varepsilon[\Theta(\sigma + \delta - r) - \Theta(\sigma - r)]$, the solid–solid transition is stable when $\delta/\sigma < 0.07$. This kind of interaction potential is not characteristic of atomic systems for which the range of the attractive part is greater than that of the repulsive part (see Fig. 8.7 and Table 8.4) and the transition will not be observed (it is metastable). However, the transition could be observed experimentally in colloidal suspensions. Assume, for example, a mixture of hard spheres of diameters σ and σ' ($\sigma' \ll \sigma$). If the density of the large spheres is low, the osmotic pressure exerted by the fluid of small spheres on each large sphere is isotropic. At high density, in contrast, if the distance r between the centers of two large spheres is such that $r < \sigma + \sigma'$, the small spheres cannot penetrate inside the region between the two large spheres, and since the osmotic pressure is unbalanced, this leads to an effective attraction of range σ' between the two large spheres. For this reason, in a colloidal suspension of particles of two different sizes, the transition might be observed upon reduction of the size of the small particles (see Sect. 8.3.3).

9.6.4 Consolute Critical Point

Many mixtures have a liquid–vapor critical point similar to that of a one-component system. Furthermore, in mixtures a demixing phase separation can also take place and lead to a critical point known as the consolute critical point. In order to simplify the study of this transition, consider the regular solution theory of Sect. 8.3.1 applied to a binary mixture of particles of the same mass $m_1 = m_2$ ($\Lambda_1 = \Lambda_2 \equiv \Lambda$). Equation (8.90) may then be written as

$$\beta f(T, \rho, x) = \ln(\rho \Lambda^3) - 1 + \beta f_0(x) + \rho B(x), \tag{9.80}$$

where

$$\beta f_0(x) = x \ln x + (1 - x) \ln(1 - x), \tag{9.81}$$

and

$$B(x) = x^2 b_{11} + 2x(1 - x)b_{12} + (1 - x)^2 b_{22}. \tag{9.82}$$

Since normally the phase separation is studied at constant pressure (the atmospheric pressure), the adequate potential, instead of $f(T, \rho, x)$, is the Gibbs free energy per particle $g(T, p, x)$ defined as the Legendre transform:

$$g(T, p, x) = f(T, \rho, x) + \frac{p}{\rho}, \tag{9.83}$$

where in (9.83) $\rho = \rho(T, p, x)$ is obtained by solving the implicit equation $p = \rho^2 \partial f(T, \rho, x)/\partial \rho$, namely

$$\beta p = \rho + \rho^2 B(x), \tag{9.84}$$

whose positive solution is

$$\rho(T, p, x) = \frac{1}{2B(x)}(\sqrt{1 + 4\beta p B(x)} - 1). \tag{9.85}$$

From (9.80) and (9.83–9.85), it follows that

$$\beta g(T, p, x) = \ln\left(\frac{\Lambda^3}{2B(x)}\right) + \beta f_0(x) + \ln(\sqrt{1 + 4\beta p B(x)} - 1)$$
$$+ \sqrt{1 + 4\beta p B(x)} - 1. \tag{9.86}$$

Note that, in a binary mixture, the Gibbs free energy is given by

$$g(T, p, x) = x\mu_1(T, p, x) + (1 - x)\mu_2(T, p, 1 - x), \tag{9.87}$$

where μ_1 and μ_2 are the chemical potentials of the two species. On the other hand, the Gibbs–Duhem relation (2.16) at constant pressure and temperature reads

$$x\frac{\partial \mu_1(T, p, x)}{\partial x} + (1 - x)\frac{\partial \mu_2(T, p, 1 - x)}{\partial(1 - x)} = 0, \tag{9.88}$$

and so from (9.87) and (9.88) one obtains

$$\frac{\partial g(T, p, x)}{\partial x} = \mu_1(T, p, x) - \mu_2(T, p, 1 - x). \tag{9.89}$$

If x' and x'' denote the number fractions of species 1 at coexistence, the conditions for chemical equilibrium are

$$\mu_1(T, p, x') = \mu_1(T, p, x''), \quad \mu_2(T, p, x') = \mu_2(T, p, x''), \tag{9.90}$$

where, after (9.87) and (9.89),

$$\mu_1(T, p, x) = g(T, p, x) + (1 - x)\frac{\partial g(T, p, x)}{\partial x}, \tag{9.91}$$

and

$$\mu_2(T, p, x) = g(T, p, x) - x\frac{\partial g(T, p, x)}{\partial x}. \tag{9.92}$$

Note that if $\beta p B(x) \ll 1$ from (9.85) it follows that, as a first approximation, $\rho(T, p, x) \simeq \beta p$, which does not depend on x, in which case from (9.80) and (9.83) one finds

$$\begin{aligned}
\beta g(T, p, x) &= \ln(\beta p \Lambda^3) + \beta f_0(x) + \beta p B(x) \\
&= \ln(\beta p \Lambda^3) + \beta p(x b_{11} + (1 - x)b_{22}) \\
&\quad + \beta f^{\text{mix}}(T, \beta p, x),
\end{aligned} \tag{9.93}$$

where $f^{\text{mix}}(T, \beta p, x)$ is given by (8.92), in which ρ has been replaced by βp. From (9.93) it follows that (9.90) reads

$$\ln\left(\frac{x'(1 - x'')}{x''(1 - x')}\right) = 4\beta p b(x' - x''), \tag{9.94}$$

and

$$\ln\left(\frac{1 - x'}{1 - x''}\right) = -2\beta p b(x' - x'')(x' + x''), \tag{9.95}$$

where the coefficient b is given by (8.93).

The phase separation diagram of a binary mixture in the $(x, \beta p b)$ plane is shown in Fig. 9.11. The consolute critical point $(x_c, \beta_c p_c b)$ is obtained, by the same arguments used to derive (9.66) now particularized to the Gibbs free energy per particle, from the following system of equations:

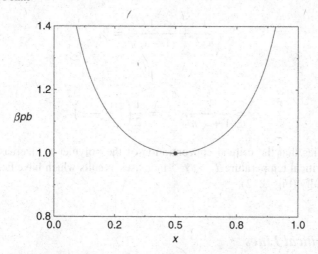

Fig. 9.11 Phase separation diagram of a binary mixture in the $(x, \beta pb)$ plane, where x is the number fraction of one of the components and βpb is the reduced pressure. The coordinates of the consolute critical point are $x_c = 1/2$ and $\beta_c p_c b = 1$

$$\left(\frac{\partial^2 g(T, p, x)}{\partial x^2}\right)_c = 0, \quad \left(\frac{\partial^3 g(T, p, x)}{\partial x^3}\right)_c = 0, \tag{9.96}$$

i.e.,

$$\frac{1}{x_c} + \frac{1}{1 - x_c} - 4\beta_c p_c b = 0, \tag{9.97}$$

$$-\frac{1}{x_c^2} + \frac{1}{(1 - x_c)^2} = 0, \tag{9.98}$$

yielding

$$x_c = \frac{1}{2}, \quad \beta_c p_c b = 1, \tag{9.99}$$

which implies that b has to be positive, namely $2b_{12} > b_{11} + b_{22}$.

Note that, according to (9.93), the consolute critical point (9.96) is obtained with the approximation $\rho(T, p, x) \simeq \beta p$ from the free energy of mixing $f^{\text{mix}}(T, \beta p, x)$. For that reason, applying the same kind of study to a polymer solution in the Flory–Huggins theory, from (8.153) it follows that

$$\frac{1}{nx_c} + \frac{1}{1 - x_c} - 2\chi_c = 0, \tag{9.100}$$

and

$$-\frac{1}{nx_c^2} + \frac{1}{(1-x_c)^2} = 0, \tag{9.101}$$

i.e.,

$$x_c = \frac{1}{1+\sqrt{n}}, \quad \chi_c = \frac{1}{2}\left(1+\frac{1}{\sqrt{n}}\right)^2, \tag{9.102}$$

which implies that the critical concentration of the polymer x_c decreases with n, while the critical temperature $T_c \sim \chi_c^{-1}$ increases, results which have been verified experimentally (Fig. 9.12).

9.6.5 Critical Lines

As has been analyzed previously, in phase transitions with symmetry breaking, two cases may occur. In the case of a discontinuous transition, i.e., when the free energy is not an even function of the order parameter, the transition is always of first order since the order parameter changes from $q_l = 0$ to $q_l \neq 0$ at the transition (see (9.54)). On the contrary, when the free energy is an even function of the order parameter, q_l is continuous at the transition (which is thus called a continuous transition) and in such a transition critical points may exist, i.e., points of coexistence for which $\Delta s = 0$ and $\Delta v = 0$. To see this, consider once more the Landau theory or the bifurcation theory of Sects. 9.4 and 9.5. In a continuous transition the contribution of the order parameter to the Helmholtz free energy difference $\Delta f_l(T, \rho)$ (see (9.39) and (9.47)) is given by

$$\beta \Delta f_l(T, \rho) = -\frac{1}{2}\frac{a_2^2(l)}{a_4(l)}, \tag{9.103}$$

with $l = 2n + 1$. Since at the coexistence between the ordered and the disordered phases $a_2(l) = 0$, one has that $\Delta f_l(T, \rho)$ and its first derivatives (with respect to ρ or T), which are denoted indistinctly as $\Delta f_l'(T, \rho)$, $\beta \Delta f_l'(T, \rho) = -a_2(l)a_2'(l)/a_4(l)$ (since $a_4(l)$ is a constant) vanish at coexistence. Therefore, $\Delta f_l(T, \rho)$ does not contribute neither to Δs nor to Δp and coexistence is produced at the critical point ($\Delta s = 0$ and $\Delta v = 0$), although the second derivatives may be discontinuous (like, for instance, c_V). The transition is, at least, of second order in the Ehrenfest sense.

At the critical point, the continuous transitions have a similar behavior to that of the isostructural transitions. For instance, the order parameter $q_{2n+1}^{\pm}$ tends to zero as

$$q_{2n+1}^{\pm} \sim \sqrt{\varepsilon_{2n+1} - 2n - \frac{3}{2}}, \tag{9.104}$$

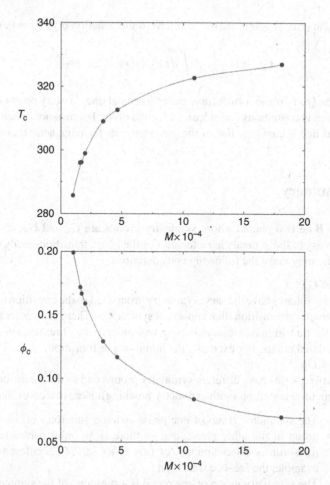

Fig. 9.12 Experimental values (*indicated by points*) of the critical temperature T_c (K) and critical concentration of polystyrene ϕ_c (volume fraction) of a polystyrene–methylcyclohexane solution as a function of the molecular weight M of the polymer (in units of the molecular weight of the monomer). The *lines* have been drawn to guide the eye. *Source* T. Dobashi, M. Nakata and M. Kaneko, J. Chem. Phys. **72**, 6685 (1980)

when one approaches the transition from the ordered phase, although the Landau theory (as all mean field theories) does not adequately describe the critical exponent in (9.104). The main difference in critical behavior between phase transitions with or without symmetry breaking is appreciated in a (ρ, T) phase diagram. For the transitions that conserve the symmetry, the critical behavior in the (ρ, T) plane usually takes place at an isolated point of the plane like, for example, in the liquid–vapor transition and in the isostructural solid–solid transition (see Figs. 9.6 and 9.7). The same occurs in the (T, p) plane (see Fig. 2.15). The transition is continuous at the critical point and discontinuous everywhere else. For the transitions with symmetry breaking, the critical points are the solutions to the equation $q(\rho, T) = 0$, namely

$$\varepsilon_{2n+1} \equiv -\frac{1}{2}\rho\beta \int d\mathbf{r} V_{2n+1}(r) = 2n + \frac{3}{2}, \qquad (9.105)$$

which in the (ρ, T) plane form a curve called a critical line. In every point of this line, the transition is continuous and at least of second order. For instance, when $\varepsilon_1 = 3/2$, this critical line is the Curie line of the paramagnetic–ferromagnetic transition.

9.7 Summary

Let A and B be two phases whose symmetry groups are G_A and G_B, respectively. As a summary of the general characteristics of the phase transitions analyzed in this chapter, one may make the following considerations.

(1) $G_A \equiv G_B$

The two phases have the same symmetry group, and so the transition is called an isostructural transition (that conserves symmetry). There is no order parameter and the transition is discontinuous (of first order in the Ehrenfest sense), except at isolated points. For example, the liquid–vapor transition.

(2) $G_A \neq G_B$

The two phases have different symmetry groups and so the transition is called a structural transition (with symmetry breaking). Here two cases may occur.

(a) The symmetry group of one phase is not a subgroup of the symmetry group of the other phase, and so there is no order parameter and the transition is discontinuous (of first order in the Ehrenfest sense), for example, the fcc–bcc transition.

(b) The symmetry group of one phase is a subgroup of the symmetry group of the other phase, $G_A \subset G_B$. The transition may be described with an order parameter q, and it is

(b1) discontinuous (of first order in the Ehrenfest sense) if the free energy is not an even function of q, for example, the isotropic–nematic transition.

(b2) continuous (at least of second order in the Ehrenfest sense) if the free energy is an even function of q, for example, the paramagnetic–ferromagnetic transition.

9.8 Triple Points

Up to this point, the coexistence between two phases and the transition from one phase to the other have been considered. Since a system may find itself in more than two phases, the derivation of the resulting phase diagrams requires the study of all the binary coexistences.

9.8.1 Ordinary Triple Point

Consider a system that may be found in three different phases whose Gibbs free energies per particle are $\mu(T, p)$ $(i = 1, 2, 3)$. There are, therefore, three possible transitions described by the following coexistence lines in the (T, p) plane: $\mu_1(T, p) = \mu_2(T, p)$, $\mu_2(T, p) = \mu_3(T, p)$, and $\mu_3(T, p) = \mu_1(T, p)$. It is possible that these three lines intersect at a point (T_t, p_t),

$$\mu_1(T_t, p_t) = \mu_2(T_t, p_t) = \mu_3(T_t, p_t), \tag{9.106}$$

which is called the triple point and in which the three phases coexist. Assume, first, that the three transitions are discontinuous in the vicinity of (T_t, p_t).

The intersection of these lines is shown in Fig. 9.13. On crossing the triple point, each line changes from being stable (continuous line) to being metastable (broken line). If, as usual, only the stable parts are represented, the angles formed by the coexistence lines are less than π.

The phase diagram of an ordinary triple point (the three transitions are discontinuous) in the (v, T) plane is represented in Fig. 9.14. The specific volumes at the triple point have been denoted by v_1, v_2, and v_3. Examples of ordinary triple points are the solid–liquid–vapor triple point (see Fig. 9.16) and the triple point fluid-bcc solid-fcc solid in charged colloidal particles interacting with a repulsive Yukawa potential.

9.8.2 Critical Endpoint

Assume now that in the vicinity of the triple point the transitions 1–2 and 1–3 are discontinuous while the transition 2–3 is continuous. The resulting phase diagram in the (v, T) plane is shown in Fig. 9.15, and the corresponding triple point is called a critical endpoint $(v_2 = v_3)$. An example of a critical end point is the paramagnetic liquid–ferromagnetic liquid transition in a fluid in which the attractive part of the isotropic interaction dominates over the anisotropic Heisenberg interaction.

9.8.3 Bicritical Point

Another special case is obtained when two of the three transitions are continuous.

If the transition 1–3 is discontinuous, the phase diagram in the (v, T) plane is shown in Fig. 9.16 and such a triple point is called a bicritical point $(v_1 = v_2 =$

Fig. 9.13 Stable parts (*continuous lines*) and metastable parts (*broken lines*) of the coexistence lines 1–2, 2–3, and 3–1 in the (T, p) plane, intersecting at the triple point (T_t, p_t). Since the coexistence lines are obtained from the intersection of convex surfaces $\mu_i = \mu_i(T, p)$, the angles θ_i formed by the stable parts are such that $\theta_i < \pi$ ($i = 1, 2, 3$)

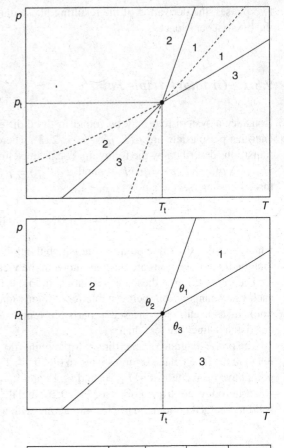

Fig. 9.14 Phase diagram in the (v, T) plane of an ordinary triple point in which three phases coexist (1, 2, and 3) at the specific volumes $v_1 \neq v_2 \neq v_3$. All the transitions are discontinuous. Note the discontinuity in the slopes of the binodals for $T = T_t$

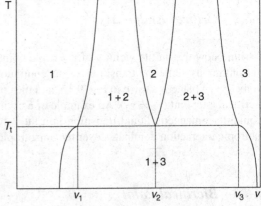

Fig. 9.15 Phase diagram in the (v, T) plane with a critical endpoint in which three phases coexist (1, 2, and 3) at the specific volumes v_1 and $v_2 = v_3$. The transitions 1–2 and 1–3 are discontinuous while the transition 2–3 is continuous. Note the change in curvature of the binodals for $T = T_t$

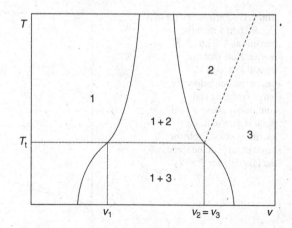

Fig. 9.16 Phase diagram in the (v, T) plane with a bicritical point at which three phases coexist (1, 2, and 3) at the specific volumes $v_1 = v_2 = v_3$. The transitions 1–2 and 2–3 are continuous while the transition 1–3 is discontinuous. Note the continuity of the slopes of the binodals for $T = T_t$

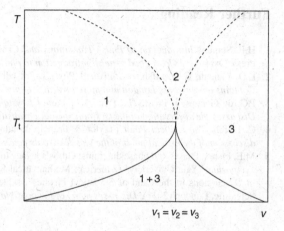

v_3). An example of a bicritical point is the paramagnetic liquid–ferromagnetic liquid transition in a fluid in which the anisotropic Heisenberg interaction dominates over the attractive part of the isotropic interaction.

9.8.4   Tricritical Point

Finally, it may occur that the continuous transitions 1–2 and 2–3 merge in the vicinity of the triple point, as is shown in Fig. 9.17. In this case, the transition 1–3 changes from being discontinuous to being continuous at the triple point, which is then called a tricritical point.

Fig. 9.17 Particular case of
Fig. 9.16 in which the
transitions 1–2 and 2–3
merge. Note that the
transition 1–3 is
discontinuous below the
temperature T_t and
continuous above it. This
point is then called a
tricritical point. Note the
discontinuity in the slopes of
the binodals for $T = T_t$

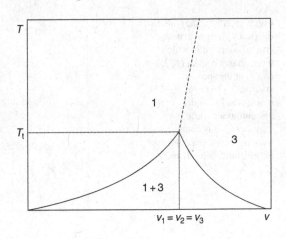

Further Reading

1. H.E. Stanley, *Introduction to Phase Transitions and Critical Phenomena* (Oxford University Press, Oxford, 1987). *A good general introduction to phase transitions.*
2. L.D. Landau, E.M. Lifshitz, *Statistical Physics*, 3rd edn. (Pergamon Press, London, 1989). *Contains the theory of Landau in a more abstract setting.*
3. P.G. de Gennes, J. Prost, *The Physics of Liquid Crystals* (Clarendon Press, Oxford, 1993). *Discusses the observation of the isotropic–nematic transition.*
4. C. Domb, *The Critical Point* (Taylor & Francis, London, 1996). *Contains a very detailed discussion of the critical point of the van der Waals equation.*
5. M.E. Fisher, Phases and phase diagrams: Gibbs's legacy today, in *The Proceedings of the Gibbs Symposium* (Yale University) (American Mathematical Society, 1990), p. 39. Now reprinted in "Excursions in the Land of Statistical Physics", selected reviews by M.E. Fisher, World Scientific, London (2017). *Discusses many of the possible complexities of phase diagrams.*

Part IV
More Advanced Topics

Chapter 10
Critical Phenomena

Abstract As was already seen in Chap. 9, there may exist particular thermodynamical states where a discontinuous (first order) phase transition becomes continuous (second order) or where the order parameter of a continuous phase transition vanishes (Fig. 10.1). Such a thermodynamical state is called a critical state, or a critical point (CP), and its variables will be distinguished here by a subscript c. In the immediate vicinity of a CP, there occur a number of peculiar (static and dynamic) phenomena which are designed collectively as "critical phenomena." During the last four decades considerable progress has been made in the understanding of these phenomena. In this chapter some of the new ideas which emerged from these studies are summarized and explained in the simpler context of the present textbook. More complete studies can be found in the Further Reading.

10.1 Classical Theory

Consider the behavior of the thermodynamic potentials, e.g., the Helmholtz free energy per particle $f(T, v)$, in the vicinity of a critical point. Until now it has been always assumed that the thermodynamic potentials are analytical, i.e., that they can be expanded in a power series around any thermodynamic state, including the critical state. The expansion of $f(T, v)$ around the CP (T_c, v_c) will then read

$$f(T, v) = \sum_{n=0}^{\infty} \sum_{k=0}^{\infty} \frac{1}{n!k!} \left(\frac{\partial^{n+k} f(T, v)}{\partial v^n \partial T^k} \right)_c (\delta v)^n (\delta T)^k, \tag{10.1}$$

where $\delta v = v - v_c$ and $\delta T = T - T_c$ measure the distance between the thermodynamic state and the CP defined by

$$\left(\frac{\partial^2 f(T, v)}{\partial v^2} \right)_c = 0, \quad \left(\frac{\partial^3 f(T, v)}{\partial v^3} \right)_c = 0. \tag{10.2}$$

© The Author(s), under exclusive license to Springer Nature Switzerland AG 2021   283
M. Baus and C. F. Tejero, *Equilibrium Statistical Physics*,
https://doi.org/10.1007/978-3-030-75432-7_10

It is easily shown that these equations are equivalent to the more familiar definition

$$\left(\frac{\partial p(T, v)}{\partial v}\right)_c = 0, \quad \left(\frac{\partial^2 p(T, v)}{\partial v^2}\right)_c = 0, \tag{10.3}$$

in terms of the pressure $p(T, v) = -\partial f(T, v)/\partial v$, or

$$\left(\frac{\partial \mu(T, v)}{\partial v}\right)_c = 0, \quad \left(\frac{\partial^2 \mu(T, v)}{\partial v^2}\right)_c = 0, \tag{10.4}$$

in terms of the chemical potential $\mu(T, v) = f(T, v) - v\partial f(T, v)/\partial v$. As in most parts of this textbook, the language of the one-component fluid is used for illustration purposes, but the same discussion can be performed for any other system (mixtures, magnetic systems, etc.) exhibiting a CP although, in fact, most of the CP literature (cf. Further Reading) is usually presented in the language of the magnetic systems.

10.2 Critical Exponents

The CP can be approached along different thermodynamical paths, e.g., along the coexistence curve, along an isotherm or along an isochore (Fig. 10.2).

If $T = T(v)$ denotes the coexistence curve (binodals) in the (v, T)-plane (Fig. 10.1), then the CP, $T_c = T(v_c)$, is found by following this curve until its extremum is reached,

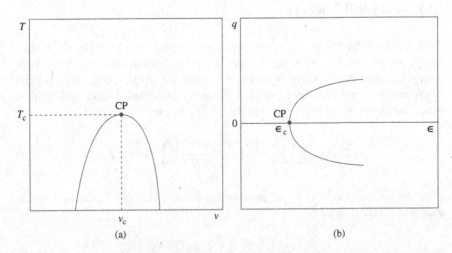

Fig. 10.1 a Upper CP terminating the coexistence curves (binodals), $T = T(v)$, of a discontinuous phase transition, such as the L-V transition. The coordinates of this CP in the (v, T) plane will be denoted by (v_c, T_c). **b** Behavior of the order parameter q versus a control parameter $\varepsilon = \varepsilon(T, v)$ for a continuous phase transition (see Sect. 9.4.1). The value $\varepsilon_c = \varepsilon(T_c, v_c)$ for which q vanishes corresponds to a CP

Fig. 10.2 Different ways to approach the CP: along the critical isotherm $T = T_c$, along the critical isochore $v = v_c$, along the coexistence curve $T = T(v)$

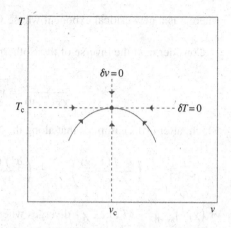

i.e., when $(\partial T(v)/\partial v)_c = 0$. Therefore, in the vicinity of the CP one finds

$$T(v) = T(v_c) + \frac{1}{2!}\left(\frac{\partial^2 T(v)}{\partial v^2}\right)_c (\delta v)^2 + O\big((\delta v)^3\big), \qquad (10.5)$$

a behavior that will be rewritten compactly as $(\delta T)_{coex} \sim (\delta v)^2$, where δT and δv are the distances to the critical point and the subscript reminds that $(ST)_{coex}$ is measured here along the coexistence curve $T = T(v)$, hence $(\delta T)_{coex} = T(v) - T(v_c)$. The relation $(\delta T)_{coex} \sim (\delta v)^2$ states that in the classical theory the shape of the coexistence curve is parabolic in the vicinity of the CP. For later use, this relation will be rewritten more generally as

$$(\delta T)_{coex} \sim (\delta v)^{1/\beta}, \qquad (10.6)$$

where β is the so-called critical exponent, which in the above classical theory takes on the value $\beta = 1/2$.

If, similarly,

$$\delta p = p(T, v) - p(T_c, v_c)$$

is followed along the critical isotherm $\delta T = 0$, (10.3) implies

$$p(T_c, v) = p(T_c, v_c) + \frac{1}{3!}\left(\frac{\partial^3 p(T, v)}{\partial v^3}\right)_c (\delta v)^3 + O\big((\delta v)^4\big), \qquad (10.7)$$

or, $(\delta P)_{\delta t=0} \sim (\delta v)^3$. This relation will be rewritten more generally as

$$(\delta p)_{\delta T=0} \sim (\delta v)^{\delta}, \qquad (10.8)$$

where δ is a new critical exponent whose value in the above classical theory is $\delta = 3$.

Consider next the inverse of the isothermal compressibility coefficient:

$$\frac{1}{\chi_T(T, v)} = v \frac{\partial^2 f(T, v)}{\partial v^2}, \tag{10.9}$$

which, after (10.2), implies that along the critical isochore $\delta v = 0$ one has

$$\left(\frac{\partial^2 f(T, v)}{\partial v^2}\right)_{v=v_c} = \left(\frac{\partial^3 f(T, v)}{\partial v^2 \partial T}\right)_c \delta T + O\left((\delta T)^2\right), \tag{10.10}$$

or $\left(\chi_T^{-1}\right)_{\delta v=0} \sim \delta T$, i.e., χ_T diverges when the CP is approached along the critical isochore $\delta v = 0$. This relation can be rewritten as

$$\left(\chi_T^{-1}\right)_{\delta v=0} \sim (\delta T)^\gamma = \begin{cases} +|\delta T|^{\gamma_+}, & \delta T > 0 \\ -|\delta T|^{\gamma_-}, & \delta T < 0 \end{cases} \tag{10.11}$$

where γ is yet another critical exponent whose classical value is $\gamma = 1$. Note that to write (10.11) it has been taken into account that this time the value of the exponent γ could depend on the sign of δT since for $\delta v = 0$ and $\delta T > 0$ the fluid is in a one-phase region, whereas for $\delta v = 0$ and $\delta T < 0$ the fluid is in a two-phase region, while in (10.8) one always remains in the one-phase region when $\delta T = 0$. However, in order to simplify the notation, in what follows the set of two exponents (γ_+, γ_-) will be designed simply as γ. More critical exponents can be introduced but since ultimately only two of them will turn out to be independent, in what follows only (10.6), (10.8), and (10.11) will be considered and be rewritten compactly as

$$(\delta T)_{\text{coex}} \sim (\delta v)^{1/\beta}, \quad (\delta p)_{\delta T=0} \sim (\delta v)^\delta, \quad \left(\chi_T^{-1}\right)_{\delta v=0} = (\delta T)^\gamma. \tag{10.12}$$

Instead of relying on the above theoretical considerations for the determination of the critical exponents, one can of course measure them experimentally. One then finds that, when one is not too close to the CP, the above classical results ($\beta = 1/2$, $\delta = 3$, $\gamma = 1$) appear to be adequate but that they become inadequate the closer one gets to the CP. Clearly, to really measure the critical exponents, a limiting procedure is required. For instance, to measure β of (10.6) one can plot $\ln (\delta T)_{\text{coex}}$ versus $\ln (\delta v)$ and measure the slope $1/\beta$ of this relation as one crosses the CP. In this way a vast body of experimental evidence has been gathered during the past decades which clearly indicates that the above theoretical values of the critical exponents are in fact incorrect (cf. Fig. 10.3), justifying hereby our rewriting their definition in the more general form of (10.12).

The experimental evidence shows clearly that the values of $1/\beta$, δ, and γ cannot be represented by integers. The very reason why in the above classical theory the values of these critical exponents did turn out to be integers can clearly be traced down to the

Fig. 10.3 Measurements of the coexistence curve of helium in the neighborhood of its critical point (circles = vapor ($v > v_c$), triangles = liquid ($v < v_c$)). The critical exponent P has a value 0.354. *Source* H. E. Stanley, *Introduction to Phase Transitions and Critical Phenomena*, Clarendon Press, Oxford (1971)

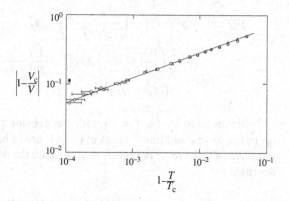

fact that in the classical theory it has been assumed that $f(T, v)$ is an analytic function of δT and δv in the vicinity of the CP. Another, more positive, experimental finding has been that a given critical exponent turns out to have the same numerical value for physically very different systems, e.g., fluids, magnetic systems, etc. This is usually expressed by saying that while the critical exponents take on non-classical values, their values are "universal." Such a universality is typical of critical phenomena, or of continuous (second order) phase transitions, and is in clear contradistinction with the results obtained for discontinuous (first order) phase transitions which are strongly system-dependent. To bring the theory of the CP behavior in agreement with these experimental observations has been a major enterprise of the statistical physics of the last few decades. Here only the major steps in the resolution of this enigma are summarized. More details can be found in the specialized literature quoted in the Further Reading.

10.3 Scaling Hypothesis

The first major step forward in bringing the theory of the CP behavior in agreement with the experimental findings was the introduction by B. Widom in 1965 of a scaled equation of state. To understand how this comes about, consider the behavior of $p(T, v)$ in the vicinity of the CP. When the CP is approached along an isotherm, the classical theory predicts that

$$p(T, v) = p(T, v_c) + \sum_{n=1}^{\infty} \frac{1}{n!} \left(\frac{\partial^n p(T, v)}{\partial v^n} \right)_{v=v_c} (\delta v)^n, \tag{10.13}$$

or when the isotherm is close to the critical isotherm, so that δT is small, (10.13) becomes up to fourth order:

$$p(T, v) = p(T, v_c) + \left(\frac{\partial^2 p(T, v)}{\partial v \partial T}\right)_c \delta v \delta T + \frac{1}{2!}\left(\frac{\partial^3 p(T, v)}{\partial v^2 \partial T}\right)_c (\delta v)^2 \delta T$$

$$+ \frac{1}{3!}\left(\frac{\partial^3 p(T, v)}{\partial v^3}\right)_c (\delta v)^3 + \frac{1}{2!}\left(\frac{\partial^3 p(T, v)}{\partial v \partial T^2}\right)_c \delta v (\delta T)^2$$

$$+ O\left((\delta v)^4, (\delta v)^2 (\delta T)^2, (\delta v)^3 \delta T\right). \tag{10.14}$$

Evaluating (10.14) for $T = T(v)$ it is seen, using (10.5), that, because $(\delta T)_{\text{coex}} \sim (\delta v)^2$, the second and fourth terms of (10.14) are of order $(\delta v)^3$ while the remaining terms are of order $(\delta v)^4$ or higher. Keeping only the dominant terms, (10.14) may be rewritten as

$$\Delta p \equiv H(\delta T, \delta v) = a_c \delta v \delta T + b_c (\delta v)^3, \tag{10.15}$$

where

$$\Delta p \equiv p(T, v) - p(T, v_c),$$

$$a_c \equiv \left(\frac{\partial^2 p(T, v)}{\partial v \partial T}\right)_c, \quad b_c \equiv \frac{1}{3!}\left(\frac{\partial^3 p(T, v)}{\partial v^3}\right)_c.$$

Consider now $H(\delta v, \delta T)$ of (10.15). In the vicinity of the CP this function obeys a relation of the form:

$$H\left(\lambda^2 \delta T, \lambda \delta v\right) = \lambda^3 H(\delta T, \delta v), \tag{10.16}$$

for any $\lambda > 0$. Such a relation is called a scaling relation because it states that when rescaling the variables of $H(\delta T, \delta v)$ as $\delta v \to \lambda \, \delta v$ and $\delta T \to \lambda^2 \delta T$, the value of $H(\delta T, \delta v)$ is restored up to a scaling factor λ^3. The mathematical property of (10.16) implies that the function of two variables $H(\delta T, \delta v)$ can always be written in the simpler form

$$H(\delta v, \delta T) = (\delta v)^3 f_0\left(\frac{\delta T}{(\delta v)^2}\right), \tag{10.17}$$

where $f_0(x)$ is a function of a single but rescaled variable $x = \delta T/(\delta v)^2$, which remains invariant under the above rescaling transformation. To prove (10.17), it is indeed sufficient to take $\lambda = 1/\delta v$ in (10.16). Moreover, from (10.15) it is seen that in the classical theory $f_0(x) = b_c + a_c x$. Suppose now that, in order to obtain non-classical results, one modifies (10.15–10.17) empirically as

$$\Delta p = (\delta v)^\delta f\left(\frac{\delta T}{(\delta v)^{1/\beta}}\right), \tag{10.18}$$

where δ and β are some (still undetermined) critical exponents and $f(x)$ a still arbitrary function of the rescaled variable $x = \delta T/(\delta v)^{1/\beta}$. Equation (10.18) is equivalent to the scaling relation:

$$H\left(\lambda^{1/\beta}\delta T, \lambda\delta v\right) = \lambda^{\delta} H(\delta T, \delta v), \tag{10.19}$$

which generalizes (10.16). Since $(\Delta p)_{\delta T=0} = (\delta p)_{\delta T=0}$, it follows that (10.18) directly implies (10.8), provided $f(0)$ is finite. Moreover, evaluating $\Delta p/\delta v$ from (10.8), and taking into account that as $\delta v \to 0$ one has

$$\left(\frac{\Delta p}{\delta v}\right)_{\delta v \to 0} = \left(\frac{\partial p(T, v)}{\partial v}\right)_{v=v_c} = \left(\frac{1}{v\chi_T(T, v)}\right)_{v=v_c}, \tag{10.20}$$

so that

$$\frac{1}{v_c \chi_T(T, v_c)} = \left[(\delta v)^{\delta-1} f\left(\frac{\delta T}{(\delta v)^{1/\beta}}\right)\right]_{\delta v \to 0}. \tag{10.21}$$

Evaluating now (10.21) along the coexistence curve $T = T(v)$, where according to (10.6) $\delta v \sim (\delta T)^{\beta}$, it is seen that (10.21) is equivalent to

$$\frac{1}{\chi_T(T, v_c)} \overset{.}{\sim} (\delta T)^{\beta(\delta-1)} f(1) \sim (\delta T)^{\beta(\delta-1)}, \tag{10.22}$$

provided $f(1)$ is finite. Therefore, the scaled equation of state (10.18) is fully consistent with the experimental observations under some fairly weak conditions on $f(x)$. Moreover, comparing (10.22) with (10.12) yields

$$\gamma = \beta(\delta - 1), \tag{10.23}$$

showing that only two of the critical exponents (β, γ, δ) are independent. The surmised non-analyticity of the equation of state in the vicinity of the CP can, according to (10.18), now be traced back to the fact that $f(x)$ is a function of the rescaled variable $x = \delta T/(\delta v)^{1/\beta}$, i.e., for $\delta T \neq 0$, $f(x)$ is a non-analytical function of δv. Note, however, that the conditions imposed on $f(x)$ are compatible with $f(x)$ being an analytic function of x. Therefore, the complete non-classical thermodynamic behavior observed in the vicinity of the CP can be deduced from the scaling relation (10.19), itself an empirical modification of the classical scaling relation (10.16). There remains now to understand why a scaling behavior such as described by (10.19) should hold, i.e., what is the physical reason behind this scale invariance of the thermodynamics in the region of the CP?

10.4   Correlation Length and Universality

Scale invariance is rather unusual since specific physical systems usually have specific scales on which its variables change. Therefore, in the vicinity of the CP these system-specific scales must in some sense become irrelevant. Of particular interest here is the fact, first pointed out by Ornstein and Zernike in 1916, that the spatial scale of the density fluctuations will change dramatically near a CP and become observable as critical opalescence. As already seen in Sect. 6.7, at the CP the density fluctuations will diverge (cf. (6.91)) and the total correlation function $h(r; T_c, \rho_c)$ will become long ranged (cf. (6.104)) even for a short-ranged pair potential. In other words, near the CP it is the spatial scale of $h(r; T, \rho)$ which is important in order to understand the phenomenon of critical opalescence, whereas the spatial scale of the system-specific pair potential becomes irrelevant. This suggested Ornstein and Zernike to introduce a new function, the direct correlation function $c(r; T, \rho)$ defined by (8.52) and (8.53) (see also Sect. 7.4), which remains short ranged at the CP (cf. (8.55)). If $c(r; T, \rho)$ is short ranged, then its Fourier transform $\tilde{c}(k; T, \rho)$ can be expanded as

$$\tilde{c}(k; T, \rho) = \int d\mathbf{r} \, e^{-i\mathbf{k}\cdot\mathbf{r}} c(r; T, \rho) = 4\pi \int_0^\infty dr \, r^2 c(r; T, \rho) \frac{\sin(kr)}{kr}$$

$$= \tilde{c}(0; T, \rho) + \tilde{c}_2(T, \rho)k^2 + O(k^4), \tag{10.24}$$

where $\rho\tilde{c}(k; T, \rho)$ is given by (8.55), i.e.,

$$1 - \rho\tilde{c}(0; T, \rho) = \frac{1}{\rho k_B T \chi_T(T, \rho)}, \tag{10.25}$$

and

$$\rho\tilde{c}_2(T, \rho) = -\frac{2\pi}{3}\rho \int_0^\infty dr \, r^4 c(r; T, \rho) \equiv -R^2(T, \rho), \tag{10.26}$$

where $R(T, \rho)$ is a system-specific length scale. Rewrite (8.53) as

$$1 + \rho\tilde{h}(k; T, \rho) = \frac{1}{1 - \rho\tilde{c}(k; T, \rho)}$$

$$= \frac{1}{1 - \rho\tilde{c}(0; T, \rho) - \rho\tilde{c}_2(T, \rho)k^2 + O(k^4)}. \tag{10.27}$$

From (10.25) and (10.26) one obtains for the small-k (large-r) asymptotic expansion of $\tilde{h}(k; T, \rho)$:

$$1 + \rho \tilde{h}(k; T, \rho) = \frac{R^{-2}(T, \rho)}{\xi^{-2}(T, \rho) + k^2},\tag{10.28}$$

where the $O(k^4)$ terms in (10.27) have been dropped, which is meaningful provided the range of $c(r; T, \rho)$ is finite, and introduced a new length scale:

$$\xi(T, \rho) = R(T, \rho)\sqrt{\rho k_B T \chi_T(T, \rho)}.\tag{10.29}$$

From (10.28) evaluating the inverse Fourier transform using, e.g., the calculus of residues yields

$$\rho h(r; T, \rho) = \frac{1}{4\pi R^2(T, \rho)} \frac{e^{-r/\xi(T,\rho)}}{r},\tag{10.30}$$

where an irrelevant term proportional to $\delta(r)$ has been deleted. It is seen from (10.30) that $\xi(T, \rho)$ determines the spatial range of $h(r; T, \rho)$. Since at the CP, $R(T, \rho)$ will remain finite but $\chi_T(T, \rho)$ will diverge like $(\delta T)^{-\gamma}$ (cf. (10.12)), one finds from (10.29) that $\xi(T, \rho)$ will diverge like $(\delta T)^{-\nu}$ with $\nu = \gamma/2$, i.e.,

$$\xi \sim (\delta T)^{-\nu}, \quad \nu = \frac{\gamma}{2}.\tag{10.31}$$

Therefore, near a CP, the spatial scale determined by the correlation length $\xi(T, \rho)$ of $h(r; T, \rho)$ will dominate all system-specific length scales, e.g., those related to the pair potential such as $R(T, \rho)$ of (10.26). This, together with a vast body of similar results (on various fluids, binary mixtures, magnetic systems, etc.) suggested to L. Kadanoff in 1968 that the reason why one witnesses scale-invariant properties such as (10.19) is that in the vicinity of a CP only those properties which change on the largest possible scale, namely $\xi(T, \rho)$, are important, i.e., $\xi(T, \rho)$ is the only relevant spatial scale. Since the divergence of $\xi(T, \rho)$ is system independent, it only depends on the existence of a CP, the CP properties must exhibit system-independent or universal features. This observation led Kadanoff to formulate the "universality hypothesis" according to which the value of the critical exponents will only depend on (a) d, the dimensionality of the working space, i.e., they are different for bulk systems ($d = 3$) and for surfaces ($d = 2$) and (b) $\bar{n}$, the number of symmetry-unrelated components of the order parameter characterizing the continuous (second order) phase transition, or if there is no order parameter, the number of diverging length scales. In other words, each pair $(d, \bar{n})$ defines a "universality class" of systems having the same critical exponents.

With respect to the first point (d), it has to be noted that in the above only ($d = 3$) bulk systems have been considered, e.g., when a d-dimensional Fourier transform is used (10.30) will become

$$h(r; T, \rho) \sim \frac{e^{-r/\xi(T,\rho)}}{r^{(d-1)/2}}.\tag{10.32}$$

whereas for the second point $(\bar{n})$, in the above, fluid systems with only one diverging length scale ξ related to the density fluctuations were considered, hence $\bar{n} = 1$. When an order parameter exists, one has to consider instead the correlation lengths of the fluctuations of the independent components of the order parameter. Therefore, one often assimilates, in this context, a fluid system to a system having an "effective" scalar order parameter related to the density difference between the coexisting fluid phases. More details can be found in the specialized literature given in the Further Reading.

10.5   Renormalization Group (RG) Idea

The scaling hypothesis and the universality hypothesis do bring the theory of the CP into agreement with a vast body of theoretical and experimental results. However, they are formulated in terms of some abstract non-classical critical exponents whose value remains undetermined. What is still lacking is a theoretical scheme to calculate their value. Such a calculational technique was introduced in 1971 by K. G. Wilson and is known as the "Renormalization Group"(RG) because it is similar to an earlier technique bearing this name and used in field theory to compute the renormalized coupling constants. This technique not only allows for the critical exponents to be computed but also provides a more profound justification for the scaling and universality assumptions. It can thus be said to "solve" the CP enigma. Unfortunately, although the physical ideas involved are simple, the calculational details are very complex. The basic idea consists in "decimating" successively all the physical information contained in the irrelevant microscopic scales until the relevant macroscopic scale (l) is reached. To this end the original microscopic system is, at each successive step, transformed into a new system with less irrelevant scales. These successive transformations form a mathematical (semi-) group which is the RG. Each such transformation will modify the system's parameters (the parameters setting the scales), and this flow of parameters culminates in a set of differential equations for the relevant parameters which is characteristic of the given RG. Evaluating finally the "fixed points" of these differential equations allows one to describe ultimately the system's behavior at the macroscopically relevant scale. It will be obvious that this technique, as formulated here, is very general and not limited to, although very well adapted for, the study of the CP. In fact, the RG technique can be used to study any problem for which the physically relevant information is independent of the system's microscopic scales. This generality will be used here to illustrate the RG technique on a physical problem closely related, but not identical, to the calculation of a critical exponent. The problem chosen to illustrate the RG technique concerns the calculation of the Flory exponent introduced in (8.142). Many of the steps used in this example can easily be translated into the CP language, but the calculational details turn out to be much simpler.

10.6 Critical Exponents and the Flory Exponent

The basic reason why the critical exponents (β, γ, δ) are non-classical is that the correlation length of the density fluctuations ξ diverges as $\xi \sim (\delta T)^{-\nu}$ $(\nu = \gamma/2)$ near the CP. Therefore, ξ is the only relevant spatial scale for the study of critical phenomena. Any theory which neglects fluctuations on this scale will always yield classical exponents.

A very similar observation can be made with respect to the Flory exponent ν introduced in (8.142) and characterizing the size R_0 of a linear chain of n segments. In Chap. 8 the classical value $\nu = 1/2$ of the Flory exponent was obtained but its evaluation did neglect the excluded volume interactions between the segments. When measuring the distance along the chain of segments, these excluded volume inter-actions will operate not only between nearby segments, i.e., at short distances, but also between any two segments whose distance (along the chain) is comparable to the total length of the chain (cf. Fig. 10.4). The presence of these excluded volume interactions will lead thus to correlations between the position of the segments with a correlation length comparable to the total length of the chain. In the asymptotic regime, where the number of segments tends to infinity, this correlation length will hence diverge. In other words, the relevant scale is not the length of a segment but the total length of the chain. Rewriting the Flory relation, $R_0 \sim n^\nu$ for $n \to \infty$, in the form $R_0 \sim (1/n)^{-\nu}$ for $(1/n) \to 0$, it becomes obvious that the evaluation of the Flory exponent ν is analogous to the evaluation of the critical exponent of $\xi \sim (\delta T)^{-\nu}$ $(\nu = \gamma/2)$ for $\delta T \to 0$, with $1/n$ playing the role of δT. It is indeed found experi-mentally that the Flory exponent ν, although non-classical $(\nu \neq 1/2)$, is universal in the sense that its value does not depend on the specific nature of the segments of the linear chain. The evaluation of the Flory exponent is, thus, a good candidate for illustrating the RG technique. Before doing so, one should nevertheless also point out a difference between the Flory exponent, $R_0 \sim (1/n)^{-\nu}$, and the critical exponent, $\xi \sim (\delta T)^{-\nu}$, because, although the two values of ν are very close and often represented by the same symbol (as here) they are not strictly equal. It can, in fact, be shown that these two problems do not belong to the same universality class which, as stated, is determined by the space dimensionality d and the number of components $\bar{n}$ (not to be confused here with the number of segments n) of the order parameter. Indeed,

Fig. 10.4 Two configurations of a chain of ten segments with an excluded volume interaction between segments: **a** 2 and 5. **b** 1 and 10

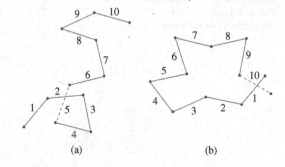

(a)   (b)

in both cases one is interested in $d = 3$, but while ξ corresponds to a scalar (one component) order parameter it can be shown that the evaluation of R_0 corresponds, formally, to an order parameter whose number of components tends to zero.

10.7   RG Calculation of the Flory Exponent

As in Chap. 8, one considers a linear chain of $n - 1 \simeq n$ ($n \gg 1$) identical segments (homopolymer), each of length b, and wants to determine its average size R_0 (or volume $4\pi R_0^3/3$) in terms of the fluctuations, namely $\langle \mathbf{R} \rangle = 0$, $\langle \mathbf{R}^2 \rangle = R_0^2$, of the end-to-end vector $\mathbf{R}$ but averaged now over a statistical ensemble which includes excluded volume interactions between any pair of segments. Let thus v (not to be confused with the specific volume) denote here the strength of the excluded volume interaction between two segments. Moreover, since the strength of the fluctuations of $\mathbf{R}$ becomes weaker when the dimensionality d of the working space increases (see (10.32) for a similar result), a linear chain in a d-dimensional space will henceforth be considered since one can always return to $d = 3$ at the end. To perform the RG calculation an original method proposed by P. G. de Gennes in 1972 is followed.

10.7.1   RG Transformations

Starting from the original chain characterized by the triplet $\{n, b, v\}$, one first tries to eliminate ("decimate") the irrelevant (microscopic) scale b by regrouping formally g successive segments, g being an integer larger than one, into a new "renormalized" segment also called a blob (cf. Fig. 10.5). This transformation, say $\mathcal{T}$, of the original chain $\{n, b, v\}$ will produce a new chain characterized by the triplet $\{n_1, b_1, v_1\}$, where n_1 is the number of new segments, b_1 their length, and v_1 the strength of their excluded volume interaction. One has $n_1 = n/g < n$, while b_1 and v_1 are unknown

Fig. 10.5 Transformation of a chain of six original segments into a renormalized chain of three blobs, each consisting of two original segments

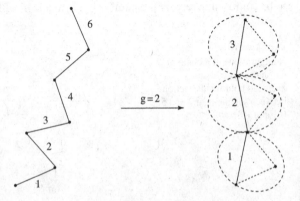

functions of b, v, and g. Without loss of generality, these two functions will be written in the following suggestive form:

$$b_1 = bg^{1/2}[1 + h(g, u)], \quad v_1 = vg^2[1 - q(g, u)], \tag{10.33}$$

where $h = h(g, u)$ and $q = q(g, u)$ are two dimensionless functions of the two dimensionless variables, g and $u = v/b^d$, which can be formed with $\{g, b, v\}$ when the space dimensionality is d, the strength v having the dimension of the volume excluded to a second segment due to the presence of a segment. The suggestive forms used here are such that all the excluded volume effects are contained in $h(g, u)$ and $q(g, u)$. Indeed, when $h(g, u) = 0$ the new chain is expected to behave classically, i.e., $b_1 \sim bg^{1/2}$, while when $h(g, u) = 0$ one expects the excluded volume effects to increase the size of the chain, hence $h(g, u) > 0$. Similarly, since the new segments contain $g(g - 1)/2 \simeq g^2$ pairs of old segments, it should be expected that $v_1 \sim vg^2$, although this probably overestimates the effect, hence one writes $v_1 = vg^2 [1 - q(g, u)]$ with $q(g, u) > 0$.

Applying now the same transformation $\mathcal{T}$ to the new chain $\{n_1, b_1, v_1\}$, a twice renormalized chain $\{n_2, b_2, v_2\}$ is obtained, and iterating $\mathcal{T}$ k-times, i.e., applying $\mathcal{T}^k$ to the original chain $\{n, b, v\}$, one obtains a chain characterized by $\{n_k, b_k, v_k\}$ with $n_k = n_{k-1}/g$,

$$b_k = b_{k-1} g^{1/2}[1 + h(g, u_{k-1})], \quad v_k = v_{k-1} g^2[1 - q(g, u_{k-1})], \tag{10.34}$$

and $k = 1, 2, \ldots, n_0 = n$, $b_0 = b$, $v_0 = v$, together with $u_k = v_k/b_k^d$. The successive transformations, $\mathcal{T}$, form a mathematical semigroup, namely $\mathcal{T}^k \mathcal{T}^l = \mathcal{T}^{k+l}$, which fully characterizes the renormalization operations. Note that this semigroup cannot be completed into a group because the inverse of $\mathcal{T}$ cannot be defined. Indeed, at each transformation the smaller scale b_{k-1} is eliminated (decimated) in favor of the larger ($g > 1$) scale b_k and this process cannot be reversed. This semigroup will nevertheless be designed briefly as the renormalization group (RG).

10.7.2 Fixed Points of the RG

The basic assumption underlying the RG equations (10.34) is that at each step of the renormalization scheme the new chain behaves in the same way as the original chain. This is expressed by the fact that in (10.34) the functions $h(g, u)$ and $q(g, u)$ do not depend explicitly on k, only implicitly via u_{k-1}. This implies a scale invariance of the type expected at a CP. To see this more clearly consider now the "flow" of the dimensionless parameter u_k as k increases. From (10.34) one obtains for $u_k = v_k/b_k^d$:

$$u_k = u_{k-1} g^{\varepsilon/2}[1 - l(g, u_{k-1})], \tag{10.35}$$

where $\varepsilon \equiv 4 - d$, together with

$$1 - l(g, u) = \frac{1 - q(g, u)}{[1 + h(g, u)]^d}, \tag{10.36}$$

so that, for a given value of $d = 4 - \varepsilon$ and g, (10.35) determines the flow $u_{k-1} \to u_k$ in terms of the function $l(g, u)$ defined by (10.36). When k increases, n_k decreases while b_k and v_k increase, so that $u_k = v_k/b_k^d$ may either increase indefinitely or tend to a finite value u^*. Physically, one expects the latter to be the case. The limiting value u^* must then obey (10.35),

$$u^* = u^* g^{\varepsilon/2}\big[1 - l(g, u^*)\big], \tag{10.37}$$

for any g, i.e., when $u_k \to u^*$ the system will have reached a state which is independent of the scale g. Note, however, that k cannot increase beyond the value k^* for which $n_{k^*} = 1$. Nevertheless, since $n \gg 1$ one expects $k^* \gg 1$, and one often assimilates u^* to $\lim_{k \to \infty} u_k$.

Any solution, $u = u^*$, of (10.37) is called a fixed point of the RG. By construction one has $l(g, 0) = h(g, 0) = q(g, 0) = 0$, and, therefore, (10.37) always admits the trivial fixed point $u^* = 0$, corresponding to the classical theory ($v = 1/2$). Any non-trivial fixed point $u^* \neq 0$ must, therefore, be a solution of

$$1 = g^{\varepsilon/2}\big[1 - l(g, u^*)\big], \tag{10.38}$$

and, henceforth, such a solution will be assumed to exist.

10.7.3 Non-classical Flory Exponent

The problem one originally did set out to solve with the aid of the RG technique did concern the relation between R_0 and n for $n \gg 1$, namely $R_0 \sim n^v$. This relation may be formally written as $R_0 = bf(n, u)$, where $f(n, u)$ is unknown except that $f(n, 0) \sim n^{1/2}$, since $u = 0$ corresponds to the classical theory. As the value of R_0 is not affected by the renormalization of the segments, one must have

$$R_0 = bf(n, u) = b_1 f(n_1, u_1) = \ldots = b_k f(n_k, u_k), \tag{10.39}$$

and, therefore,

$$b_{k-1} f(n_{k-1}, u_{k-1}) = b_k f(n_k, u_k), \tag{10.40}$$

which, using (10.34) to evaluate b_k/b_{k-1}, can be transformed into

$$f\left(\frac{n}{g^k}, u_k\right) = f\left(\frac{n}{g^{k-1}}, u_{k-1}\right)\frac{1}{g^{1/2}[1 + h(g, u_{k-1})]}. \tag{10.41}$$

Assume now that k is large enough ($k \gg 1$) so that one may approximate the solution of u_k of (10.35) by the fixed point value u^*. Then (10.41) reduces to

$$f\left(\frac{n}{g^k}, u^*\right) = f\left(\frac{n}{g^{k-1}}, u^*\right)\frac{1}{g^{1/2}[1 + h(g, u^*)]}. \tag{10.42}$$

To solve (10.42) one may write $f(n_k, u^*) = C(u^*)n_k^\nu$, where ν is the Flory exponent. Substituting this power law into (10.42) yields

$$C(u^*)\frac{n^\nu}{g^{k\nu}} = C(u^*)\frac{n^\nu}{g^{(k-1)\nu}}\frac{1}{g^{1/2}[1 + h(g, u^*)]}, \tag{10.43}$$

or

$$\nu = \frac{1}{2} + \frac{\ln[1 + h(g, u^*)]}{\ln g}. \tag{10.44}$$

It is seen from (10.44) that for any non-trivial fixed point $u^* \neq 0$, the Flory exponent ν will take on a non-classical value ($\nu \neq 1/2$), whereas for $u^* = 0$ one recovers the classical value $\nu = 1/2$, since $h(g, 0) = 0$.

10.7.4 Critical Dimension

The search for a non-classical value of ν is thus reduced to the search for a non-trivial fixed point $u^* \neq 0$. The very reason why one did use a d-dimensional presentation is that there exists a particular value of d for which the solution of the fixed point equation (10.38) becomes trivial. Indeed, for $\varepsilon = 0$, or equivalently $d = 4$, (10.38) reduces to $l(g, u^*) = 0$, which implies $u^* = 0$. Hence for the particular value $d = 4$ the only fixed point is the trivial one $u^* = 0$, and the Flory exponent will always be classical. This value of d is called the critical dimension of the present RG. Of course, experiments in $d = 4$ are not possible but, following a suggestion of K. Wilson and M. Fischer (1971), one may try to evaluate ν or u^* by treating $d = 4 - \varepsilon$ as a continuous variable and by performing a perturbation expansion for small ε-values, i.e., by expanding ν in a power series in ε. Theoretically there is, of course, no difficulty in treating d as a continuous variable.

10.7.5 ε-Expansion

Since for $\varepsilon = 0$ the solution of (10.38) is $u^* = 0$, one expects that for $\varepsilon = 0$ but $|\varepsilon| \ll 1$ (10.38) will have a solution such that $|u^*| \ll 1$. For small-ε one may thus expand the basic unknowns, $h(g, u)$, $q(g, u)$, and $l(g, u)$ in a power series in u and retain only the dominant terms, namely

$$
\begin{aligned}
h(g, u) &= C_h u + O(u^2), \\
q(g, u) &= C_q u + O(u^2), \\
l(g, u) &= C_l u + O(u^2),
\end{aligned}
\tag{10.45}
$$

where C_h, C_q, and C_l are constants depending on g and ε but not on u. Note that from (10.36) one has $C_l = C_h + dC_q$, while (10.38) becomes now to dominant order

$$
1 = g^{\varepsilon/2}(1 - C_l u^*).
\tag{10.46}
$$

Rewriting (10.46) as

$$
g^{\varepsilon/2} C_l u^* = g^{\varepsilon/2} - 1
\tag{10.47}
$$

where $g^{\varepsilon/2} - 1$ is a small quantity which may be rewritten as

$$
g^{\varepsilon/2} = 1 + \frac{\varepsilon}{2} \ln g + O(\varepsilon^2),
\tag{10.48}
$$

so that the solution of (10.47) will read to dominant order in ε

$$
u^* = \frac{\varepsilon \ln g}{2 C_l}.
\tag{10.49}
$$

Substituting (10.49) into (10.44) yields then

$$
v = \frac{1}{2} + \frac{\ln(1 + C_h u^*)}{\ln g} \simeq \frac{1}{2} + \frac{C_h u^*}{\ln g} = \frac{1}{2} + \frac{\varepsilon}{2}\left(\frac{C_h}{C_l}\right)_{\varepsilon=0},
\tag{10.50}
$$

which clearly shows that v is non-classical when $\varepsilon = 0$, and hence one expects v to be non-classical also for the physical value $\varepsilon = 1$ or $d = 3$, although it is not obvious that the small ε-expansion (10.50) can still be used when $\varepsilon \to 1$. It is known, in fact, that the perturbation expansion (10.45) diverges when $\varepsilon = 1$, i.e., the excluded volume effects are not small in $d = 3$, so why should things improve when ε is small? To understand why this is the case, some qualitative properties of the RG equations (10.34) are now considered.

10.7.6   Differential RG Equations

In order to study the two basic RG equations, namely

$$b_{k+1} = b_k g^{1/2}[1 + h(g, u_k)], \quad u_{k+1} = u_k g^{\varepsilon/2}[1 - q(g, u_k)], \tag{10.51}$$

in more detail, one first transforms these finite difference equations into differential equations. Although, physically, k and g are integers, the only mathematical condition used until now is $g > 1$. One replaces, therefore, the (finite) elementary step ($k \to k + 1$) of the RG transformation by an infinitesimal one by treating now also g as a continuous variable and writing $g = 1 + \Delta g$, with $\Delta g \ll 1$. If at the step k the length of the renormalized segment was s, then at the next step $k + 1$ this length will increase to $s + \Delta s$. One will thus also replace the discrete variable k by the continuous variable s, namely $b_k \to b(s)$, and $u_k \to u(s)$, and the discrete transformation by $s + \Delta s = \Delta s = (1 + \Delta g) s$ so that $\Delta g = \Delta s / s$ represents now the infinitesimal relative increase in the length s of the segments. In this representation, the original transformation, $b_k \to b_{k+1}$, will thus be replaced by $b(s) \to b(s + \Delta s)$ with $\Delta s = s \Delta g$. Now first rewrite (10.51) as follows:

$$b_{k+1} - b_k = b_k\left(g^{1/2} - 1\right) + b_k g^{\varepsilon/2} h(g, u_k), \tag{10.52}$$

and

$$u_{k+1} - u_k = u_k\left(g^{\varepsilon/2} - 1\right) - u_k g^{\varepsilon/2} q(g, u_k) \tag{10.53}$$

and consider the limit $\Delta g \to 0$. Since

$$b_{k+1} - b_k \to b(s + \Delta s) - b(s) = \Delta s \frac{db(s)}{ds} + O\left((\Delta s)^2\right),$$

the LHS of (10.52) and (10.53) will be first order in Δs. The first term in the RHS of (10.52) and (10.53) will also be first order in Δs, namely

$$b_k\left(g^{1/2} - 1\right) \to b(s)\left((1 + \Delta g)^{1/2} - 1\right) = b(s)\frac{\Delta g}{2} + O\left((\Delta g)^2\right),$$

since $\Delta g = \Delta s / s$. The limit $\Delta g \to 0$ of the second term can, however, not be taken as such. To perform this limit first observe that for small-ε one has $l(g, u_k) = C_l u_k + O\left(u_k^2\right)$, but that according to (10.47) u^* will remain finite for small ε only if one has $C_l \sim g^{1/2} - 1$. Now introduce the new constants $\overline{C}_h$, etc., according to

$$C_i = \left(g^{1/2} - 1\right)\overline{C}_i, \quad (i = h, q, l), \tag{10.54}$$

instead of the constants C_h, etc. of (10.45). Note that in the present notation one has $u^* = 1/\overline{C}_l$, cf. (10.49), $C_h/C_l = \overline{C}_h/\overline{C}_l$, cf. (10.50). Returning now to (10.52) and

(10.53), it is seen that the second term is also first order in Δg because

$$b_k g^{1/2} h(g, u_k) \rightarrow b(s) g^{1/2} \left(g^{\varepsilon/2} - 1\right) \overline{C}_h u$$

and

$$g^{\varepsilon/2} - 1 = \frac{\varepsilon}{2} \Delta g + O\left((\Delta g)^2\right)$$

Therefore, dividing (10.52) and (10.53) by $\Delta s = s\Delta g$ and taking the limit $\Delta g \rightarrow 0$, one obtains finally

$$\frac{db(s)}{ds} = \frac{b(s)}{2s}\left(1 + \varepsilon \overline{C}_h u(s)\right), \tag{10.55}$$

and

$$\frac{du(s)}{ds} = \varepsilon \frac{u(s)}{2s}\left(1 - \overline{C}_l u(s)\right), \tag{10.56}$$

i.e., two differential equations which, for $g - 1$ and e small, are equivalent to the original RG equations (10.51), except that the finite transformation g has been replaced here by an infinitesimal transformation $g = 1 + \Delta g$. Since (10.55) is linear in $b(s)$, it can be easily integrated in terms of $u(s)$. The basic equation is thus (10.56) which is nonlinear in $u(s)$.

10.7.7   Stability of the Fixed Points

Using (10.49) and (10.54) one may rewrite (10.56) as

$$\frac{du(s)}{ds} = \varepsilon \frac{u(s)}{2s}\left(1 - \frac{u(s)}{u^*}\right), \tag{10.57}$$

whose qualitative properties are now considered. The two stationary points of the differential equation (10.57) are seen to coincide with the two fixed points, $u(s) = 0$ and $u(s) = u^*$ of the RG. When u is not a fixed point, $u'(s) \equiv du(s)/ds$ will have the same sign as $\varepsilon u(s)(1 - u(s)/u^*)$ because $s > 0$, while for $s \rightarrow \infty$, $u'(s)$ will decrease as $1/s$. This qualitative relation between $u'(s)$ and $u(s)$, i.e., the so-called phase portrait of the differential equation (10.57), can thus be represented as shown in Fig. 10.6. As seen from the figure, $u'(s)$ changes sign when $u(s)$ is close to a fixed point but the detailed behavior still depends on the sign of ε. Assume, for example, that one starts initially from a value $u(s) = u_0$ close to u^* and considers the case $\varepsilon > 0$. If one increases s from s_0 to s_1 $(s_1 > s_0)$, $u(s)$ will change from $u_0 \equiv u(s_0)$ to $u_1 = u(s_1)$. From the figure it is seen that if $u_1 > u^*$ then $u'(s_1) < 0$ and the differential equation

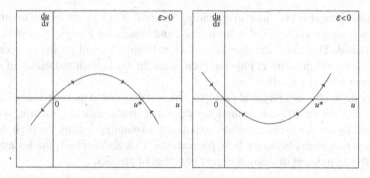

Fig. 10.6 Phase portrait of the differential equation (10.57). The arrows indicate the flow of u in the vicinity of the two fixed points

(10.57) counteracts the further increase of $u(s)$. Similarly, if $u_1 < u^*$, then $u'(s_1) > 0$ and (10.57) also counteracts this decrease of $u(s)$. Therefore, the differential equation (10.57) is such as to prevent the solution $u(s)$ to run away from $u = u^*$. Hence, all the trajectories $u(s)$, i.e., all the solutions of (10.57) corresponding to different initial values u_0, will ultimately tend to the point $u = u^*$. Such a point is called a stable fixed point or an "attractor" of the differential equation (10.57). Note that when $\varepsilon < 0$ this behavior is reversed, i.e., all the trajectories $u(s)$ now run away from $u = u^*$. Such a point is called an unstable fixed point. One can thus say that the fixed point $u = u^*$ is stable when $\varepsilon > 0$ and unstable for $\varepsilon < 0$ while, conversely, the fixed point $u = 0$ is unstable when $\varepsilon > 0$ and stable for $\varepsilon < 0$. This same behavior is also often represented in the flow diagram of $u(s)$ versus s as shown in Fig. 10.7 which represents a few trajectories $u(s)$ here shown versus $1/s$ so that asymptotically $1/s \to 0$. From this figure it is clearly seen that the solutions of (10.57) come in two families separated by a separatrix. When $\varepsilon > 0$, this separatrix is $u = 0$, while it is $u = u^*$ for $\varepsilon < 0$. Translated in the present context this implies that for the physically relevant values of e, i.e., $\varepsilon > 0$, one may start from any initial value, $u_0 > 0$, the

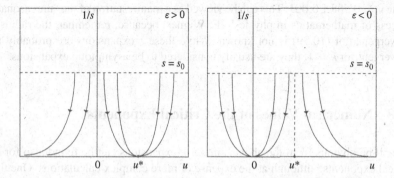

Fig. 10.7 Solutions $u(s)$ of (10.57) for different values $u_0 = u(s_0)$. For convenience $u(s)$ is plotted versus $1/s$ so that the asymptotic state ($s \to \infty$) corresponds here to $1/s \to 0$

asymptotic properties of the corresponding solution $u(s)$ will always be controlled by the non-trivial stable fixed point $u = u^*$, and hence, the Flory exponent will be non-classical. Therefore, the above calculation, based on small positive values of ε and u^*, describes the correct physics, even when the final (positive) values of ε and u^* turn out not to be small.

More generally, any physical problem which is dominated by long-ranged correlations, and whose RG equations have the same mathematical structure, will be described by the same characteristic exponents, explaining hereby the high degree of universality found between these phenomena. This also explains the widespread use of RG techniques in many different branches of physics.

10.7.8 Numerical Value of the Flory Exponent

As one is insured now that (10.50) contains the correct physics of this problem, it is tempting to evaluate it for the physically relevant value $\varepsilon = 1$. This can be done provided one knows $C_h/C_l = \overline{C}_h/\overline{C}_l$ for $g = 1$ and $\varepsilon \to 0$. Since these constants result from the perturbation expansion (10.45), their evaluation is not more difficult than, say, the evaluation of the second and the third virial coefficient for this, or a related, problem. One then finds that $(C_h/C_l)_{\varepsilon=0} = 1/8$, so that (10.50) becomes

$$v = \frac{1}{2} + \frac{1}{16}\varepsilon + O(\varepsilon^2) \tag{10.58}$$

or, if the expansion is extended to second order in ε,

$$v = \frac{1}{2} + \frac{1}{16}\varepsilon + \frac{15}{512}\varepsilon^2 + O(\varepsilon^3). \tag{10.59}$$

Evaluating now the Flory exponent v for $\varepsilon = 1$ ($d = 3$) one finds from (10.58) $v \simeq 0.5625$ and from (10.59) $v \simeq 0.592$, which compare well with the experimental value $v = 0.588 \pm 0.001$. This can be viewed as a manifestation of the "unreasonable success of mathematics in physics" (E. Wigner) because, remember, the radius of convergence of (10.59) is not known. While these ε-expansions are probably not convergent for $\varepsilon = 1$, they are usually considered to be asymptotic expansions.

10.8 Numerical Values of the Critical Exponents

As will be obvious from the above, similar ε-expansions can be found also for the critical exponents, although at the expense of more complex calculations. One then finds to first order in ε that

$$\beta = \frac{1}{2} - \frac{3}{2(\bar{n}+8)}\varepsilon + O\left(\varepsilon^2\right), \tag{10.60}$$

$$\gamma = 1 + \frac{\bar{n}+2}{2(\bar{n}+8)}\varepsilon + O\left(\varepsilon^2\right), \tag{10.61}$$

$$\delta = 3 + \varepsilon + O\left(\varepsilon^2\right), \tag{10.62}$$

for a system which is $d = 4 - \varepsilon$ dimensional and has a CP described by a $\bar{n}$-component order parameter, i.e., for the universality class $(d, \bar{n})$. For instance, for the liquid–vapor CP one obtains by substituting $\varepsilon = 1$ and $\bar{n} = 1$ in (10.60–10.62) that $\beta \simeq 1/3, \gamma \simeq 7/6$, and $\delta \simeq 4$, which are in fair agreement with the experimental values $\beta \simeq 0.3265, \gamma \simeq 1.239$, and $\delta \simeq 4.5$. Note also that the Flory exponent expression of (10.59) corresponds to (10.61), namely $\nu = \gamma/2$, for $\bar{n} \to 0$, a fact first pointed out by P. G. de Gennes (1972). Observe also that (10.60–10.62) satisfy (10.23).

The RG idea can thus be considered to solve the CP enigma in a satisfactory manner although, of course, the very reason of the quantitative success of the expansions (10.60–10.62) for $\varepsilon = 1$ remains somewhat obscure.

Further Reading

1. H.E. Stanley, *Introduction to Phase Transitions and Critical Phenomena* (Clarendon Press, Oxford, 1971). *A classic introduction to critical phenomena.*
2. P.G. de Gennes, *Scaling Concepts in Polymer Physics* (Cornell University Press, New York, 1979). *Contains the RG-method used in the main text as well as many other applications of the RG-idea to polymer physics.*
3. C. Domb, *The Critical Point* (Taylor & Francis, London, 1996). *Provides a complete history of the study of the CP.*
4. F. Schwabl, *Statistical Mechanics* (Springer-Verlag, Berlin, 2002). *Contains alternative RG-theories.*

Chapter 11
Interfaces

Abstract As has been analyzed in Chap. 9, in a discontinuous phase transition two phases of different average density may coexist in equilibrium. In the various examples of coexistence considered in Sects. 9.1 and 9.2, the average densities of the phases are uniform, i.e., the conditions of mechanical equilibrium and chemical equilibrium at a given temperature have been derived from the free energy of each bulk phase (in the thermodynamic limit), and so the density at coexistence changes discontinuously from one phase to the other. For instance, in the liquid–vapor transition the densities at coexistence are uniform, $\rho_L(T)$ and $\rho_V(T)$, according to (9.18). This is only an approximation since, as may be observed experimentally, there are two kinds of surface effects to be considered. The first one is the effect of the walls of the container, which is not specific to the system and which gives rise to the well-known meniscus effects, which may be neglected at points far away from the wall. The second effect is that, due to the gravitational field, the denser phase (the liquid) occupies the lower part of the container, the less dense phase (the vapor) the upper part, and the local equilibrium density $\rho_1(\mathbf{r}) = \rho_1(z)$ will change continuously (albeit abruptly) from one phase to the other. This profile and the discontinuous profile of Chap. 9 are shown schematically in Fig. 11.1 Note that for the continuous profile $\rho_1(z) \simeq \rho_L(T)(z < z_0)$, $\rho_1(z) \simeq \rho_V(T)(z > z_0)$ and in the interface ($z \simeq z_0$) the gradient of the local density is $\rho_1'(z) = d\rho_1(z)/dz \neq 0$. The consequences of the existence of this transition region or interface between the two coexisting phases are studied in this chapter.

11.1 Non-uniform Systems

A system like the one just described is an example of a non-uniform fluid in which the local density of particles at equilibrium $\rho_1(\mathbf{r}) > 0$ is a function of $\mathbf{r}$ normalized as

$$\int d\mathbf{r}\rho_1(\mathbf{r}) = N, \tag{11.1}$$

Fig. 11.1 Schematic representation of the liquid–vapor local density, $\rho_1(z)$, as a function of the height z. The thermodynamic density changes discontinuously at $z = z_0$ from ρ_L to ρ_V (*broken line*) while the local density changes continuously from ρ_L to ρ_V through the interface $(\rho_1'(z) \neq 0)$ (*continuous line*)

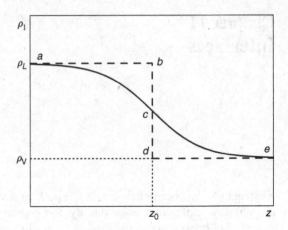

where N is the number of particles. Other examples of non-uniform systems (crystals, quasicrystals, and liquid crystals) have been analyzed in Chap. 8. In non-uniform systems, the intrinsic Helmholtz free energy is a functional of the local density of particles, which is denoted by $\mathcal{F}(T, V, N; [\rho_1])$ or, in abbreviated form, $\mathcal{F}[\rho_1]$ (see Appendix C). According to the mean field theory of Sect. 7.6, if the interaction potential may be written as $V(r) = V_R(r) + V_A(r)$, where $V_R(r)$ and $V_A(r)$ are the repulsive and attractive parts of the interaction, the functional $\mathcal{F}[\rho_1]$ (see (7.78)) may be written as

$$\mathcal{F}[\rho_1] = \mathcal{F}^{id}[\rho_1] + \mathcal{F}_R[\rho_1] + \mathcal{F}_A[\rho_1], \tag{11.2}$$

where

$$\mathcal{F}^{id}[\rho_1] = k_B T \int d\mathbf{r} \rho_1(\mathbf{r}) \big(\ln(\rho_1(\mathbf{r})\Lambda^3) - 1\big) \tag{11.3}$$

is the ideal part, $\mathcal{F}_R[\rho_1]$ is the contribution due to $V_R(r)$, and

$$\mathcal{F}_A[\rho_1] = \frac{1}{2} \int d\mathbf{r} \rho_1(\mathbf{r}) \int d\mathbf{r}' \rho_1(\mathbf{r}') V_A(|\mathbf{r} - \mathbf{r}'|). \tag{11.4}$$

If $V_R(r)$ is a hard-sphere potential, the van der Waals theory of Sect. 8.4.2 may be generalized to a non-uniform system (see (8.74)) and one may approximate the functional $\mathcal{F}_R[\rho_1]$ by

$$\mathcal{F}_R[\rho_1] = -k_B T \int d\mathbf{r} \rho_1(\mathbf{r}) \ln\left(1 - \frac{\rho_1(\mathbf{r})}{\rho_0}\right), \tag{11.5}$$

which implies that

$$1 - \frac{\rho_1(\mathbf{r})}{\rho_0} > 0, \tag{11.6}$$

for all $\mathbf{r}$. From (11.3) and (11.5) one has

$$\mathcal{F}^{\text{id}}[\rho_1] + \mathcal{F}_R[\rho_1] = \int d\mathbf{r} \, \bar{f}_{\text{HS}}(T, \rho_1(\mathbf{r})), \tag{11.7}$$

where

$$\bar{f}_{\text{HS}}(T, \rho) = k_B T \rho \left(\ln(\rho \Lambda^3) - 1 - \ln\left(1 - \frac{\rho}{\rho_0}\right) \right) \tag{11.8}$$

is the free energy density (Helmholtz free energy per unit volume) of a uniform hard-sphere (HS) fluid of density ρ. Observe that the integrand in (11.7) $\bar{f}_{\text{HS}}(T, \rho_1(\mathbf{r}))$ is a function of $\rho_1(\mathbf{r})$ (not a functional) obtained by replacing the uniform density ρ by the local non-uniform density $\rho_1(\mathbf{r})$ in (11.8). Since in this case $\mathcal{F}^{\text{id}}[\rho_1] + \mathcal{F}_R[\rho_1]$ is the integral of a function of the local density $\rho_1(\mathbf{r})$, one says that this approximation of the functional is "local."

Note that, in contrast, $\mathcal{F}_A[\rho_1]$ is "non-local," since it is a functional of $\rho_1(\mathbf{r})$ and $\rho_1(\mathbf{r}')$, that may be written as

$$\mathcal{F}_A[\rho_1] = \frac{1}{2} \int d\mathbf{r} \rho_1(\mathbf{r}) U(\mathbf{r}, [\rho_1]), \tag{11.9}$$

where $U(\mathbf{r}, [\rho_1])$ is the mean potential at r due to the attractive part of the potential, i.e.,

$$U(\mathbf{r}, [\rho_1]) = \int d\mathbf{r}' \rho_1(\mathbf{r}') V_A(|\mathbf{r} - \mathbf{r}'|). \tag{11.10}$$

Since this potential is the sum of a local contribution

$$U_0(\mathbf{r}, [\rho_1]) = \rho_1(\mathbf{r}) \int d\mathbf{r}' V_A(|\mathbf{r} - \mathbf{r}'|), \tag{11.11}$$

and a non-local contribution

$$\delta U(\mathbf{r}, [\rho_1]) = \int d\mathbf{r}' \{\rho_1(\mathbf{r}') - \rho_1(\mathbf{r})\} V_A(|\mathbf{r} - \mathbf{r}'|), \tag{11.12}$$

from (11.7), (11.11), and (11.12) one finally has

$$\mathcal{F}[\rho_1] = \int d\mathbf{r} \bar{f}_0(T, \rho_1(\mathbf{r})) + \frac{1}{2} \int d\mathbf{r} \rho_1(\mathbf{r}) \delta U(\mathbf{r}, [\rho_1]), \tag{11.13}$$

where

$$\bar{f}_0(T, \rho) = k_B T \rho \left(\ln(\rho \Lambda^3) - 1 - \ln\left(1 - \frac{\rho}{\rho_0}\right) \right) - a\rho^2, \qquad (11.14)$$

is the free energy density (Helmholtz free energy per unit volume) of a uniform fluid of density ρ in the van der Waals theory (see (8.81) and (8.82)).

In the study of some applications, it is useful to make a local approximation for the non-local potential $\delta U(\mathbf{r}, [\rho_1])$ of the form

$$\delta U(\mathbf{r}, [\rho_1]) = \int d\mathbf{r}' V_A(|\mathbf{r} - \mathbf{r}'|)((\mathbf{r}' - \mathbf{r}) \cdot \nabla)\rho_1(\mathbf{r})$$

$$+ \int d\mathbf{r}' V_A(|\mathbf{r} - \mathbf{r}'|) \frac{1}{2}((\mathbf{r}' - \mathbf{r}) \cdot \nabla)^2 \rho_1(\mathbf{r}) + \cdots$$

$$= -a_2 \nabla^2 \rho_1(\mathbf{r}) + \cdots, \qquad (11.15)$$

where it has been taken into account that

$$\int d\mathbf{r}'(\mathbf{r}' - \mathbf{r}) V_A(|\mathbf{r} - \mathbf{r}'|) = \int d\mathbf{r}\, \mathbf{r}\, V_A(r) = 0, \qquad (11.16)$$

and

$$\frac{1}{2} \int d\mathbf{r}'(\mathbf{r}' - \mathbf{r})(\mathbf{r}' - \mathbf{r}) V_A(|\mathbf{r} - \mathbf{r}'|) = \frac{1}{2} \int d\mathbf{r}\, \mathbf{r}\, \mathbf{r} V_A(r) = -a_2 I, \qquad (11.17)$$

where I is the unit tensor and a_2 is a positive constant:

$$a_2 = -\frac{1}{6} \int d\mathbf{r}\, r^2 V_A(r), \qquad (11.18)$$

which is finite as long as $V_A(r)$ decreases more rapidly than r^{-5} when $r \to \infty$.

Note that within the local approximation (11.15) the functional (11.13) reads

$$\mathcal{F}[\rho_1] = \int d\mathbf{r} \left(\bar{f}_0(T, \rho_1(\mathbf{r})) - \frac{1}{2} a_2 \rho_1(\mathbf{r}) \nabla^2 \rho_1(\mathbf{r}) \right), \qquad (11.19)$$

and so, if $\nabla \rho_1(\mathbf{r})$ vanishes at the limits of integration, upon integration by parts one has

$$\mathcal{F}[\rho_1] = \int d\mathbf{r} \left(\bar{f}_0(T, \rho_1(\mathbf{r})) + \frac{1}{2} a_2 \{\nabla \rho_1(\mathbf{r})\}^2 \right), \qquad (11.20)$$

which is called the square gradient approximation, since the correction due to the gradients of $\rho_1(\mathbf{r})$ is proportional to $\{\nabla \rho_1(\mathbf{r})\}^2$.

11.2 Density Profile

In the van der Waals theory, the functional (11.13) has to be minimized to obtain the equilibrium local density $\rho_1(\mathbf{r})$ but, due to the restriction (11.1), it is necessary to use a Lagrange multiplier to eliminate this constraint (see Chap. 7). In this way, if μ is the Lagrange multiplier, one has to minimize, now without restriction, the functional $\mathcal{F}[\bar{\rho}_1] - \mu \int d\mathbf{r}\bar{\rho}_1(\mathbf{r})$. Therefore, consider the functional (7.66),

$$A[\bar{\rho}_1] = \mathcal{F}[\bar{\rho}_1] - \mu \int d\mathbf{r}\bar{\rho}_1(\mathbf{r}) + \int d\mathbf{r}\bar{\rho}_1(\mathbf{r})\phi(\mathbf{r}), \qquad (11.21)$$

where $\phi(\mathbf{r})$ is an external potential. The extremum $\rho_1(\mathbf{r})$ of (11.21) is obtained from

$$\left(\frac{\delta\mathcal{F}[\bar{\rho}_1]}{\delta\bar{\rho}_1(\mathbf{r})}\right)_{\bar{\rho}_1(\mathbf{r})=\rho_1(\mathbf{r})} - \mu + \phi(\mathbf{r}) = 0, \qquad (11.22)$$

which using (11.12–11.14) may be written as

$$\mu = \phi(\mathbf{r}) + k_B T \left(\ln\left(\frac{\rho_1(\mathbf{r})\Lambda^3}{1 - b\rho_1(\mathbf{r})}\right) + \frac{b\rho_1(\mathbf{r})}{1 - b\rho_1(\mathbf{r})}\right) + \int d\mathbf{r}'\rho_1(\mathbf{r}')V_A(|\mathbf{r} - \mathbf{r}'|)$$

$$(11.23)$$

where $b = 1/\rho_0$.

Since

$$\mu_{\mathrm{HS}}(T, \rho) = \frac{\partial \bar{f}_{\mathrm{HS}}(T, \rho)}{\partial \rho}, \qquad (11.24)$$

and

$$\mu_0(T, \rho) = \frac{\partial \bar{f}_0(T, \rho)}{\partial \rho} = \mu_{\mathrm{HS}}(T, \rho) - 2a\rho, \qquad (11.25)$$

(11.23) for the equilibrium profile $\rho_1(\mathbf{r})$ may be expressed in different ways as

$$\mu = \phi(\mathbf{r}) + \mu_{\mathrm{HS}}(T, \rho_1(\mathbf{r})) + \int d\mathbf{r}'\rho_1(\mathbf{r}')V_A(|\mathbf{r} - \mathbf{r}'|), \qquad (11.26)$$

$$\mu = \phi(\mathbf{r}) + \mu_0(T, \rho_1(\mathbf{r})) + \delta U(\mathbf{r}, [\rho_1]), \qquad (11.27)$$

or, alternatively, in the local approximation of $\delta U(\mathbf{r}, [\rho_1])$,

$$\mu = \phi(\mathbf{r}) + \mu_0(T, \rho_1(\mathbf{r})) - a_2 \nabla^2 \rho_1(\mathbf{r}). \qquad (11.28)$$

Note that when solving any of these equations, the parameter μ must be chosen in such a way that the normalization condition (11.1) is verified.

As a simple application of a non-uniform system in equilibrium, consider the liquid–vapor coexistence with a planar interface, i.e., $\rho_1(\mathbf{r}) = \rho_1(z)$, where z is the coordinate perpendicular to the planar interface, so that the density profile behaves qualitatively as the continuous profile of Fig. 11.1. Note that although (11.26) is a nonlinear integral equation for $\rho_1(z)$ which has to be solved numerically, in the local approximation it is replaced by the differential Eq. (11.28). The potential $\phi(\mathbf{r})$ in (11.28) is the sum of two contributions. The first, $\phi_R(\mathbf{r})$, is the one that keeps the system inside the container and whose effect may be neglected at points far away from the walls. The second contribution, $\phi(z)$, is the gravitational potential responsible for the spatial separation of the phases. If this potential is weak ($\phi(z) \to 0$), it may be neglected in (11.28) and replaced by adequate boundary conditions. Therefore, the equation that determines the density profile when the interface is planar is given by

$$\mu = \mu_0(T, \rho_1(z)) - a_2 \rho_1''(z), \tag{11.29}$$

where a prime denotes a derivative with respect to the argument, which is an ordinary second-order differential equation whose solution has to satisfy the boundary conditions

$$\lim_{z \to -\infty} \rho_1(z) = \rho_L, \quad \lim_{z \to \infty} \rho_1(z) = \rho_V, \tag{11.30}$$

and

$$\lim_{z \to -\infty} \rho_1'(z) = 0, \quad \lim_{z \to \infty} \rho_1'(z) = 0, \tag{11.31}$$

which express that $\rho_1(z)$ has to smoothly turn over into the liquid ($z = -\infty$) or the vapor ($z = \infty$) bulk phase. If this is the case then, $\rho_1''(\pm\infty) = 0$, and, when $z = \pm\infty$, (11.29) reduces, according to (11.30) and (11.31), to

$$\mu = \mu_0(T, \rho_L) = \mu_0(T, \rho_V), \tag{11.32}$$

which expresses the equality of the chemical potentials of the liquid and vapor phases. Note that since μ in (11.32) is equal to the chemical potential of both phases, this parameter may be interpreted as being the chemical potential of the non-uniform system which, according to (11.29), is constant all over the system. Observe that (11.29) contains two positive constants a and a_2, given by (11.14) and (11.18), which yield a characteristic length l:

$$l^2 \equiv \frac{a_2}{a}, \tag{11.33}$$

which, together with the parameter $b = 1/\rho_0$, completely determines the van der Waals theory of planar interfaces. Introducing the dimensionless variables

$$x = \frac{z}{l}, t = \frac{k_B T b}{a}, \eta(x) = b\rho_1(z),$$
(11.34)

(11.29) may be cast in the form

$$\bar{\mu}(t, \eta(x)) - \bar{\mu}(t, \eta_{L,V}) - \frac{1}{t}\eta''(x) = 0,$$
(11.35)

where $\eta_{L,V}$ denotes indistinctly η_L or η_V and $\bar{\mu}(t, \eta)$ has been defined as

$$\bar{\mu}(t, \eta) = \ln\left(\frac{\eta}{1 - \eta}\right) + \frac{\eta}{1 - \eta} - \frac{2}{t}\eta = \frac{\partial \bar{f}(t, \eta)}{\partial \eta},$$
(11.36)

with

$$\bar{f}(t, \eta) = \eta\left(\ln\left(\frac{\eta}{1 - \eta}\right) - 1\right) - \frac{1}{t}\eta^2.$$
(11.37)

Multiplying (11.35) by $\eta'(x)$, the resulting equation may be written as

$$\frac{\partial}{\partial x}\left(\bar{f}(t, \eta(x)) - \eta(x)\bar{\mu}(t, \eta_{L,V}) - \frac{1}{2t}\{\eta'(x)\}^2\right) = 0,$$
(11.38)

i.e.,

$$\bar{f}(t, \eta(x)) - \eta(x)\bar{\mu}(t, \eta_{L,V}) - \frac{1}{2t}\{\eta'(x)\}^2 = C_{L,V},$$
(11.39)

where the constants $C_{L,V}$ are obtained from the determination of the limits $x \to \pm\infty$, namely

$$C_{L,V} = \bar{f}(t, \eta_{L,V}) - \eta_{L,V}\bar{\mu}(t, \eta_{L,V}) = -\frac{\eta_{L,V}}{1 - \eta_{L,V}} + \frac{1}{t}\eta_{L,V}^2.$$
(11.40)

Note that $C_L = C_V$, since $-\bar{f}(t, \eta) + \eta\bar{\mu}(t, \eta)$ is proportional to the pressure, which is the same in both phases. Therefore, from (11.40) it follows that the equation for the density profile (11.39) reads

$$\bar{f}(t, \eta(x)) - \bar{f}(t, \eta_{L,V}) - (\eta(x) - \eta_{L,V})\frac{\partial \bar{f}(t, \eta_{L,V})}{\partial \eta_{L,V}} - \frac{1}{2t}\{\eta'(x)\}^2 = 0.$$
(11.41)

In the vicinity of the critical point $(T < T_c)$ the variables η_L, η_V, and $\eta(x)$ do not differ much from $\eta_c = 1/3$ (see (9.68)) and the gradients of $\rho_1(z)$ are small, and hence the expansion (11.15) converges whenever $V_A(r)$ is short ranged as indicated in (11.18). Therefore, in the critical region all the terms of (11.41) are small. Note, further, that the expression

$$
\Delta \bar{f}(T, \eta(x)) = \bar{f}(t, \eta(x)) - \left(\bar{f}(t, \eta_{L,V}) + (\eta(x) - \eta_{L,V}) \frac{\partial \bar{f}(t, \eta_{L,V})}{\partial \eta_{L,V}} \right),
$$

$$(11.42)$$

that appears in (11.41) is the difference between $\bar{f}(t, \eta)$ evaluated at $\eta = \eta(x)$ and the Maxwell double tangent construction on this function between $\eta(x) = \eta_L$ and $\eta(x) = \eta_V$. Since $\bar{f}(t, x)$ is the free energy density in the van der Waals theory, this function has a loop when $T < T_c$, and so $\Delta \bar{f}(T, \eta(x)) > 0$ when $\eta(x) \neq \eta_{L,V}$ and $\Delta \bar{f}(T, \eta(x)) = 0$ at $\eta(x) = \eta_{L,V}$. These results are consistent with (11.41), which may be written as

$$
\eta'(x) = \pm \sqrt{2t \Delta \bar{f}(T, \eta(x))}.
$$

$$(11.43)$$

When $T \simeq T_c (T < T_c)$, $\Delta \bar{f}(T, \eta)$ may, therefore, be approximated, according to the comments below (11.42), by the function

$$
\Delta \bar{f}(T, \eta) = c(\eta - \eta_L)^2 (\eta - \eta_V)^2,
$$

$$(11.44)$$

where c is a positive constant. Note that with the approximation (11.44) the derivative $\partial \Delta \bar{f}(T, \eta)/\partial \eta$ vanishes at the points η_L, η_V, and $(\eta_L + \eta_V)/2$, and hence the constant c may be determined by imposing the condition that (11.44) verifies (11.41) for the particular value $\eta = (\eta_L + \eta_V)/2$. Upon substitution of (11.44) into (11.43) and subsequent integration, one finds

$$
\ln \left(\frac{\eta(x) - \eta_V}{\eta_L - \eta(x)} \right) = \pm \sqrt{2ct} (\eta_L - \eta_V)(x - x_0),
$$

$$(11.45)$$

where x_0 is a constant of integration whose value will be determined later on, so that, in terms of the variables z and $\rho_1(z)$, eliminating $\eta(x)$ in (11.45) leads to

$$
\rho_1(z) = \frac{1}{2}(\rho_L + \rho_V) - \frac{1}{2}(\rho_L - \rho_V) \tanh \left(\frac{z - z_0}{2\xi} \right),
$$

$$(11.46)$$

where ξ is a new characteristic length defined by

$$
\xi = \frac{l}{b(\rho_L - \rho_V)\sqrt{2ct}}.
$$

$$(11.47)$$

Equation (11.46) is the density profile of a planar liquid–vapor interface in the van der Waals theory. Note that, due to the local approximation (11.28), $\rho_1(z)$ decreases exponentially to ρ_L when $z \to \infty$ and to ρ_V when $z \to \infty$. On the other hand, when $\rho_1(z)$ is obtained from (11.26) or (11.27), the density profile depends on the particular form of $V_A(r)$ and not only of the moments of the attractive potential (a and a_2) as in (11.46). In spite of the approximations that have been introduced, (11.46) provides a good fit for the experimental results and its form is similar to the continuous profile of Fig. 11.1. The characteristic length ξ, which diverges at the critical point, is a measure of the width of the interface. Note that $\rho(z_0) = (\rho_L + \rho_V)/2$, although the van der Waals theory does not allow one to predict the value of z_0. In fact, since (11.29) depends on z only through the density profile, if $\rho_1(z)$ is a solution to the equation, so is $\rho_1(z - z_0)$ for any z_0. When the density profile (11.46) is replaced by a discontinuous profile, as done in thermodynamics, the discontinuity is located in a plane $z = z_0$. For this plane one generally chooses the so-called zero adsorption Gibbs surface, which is defined by the following equation,

$$\int_{-\infty}^{z_0} dz\{\rho_L - \rho_1(z)\} = \int_{z_0}^{\infty} dz\{\rho_1(z) - \rho_V\}, \tag{11.48}$$

and which implies that the areas abc and cde of Fig. 11.1 are equal, as is the case of (11.46). Note that the choice of the Gibbs surface (11.48) to locate the interface is an extra-thermodynamic hypothesis.

11.3 Pressure Profile

In thermodynamics, the liquid–vapor interface is viewed as a discontinuous surface characterized by a surface tension coefficient γ. In the case of a spherical interface of radius R (a drop of liquid), the pressure profile has been represented in Fig. 11.2 and the pressure difference $p_L - p_V$ is given by the Laplace equation $p_L - p_V = 2\gamma/R$. Note the similarity with Fig. 11.1, but with the difference that $p_L - p_V \to 0$ when $R \to \infty$, while $\rho_L - \rho_V \neq 0$ even at a planar interface (corresponding to $R = \infty$).

As will be shown in this section, the real pressure profile is not discontinuous but rather it is continuous, as shown in Fig. 11.3 (spherical and planar interfaces). Note that in the interface the local pressure becomes negative, which gives rise to a surface or interfacial "tension" (negative pressure). This occurs even at a planar interface (Fig. 11.3) and so one has $\gamma \neq 0$ even when $R = \infty$. The pressure profile of a planar interface is determined in this section while the determination of the surface tension coefficient and the analysis of the relation between the planes $z = z_0'$ and $z = z_0$ of Figs. 11.3 and 11.1 are carried out in the next section.

In order to study the pressure profile across a planar interface, consider once more the van der Waals theory of Sect. 11.1. In the first place, it is necessary to introduce the concept of local pressure in a non-uniform system for which, as usual, this field

Fig. 11.2 Schematic representation of the discontinuous local pressure $p(\mathbf{r})$ of a spherical liquid drop of radius R as a function of the distance r to the center of the sphere. p_L and p_V are the pressures of the liquid and the vapor, respectively, whose difference is given by the Laplace equation $p_L - p_V = 2\gamma/R$, where γ is the surface tension coefficient

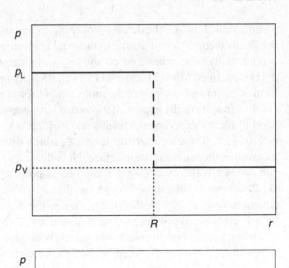

Fig. 11.3 Schematic representation of the continuous local pressure $p(r)$ of a spherical liquid drop of radius R as a function of the distance r to the center of the sphere and of the continuous local pressure $p(z)$ of a planar interface as a function of the height z. p_L and p_V are the pressures of the liquid and the vapor, respectively. Note that in both cases the local pressure becomes negative in the interfacial region

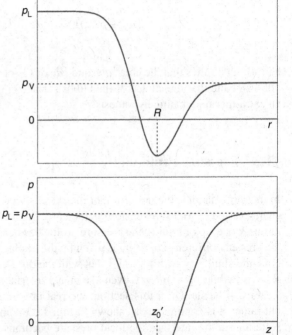

is defined through the linear relation in an equilibrium system between the "work of deformation" (or variation of the free energy) and the "deformation" variable. The latter, up to a change of sign, is the differential change of volume of the system which is generally obtained via an infinitesimal displacement of one of the boundaries of the region that contains the system (at constant T and N). Therefore,

$$(\delta\mathcal{F}[\rho_1])_{T,N} = -\int_A d\mathbf{S} \cdot \delta\mathbf{u}(S)p(S), \qquad (11.49)$$

where $(\delta\mathcal{F}[\rho_1])_{T,N}$ is the variation of the free energy, $p(S)$ is the local pressure acting on the surface S, $\delta\mathbf{u}(S)$ is the infinitesimal displacement of a surface element dS, and the integral extends over the area A of the closed surface of the container. Since $d\mathbf{S}$ is a surface element oriented in the normal direction and pointing outwards of the system, (11.49) may be written as

$$(\delta\mathcal{F}[\rho_1])_{T,N} = -\int_A dS\delta u_n(S)p(S), \qquad (11.50)$$

where $\delta u_n(S)$ is the normal component of the displacement. Note that in the case of a uniform system, $p(S) = p$ and $\delta u_n(S) = \delta u_n$ (independent of S), and so (11.50) reduces to

$$(\delta\mathcal{F}[\rho_1])_{T,N} = -p\int_A dS\delta u_n = -pA\delta u_n = -pdV, \qquad (11.51)$$

which is the thermodynamic relation (2.27). Equation (11.50) is, therefore, the generalization of (11.51) to non-uniform systems. Since $p(S)$ in (11.50) is a thermodynamic definition of the local pressure, one needs a statistical theory to determine it. Note that if the potential that keeps the particles in the closed region $\phi_R(\mathbf{r})$ changes (at constant T and N) by $\delta\phi_R(\mathbf{r})$, then upon application of (11.21) to an equilibrium system one has

$$\left(\frac{\delta\mathcal{F}[\rho_1]}{\delta\phi_R(\mathbf{r})}\right)_{T,N} = \rho_1(\mathbf{r}), \qquad (11.52)$$

and thus (see Appendix C)

$$(\delta\mathcal{F}[\rho_1])_{T,N} = \int d\mathbf{r}\rho_1(\mathbf{r})\delta\phi_R(\mathbf{r}). \qquad (11.53)$$

On the other hand, from (11.26) it follows that

$$\delta\phi_R(\mathbf{r}) = -\frac{\partial\mu_{HS}(T, \rho_1(\mathbf{r}))}{\partial\rho_1(\mathbf{r})}\delta\rho_1(\mathbf{r}) - \int d\mathbf{r}'\delta\rho_1(\mathbf{r}')V_A(|\mathbf{r} - \mathbf{r}'|), \qquad (11.54)$$

where $\delta\rho_1(\mathbf{r})$ is the variation of the equilibrium density profile corresponding to $\delta\phi_R(\mathbf{r})$. In order to determine the local pressure on the plane $x = 0$, note that a local deformation of this plane $x \to x + \delta u_n(y, z)$ produces a change in the density $\rho_1(x, y, z) \to \rho_1(x + \delta u_n(y, z), y, z)$, namely

$$\delta\rho_1(\mathbf{r}) = \frac{\partial\rho_1(\mathbf{r})}{\partial x}\delta u_n(y, z), \tag{11.55}$$

which after substitution into (11.53) and (11.54) yields

$$(\delta\mathcal{F}[\rho_1])_{T,N} = -\int d\mathbf{r}\rho_1(\mathbf{r})\left\{\frac{\partial\mu_{HS}(T, \rho_1(\mathbf{r}))}{\partial\rho_1(\mathbf{r})}\frac{\partial\rho_1(\mathbf{r})}{\partial x}\delta u_n(x, y)\right.$$
$$\left. + \int d\mathbf{r}'\frac{\partial\rho_1(\mathbf{r}')}{\partial x'}\delta u_n(y', z')V_A(|\mathbf{r} - \mathbf{r}'|)\right\}. \tag{11.56}$$

On the other hand, (11.50) reads

$$(\delta\mathcal{F}[\rho_1])_{T,N} = -\int dy\int dz\delta u_n(y, z)p(y, z), \ (x = 0), \tag{11.57}$$

and so, comparing (11.56) and (11.57), it follows that

$$p(y, z) = \int\limits_{-\infty}^{\infty} dx\rho_1(\mathbf{r})\frac{\partial\mu_{HS}(T, \rho_1(\mathbf{r}))}{\partial\rho_1(\mathbf{r})}\frac{\partial\rho_1(\mathbf{r})}{\partial x} \tag{11.58}$$

$$+ \int\limits_{-\infty}^{\infty} dx\frac{\partial\rho_1(\mathbf{r})}{\partial x}\int d\mathbf{r}'\rho_1(\mathbf{r}')V_A(|\mathbf{r} - \mathbf{r}'|), \tag{11.59}$$

or, alternatively, since

$$\rho\frac{\partial\mu_{HS}(T, \rho)}{\partial\rho} = \frac{\partial p_{HS}(T, \rho)}{\partial\rho}, \tag{11.60}$$

where $p_{HS}(T, \rho)$ is the pressure of a hard-sphere fluid, one finally obtains

$$p(y, z) = \int\limits_{-\infty}^{\infty} dx\frac{\partial p_{HS}(T, \rho_1(\mathbf{r}))}{\partial x}$$

$$+ \int\limits_{-\infty}^{\infty} dx\frac{\partial\rho_1(\mathbf{r})}{\partial x}\int d\mathbf{r}'\rho_1(\mathbf{r}')V_A(|\mathbf{r} - \mathbf{r}'|). \tag{11.61}$$

Fig. 11.4 In order to define the local pressure $p(z)$ in a liquid–vapor interface whose local density of particles is $\rho_1(z)$ one introduces a virtual plane perpendicular to the planes of constant density. This plane (the plane $x = 0$) divides the infinite system into two parts: $x < 0$ (empty half-space) and $x > 0$ (occupied half-space)

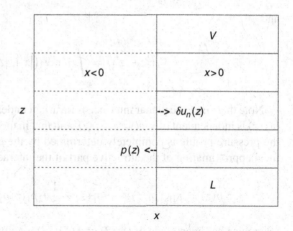

In order to apply (11.61) to a planar liquid–vapor interface one has, first, to create a boundary in the infinite system. To that end, one removes the fluid contained in the half-space $x < 0$ and the local pressure exerted by the half-space $x < 0$ on the half-space $x > 0$ is determined through (11.59) with (Fig. 11.4):

$$\rho_1(\mathbf{r}) = \rho_1(z)\Theta(x), \tag{11.62}$$

where $\Theta(x)$ is the Heaviside step function. Substituting (11.61) into (11.59), one has

$$p(z) = p_{HS}(\rho_1(z))$$
$$+ \int_{-\infty}^{\infty} dx \rho_1(z)\delta(x) \int d\mathbf{r}' \rho_1(z')\Theta(x')V_A(|\mathbf{r} - \mathbf{r}'|), \tag{11.63}$$

or, alternatively,

$$p(z) = p_{HS}(\rho_1(z)) + \rho_1(z) \int_0^{\infty} dx' \int_{-\infty}^{\infty} dz' \rho_1(z')V_2(|x'|, |z - z'|)$$

$$= p_{HS}(\rho_1(z)) + \frac{1}{2}\rho_1(z) \int_{-\infty}^{\infty} dz' \rho_1(z')V_1(|z - z'|), \tag{11.64}$$

where

$$V_2(|x'|, |z - z'|) = \int_{-\infty}^{\infty} dy\, V_A\left(\sqrt{|x'|^2 + y^2 + (z - z')^2}\right), \tag{11.65}$$

and

$$V_1(|z - z'|) = \int_{-\infty}^{\infty} dx' V_2(|x'|, |z - z'|). \tag{11.66}$$

Note that since the planar interface is invariant under translations along the y-axis, $p(y, z)$ is independent of y, $p(y, z) = p(z)$, and that in the van der Waals approximation the pressure profile is completely determined by the density profile. Finally, in the local approximation of the attractive part of the interaction, (11.64) reads

$$p(z) = p_{\mathrm{HS}}(\rho_1(z)) - a\rho_1^2(z) - \frac{1}{2}a_2\rho_1(z)\rho_1''(z) + \cdots, \tag{11.67}$$

where use has been made of (8.82) and (11.18), or, alternatively,

$$p(z) = p_0(\rho_1(z)) + \frac{1}{2}\rho_1(z)[\mu_0(\rho_{L,V}) - \mu_0(\rho_1(z))], \tag{11.68}$$

where $p_0(\rho)$ is the van der Waals pressure (8.84) and $\mu_0(\rho)$ is the chemical potential (11.25). Note that when $z \to \pm\infty$, $p(z)$ tends to $p_0(\rho_L)$ or to $p_0(\rho_V)$, while in an interval of z around the interface, $p(z)$ is negative, i.e., the "pressure" transforms into a "tension" (see Fig. 11.3).

11.4   Surface Tension

As has been analyzed in the previous sections, the planar liquid–vapor interface is characterized by the density profile $\rho_1(z)$ and by the pressure profile $p(z)$. Since $\rho_1'(z)$ and $p'(z)$ are different from zero only in the interface, whose width (far from the critical point) is of a few molecular diameters, in macroscopic theories these profiles are replaced by discontinuous profiles. In this section the relation between both descriptions is examined.

If one denotes by $\bar{\rho}_1(z)$ and $\bar{p}(z)$ the discontinuous profiles (which in macroscopic theories replace the continuous profiles $\rho_1(z)$ and $p(z)$), one may write

$$\bar{\rho}_1(z) = \rho_V\Theta(z - z_0) + \rho_L\Theta(z_0 - z) + \alpha\delta(z - z_0), \tag{11.69}$$

$$\bar{p}(z) = p_V\Theta(z - z_0') + p_L\Theta(z_0' - z) - \gamma\delta(z - z_0'), \tag{11.70}$$

where ρ_L and p_L (ρ_V and p_V) indicate the density and the pressure of the liquid (vapor) and z_0 and z_0' are the planes where the discontinuities are supposed to be located. These surfaces are usually referred to as "the dividing surface" ($z = z_0$) and "the surface of tension" ($z = z_0'$). In macroscopic theories a surface variable is associated to each of these surfaces, the adsorption coefficient α to the dividing surface and the surface tension coefficient γ to the surface of tension. If (11.70) is substituted into (11.57), one has

$$(\delta \mathcal{F}[\rho_1])_{T,N} = -p_V \int\limits_{z_0'}^{\infty} dz \int\limits_{-\infty}^{\infty} dy \delta u_n(y, z)$$

$$- p_L \int\limits_{-\infty}^{z_0'} dz \int\limits_{-\infty}^{\infty} dy \delta u_n(y, z)$$

$$+ \gamma \int\limits_{-\infty}^{\infty} dy \delta u_n(y, z_0')$$

$$= -p_V dV_V - p_L dV_L + \gamma dA, \tag{11.71}$$

where dV_V and dV_L are the volume changes of the vapor and liquid phases due to the deformation $\delta u_n(y, z)$ of the boundary surface and dA is the variation of the surface area of the interface. This equation is the generalization to a two-phase system of the thermodynamic relation $(\delta \mathcal{F}[\rho_1])_{T,N} = -pdV$. Since

$$\Theta'(x) = \delta(x), \, x\delta(x) = 0, \, x\delta'(x) = -\delta(x),$$

from (11.70) it follows that

$$\bar{p}'(z) = (p_V - p_L)\delta(z - z_0') - \gamma \delta'(z - z_0'), \tag{11.72}$$

namely

$$(z - z_0')\bar{p}'(z) = \gamma \delta(z - z_0'), \tag{11.73}$$

and

$$(z - z_0')^2 \bar{p}'(z) = 0. \tag{11.74}$$

Integrating (11.72–11.74) one has

$$\int\limits_{-\infty}^{\infty} dz \, \bar{p}'(z) = p_V - p_L = 0, \tag{11.75}$$

$$\int\limits_{-\infty}^{\infty} dz(z - z_0')\bar{p}'(z) = \gamma, \tag{11.76}$$

$$\int_{-\infty}^{\infty} dz (z - z_0')^2 \bar{p}'(z) = 0,$$ (11.77)

where in (11.75) use has been made of the fact that the pressures of the phases are equal at equilibrium. From these equations it finally follows that

$$\gamma = \int_{-\infty}^{\infty} dz \, z \, \bar{p}'(z),$$ (11.78)

and

$$2\gamma z_0' = \int_{-\infty}^{\infty} dz \, z^2 \, \bar{p}'(z),$$ (11.79)

which are the expressions of γ and of z_0' in terms of the first two moments of the pressure profile $\bar{p}(z)$.

If one performs a similar calculation with the density profile (11.69), one finds

$$\int_{-\infty}^{\infty} dz \bar{\rho}_1'(z) = \rho_V - \rho_L,$$ (11.80)

$$\int_{-\infty}^{\infty} dz (z - z_0) \bar{\rho}_1'(z) = -\alpha,$$ (11.81)

$$\int_{-\infty}^{\infty} dz (z - z_0)^2 \bar{\rho}_1'(z) = 0,$$ (11.82)

from which it follows that

$$z_0(\rho_V - \rho_L) - \alpha = \int_{-\infty}^{\infty} dz \, z \bar{\rho}_1'(z),$$ (11.83)

and

$$z_0^2(\rho_V - \rho_L) - 2z_0\alpha = \int_{-\infty}^{\infty} dz \, z^2 \, \bar{\rho}_1'(z).$$ (11.84)

In thermodynamics it is usually assumed that the molecules are either in the liquid phase or in the vapor phase so that there are no particles in the dividing surface, i.e., $\alpha = 0$. Note that in the case of a continuous profile $\rho_1(z)$, as a generalization of (11.80) and (11.83), one may define the zero adsorption ($\alpha = 0$) Gibbs surface $z = z_0$ by the equation

$$z_0 = \frac{\int_{-\infty}^{\infty} dz z \rho_1'(z)}{\int_{-\infty}^{\infty} dz \rho_1'(z)}, \qquad (11.85)$$

which is equivalent to (11.48), while, in analogy with (11.78) and (11.79), the surface tension coefficient γ and the surface of tension $z = z_0'$ are defined by the following equations:

$$\gamma = \int_{-\infty}^{\infty} dz z p'(z). \qquad (11.86)$$

$$z_0' = \frac{1}{2} \frac{\int_{-\infty}^{\infty} dz z^2 p'(z)}{\int_{-\infty}^{\infty} dz z p'(z)}. \qquad (11.87)$$

Using the van der Waals theory it may be shown that with the expressions (11.85) and (11.87) $z_0 \neq z_0'$, so that the location of the discontinuity in the macroscopic theory depends on the variable that is being considered. The difference $z_0 - z_0'$, called the Tolman length, although being small, reflects the fundamental inadequacy of the discontinuous description of an interface.

From an experimental point of view, γ may be determined from (11.71) and the result compared with the one obtained in (11.86). For instance, from (11.68) one has (since $\partial p / \partial \rho = \rho \partial \mu / \partial \rho$)

$$p'(z) = \frac{1}{2} \rho_1'(z) \left(\mu_0(\rho_{L,V}) - \mu_0(\rho_1(z)) + \frac{\partial p_0(\rho_1(z))}{\partial \rho_1(z)} \right), \qquad (11.88)$$

a result which when substituted into (11.86) allows one to obtain γ. Note that, according to (11.46), $\rho_1'(z)$ tends to zero when $\rho_L - \rho_V \to 0$. From (11.86) it follows then that γ vanishes on approaching the critical point from coexistence, as shown in Fig. 11.5.

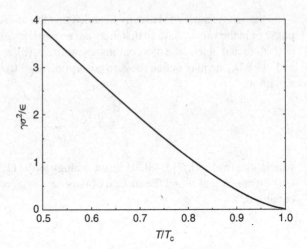

Fig. 11.5 Dimensionless variable $\gamma\sigma^2/\varepsilon$, where γ is the surface tension coefficient, as a function of the reduced temperature T/T_c of a planar liquid–vapor interface in the van der Waals theory. The interaction potential is the potential of hard spheres of diameter σ plus the attractive part of a Lennard–Jones potential of parameters ε and σ. T_c is the critical temperature

Further Reading

1. J.S. Rowlinson, B. Widom, *Molecular Theory of Capillarity* (Oxford University Press, Oxford, 1982). *A classic introduction to the theory of interfaces.*
2. R. Lovett, M. Baus, *Advances in Chemical Physics*, vol. 102 (J. Wiley, New York, 1997). *Contains the methodology used in the main text to compute the pressure profile across an interface.*

Chapter 12
Topological Defects and Topological Phase Transitions

Abstract As was seen in Chap. 8 many bulk states of matter exhibit a, partial or full, orientational and/or translational order. Until now this order has always been considered to be perfect although in practice, many bulk phases exhibit defects, i.e., domains where this order departs from the one originally postulated. Such defects are usually very stable, because the free energy cost to remove them easily exceeds the thermal energy and prevent the system to reach its true defect-free equilibrium state. In some cases they are induced by the system's boundary conditions and are hence, unavoidable; i.e., they correspond to the true equilibrium state given the applied boundary conditions.

In most cases these defects correspond to an elastic deformation of the ideal structure. In the context of soft matter, where by definition of "soft" the elastic constants are weak, the number of observable defect structures is extremely large. Defects are usually classified according to their topological nature: point defects, line defects, etc. Whereas point defects are easily visualized, this is not the case of the line defects because the latter are spatially extended structures. Point defects are often encountered in crystalline structures under the form of vacancies (a crystal node without particle) or interstitials (a particle not attached to a crystal mode). The study of point defects is presently well understood and requires no new concepts. On the contrary, the study of line defects, which are very frequent in liquid crystals, is much more difficult and not yet fully understood. Line defects produce a texture in many liquid crystals and the optical observation of this texture, which is typical of a particular type of liquid crystal, is often used as a means to identify the liquid crystal type. By "texture" one should understand here the spatial organization of the different line defects. Many different liquid crystal textures have been observed but, for simplicity, only the study of line defects, and their texture, in uniaxial nematic liquid crystals will be considered here.

Topological defects also play an important role in physical systems with a reduced spatial dimensionality such as 2D films or 1D wires. When the spatial dimensionality is reduced, the importance of the thermal fluctuations is increased and the latter will in general destroy any perfect order present in the system. The low-temperature ordered phase (cf. Chap. 8) will then be replaced by a phase with topological defects and

© The Author(s), under exclusive license to Springer Nature Switzerland AG 2021   323
M. Baus and C. F. Tejero, *Equilibrium Statistical Physics*,
https://doi.org/10.1007/978-3-030-75432-7_12

such a topological phase can then undergo a (topological) phase transition to a phase with a different topology. Such a topological phase transition has properties which can be very different from those studied in Chaps. 9 and 10 for ordinary defect-free phases. Here this will be illustrated for the particular case of the Kosterlitz–Thouless transition (2016 Physics Nobel Prize).

12.1   Topological Defects in a 3D Nematic Liquid Crystal

In Chap. 2 it was said that thermodynamics only requires that the system's volume be specified. In Chap. 3 it was stated that statistical mechanics and thermodynamics will become compatible only in the thermodynamic limit where the boundaries of the system's volume recede to infinity.

This may give the impression that the system's boundary conditions are not important.

In this section we show however that, in the presence of a broken symmetry (Chap. 9), the system's boundary conditions may influence the system's bulk behavior. In order to illustrate this we will consider a system with a simple broken rotational symmetry, namely a 3D anisotropic fluid with a uniaxial nematic order.

12.1.1   Anchoring

As already seen in Chap. 8 the orientation of molecules with a cylindrical symmetry can be indicated by a unit vector u placed along their axis of symmetry. If the intermolecular potential is such as to favor the parallelism of the molecular axis, the system may form a uniform fluid phase characterized by a non-trivial angular distribution $h(u)$. If the molecules have a top–bottom symmetry, $h(u) = h(-u)$, and such a fluid is called a nematic. If this nematic is uniaxial (cf. App. E) we have moreover, $h(u) \equiv h(u \cdot n)$, and the molecular axes are distributed around a common direction n, called the director. Suppose now that this nematic encounters a solid wall introducing hence a wall–nematic interaction potential. The latter will in general favor a certain orientation for the nematic molecules close to the wall. If this orientation is specified by the angle θ between the normal to the wall and the favored orientation (see Fig. 12.1), then we will distinguish the following types of boundary conditions for the nematic against this wall. If $\theta = \frac{\pi}{2}$ the nematic is said to have a planar anchoring against this wall; i.e., the molecular axes tend to remain parallel to the wall. As a consequence, the nematic close to such a wall will have its director parallel to the wall. Similarly, when $\theta = 0$, the nematic is said to have a homeotropic anchoring and its director will be perpendicular to the wall. For the other values of θ the anchoring is said to be mixed. Note that the type of anchoring is a characteristic of both the nematic and the wall material. To fix the ideas consider a nematic enclosed in a cylindrical vessel as shown in Fig. 12.2. It is then possible to

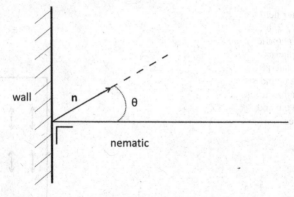

Fig. 12.1 Interactions between the molecules of the nematic and the molecules of the wall determine a characteristic orientation for the director field *n* of the nematic close to the wall. This orientation can be characterized by the angle θ between *n* and the normal to the wall. The value of θ determines then the type of anchoring between the nematic and the wall

Fig. 12.2 A cylindrical vessel filled with a perfectly ordered nematic with its director *n* parallel to the cylinder's axis. Since *n* and −*n* are equivalent directors we have indicated here their orientation by a double arrow (↕). For greater facility, in what follows we will consider the director field only in a planar section of the cylinder. Rotating this plane around the axis, the whole director filed can be reconstructed

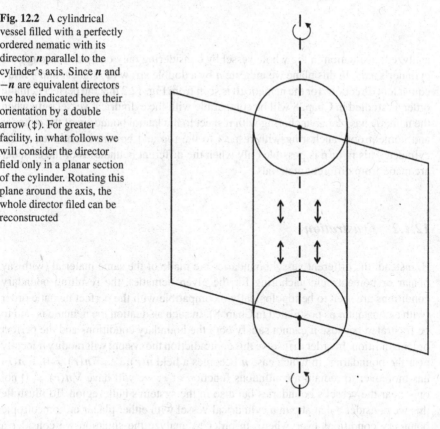

Fig. 12.3 Director field for
the case of Fig. 12.2. It is
seen that perfect nematic
order requires a planar
anchoring on the lateral
boundaries of the cylinder
together with a homeotropic
anchoring on the top and
bottom boundaries

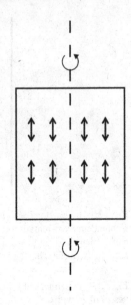

analyze the situation in the whole vessel by considering only one plane through the
cylinder's axis. In this plane we indicate n by a double arrow ($\updownarrow$) since n and $-n$ are
equivalent directions for the nematic. It is seen from Fig. 12.2 that the perfect nematic
order as studied in Chap. 8 will be compatible with the cylindrical vessel only when
the nematic has planar anchoring with respect to the lateral boundary of the cylinder
and homeotropic anchoring with respect to the top and bottom boundaries of the
cylinder. This in turn is possible only when the different boundaries of the cylinder
are made from different materials.

12.1.2   Frustration

If, instead, the different vessel boundaries are made of the same material (with say
planar or homeotropic anchoring for the given nematic), the resulting boundary
conditions are seen to be topologically incompatible with the perfect nematic order
(with a constant n as postulated in Chap. 8). In such a situation the nematic is said to
be frustrated because it cannot satisfy both the boundary conditions and the perfect
order condition. In order to resolve this contradiction the system will modify n locally
near the boundaries, in which case n becomes a field $n(r)$, i.e., $\nabla n(r) \neq 0$. If $n(r)$
has moreover to remain a continuous function of r, we will have $\nabla n(r) \neq 0$ not
only near the vessel's boundaries but also in the system's bulk region. To illustrate
this we consider a nematic in a cylindrical vessel with either planar or homeotropic
boundary conditions everywhere. In order to analyze the situation we consider a
plane through the cylinder's axes (Fig. 12.4) or a plane perpendicular to this axis
(Fig. 12.5). In these planes the field $n(r)$ will be represented by field lines, i.e.,

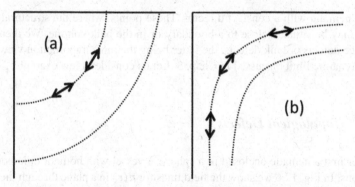

Fig. 12.4 Deformation of the director *n* between two perpendicular boundaries with **a** homeotropic anchoring, **b** planar anchoring. Both a field line (dotted line) and some directors (double arrows) are shown so as to make contact with the representation used in Fig. 12.3

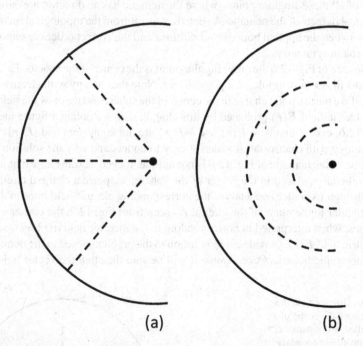

Fig. 12.5 Deformation of the director *n* along a curved boundary (such as the lateral boundary of a cylinder) for the case of **a** homeotropic anchoring, **b** planar anchoring. From here on, only field lines (dashed lines) will be shown

lines, such that in each of their points the director is tangent to the field line. A few directions are also indicated so as to get an idea of *n*(*r*) in the whole vessel. From Figs. 12.4 and 12.5 it can be deduced that, although there may exist regions of *r* where *n*(*r*) remains almost constant, in a frustrated nematic there will always exist points (or regions) where *n*(*r*) is strongly deformed, i.e., strongly departs from a perfectly ordered nematic with a constant director. These points, where the structural defect occurs, may be situated close to a boundary or in the bulk volume. We then speak of surface defects or bulk defects, the latter being the most important in view of the thermodynamic limit discussed in Chap. 3. Let us consider a few examples.

12.1.3 Topological Defects

Consider first a nematic enclosed in a spherical vessel with homeotropic boundary conditions. In Fig. 12.6 we show the field lines for $n(r)$ in a plane through the center of the sphere. It is seen there that as r approaches the center of the sphere, $n(r)$ becomes a multivalued function. We will say then that the points r, where $n(r)$ no longer indicates a direction, constitute the singular points of the vector field $n(r)$. The set of all these singular points, where the nematic has no director, are called the topological defects of the nematic. A seen they result from the topological incompatibility between the applied boundary conditions and the perfect order corresponding to the broken symmetry.

In the case of Fig. 12.6 the only singular point is the center of the sphere. This point represents thus a bulk topological point defect. Note the analogy with electrostatics. Indeed, if we place a point charge in the center of the sphere and we draw the field lines of the electric field $E(r)$ produced by this charge, then we obtain a figure identical to Fig. 12.6, except that here $E(r)$ and $-E(r)$ are not equivalent and $|E(r)|^2 \neq 1$. This analogy with electrostatics makes it easy to understand why the solution to the frustration of the nematic of Fig. 12.6 is not unique. Indeed, assume now that instead of a point charge placed in the center of the sphere, we place a charged circuit in a plane through its center and analyze the corresponding electric field lines in a plane perpendicular to the plane of the circuit. As seen from Fig. 12.7, the corresponding field lines, when interpreted as corresponding to the director field $n(r)$ instead of to the electric field $E(r)$, provide a new solution to the problem posed to the nematic by the homeotropic boundary conditions. It will be said then that the vector field $n(r)$

Fig. 12.6 Director field in a plane through the center of a sphere filled with a nematic with homeotropic boundary conditions

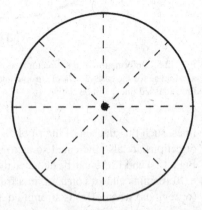

Fig. 12.7 Field lines produced by a circle of singular points located in the equatorial plane of a sphere and observed here in a plane through the poles

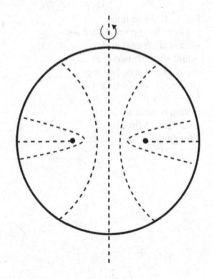

shown in Fig. 12.7 presents a topological line defect; i.e., the set of singular points forms a line which in this case is a circle.

Using this analogy with electrostatics, it becomes easy to construct a solution to any problem of frustration. For instance, consider a nematic enclosed in a cylindrical vessel with homeotropic boundary conditions on all its boundaries. A possible solution to this frustration problem is shown in Fig. 12.8 where the field lines of $n(r)$ are shown in a plane containing the cylinder's axis. We see that in this case the singular points of $n(r)$ consist of a topological point defect in the center of the cylinder (hence, a bulk defect), a topological line defect (here a circle) on the intersection of the lateral boundary of the cylinder with its upper boundary and similarly on its intersection with its lower boundary (the latter two being thus surface defects).

To any solution of a frustration problem one can thus associate an electrostatic analogy. For instance, to the solution shown in Fig. 12.8 one may associate an electrostatic model where the cylinder contains a point charge in its center and a charged loop on its upper and lower boundary. Because of this analogy, and although there are no real charges present in the nematic, it is often said that a given vector field $n(r)$ is produced by a given set of "topological" charges when this same set, but of real charges, would produce the same field lines as those of the nematic's director.

Finally, although the origin of any frustration resides in the applied boundary conditions, the resulting deformation of the director field can "propagate" into the bulk region and hence become observable, even in the thermodynamic limit, notwithstanding the fact that in this limit the system's boundaries are rejected to infinity.

Since a frustration problem has no unique solution, the question arises now of which solution will be chosen by the system. This will be the solution which minimizes the nematic's free energy. Till now we have considered (cf. Chap. 8) only the free energy of an undeformed nematic (with a constant director), whereas here the

Fig. 12.8 Field lines
produced by a point defect
located at the center of a
cylinder and by two circular
line defects located on the
intersection of the lateral
boundary with the top and
bottom boundaries and
observed here in a plane
through the cylinder's axis

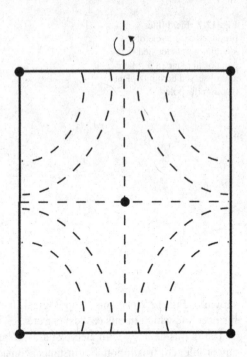

nematic has undergone an elastic deformation by locally adapting its director to the
boundary conditions. Next we will thus first study the free energy of a deformed
nematic.

12.1.4   Elastic Deformations

The Helmholtz free energy of the undeformed nematic is a function of its density (ρ)
and temperature (T) and a functional (cf. App. C) of its angular distribution $h(u)$,
viz. $\Im[h]$. The equilibrium distribution, say $h_0(u)$, being given then as a solution to
the following Euler–Lagrange equation:

$$\frac{\delta \Im[h]}{\delta h(u)} = 0. \tag{12.1}$$

For a uniaxial nematic the solution of (12.1) must have the form, $h_0 = h_0(u \cdot n)$,
with n the director of the undeformed nematic. When this nematic is deformed,
say $n \to n(r)$, the Helmholtz free energy of the deformed system can be written,
according to the theory of Frank and Oseen, as:

$$\Im_{FO}[n] = \Im\left[\tilde{h}_0\right] + \Im_{El}[n], \tag{12.2}$$

where $\tilde{h}_0 = h_0(\mathbf{u} \cdot \mathbf{n}(r))$ is the equilibrium distribution of the undeformed system but now evaluated for the local director $\mathbf{n}(r)$, and $\mathfrak{I}_{El}[\mathbf{n}]$ is the so-called free energy of elastic deformation. All these free energies do of course depend also on the thermodynamic state (ρ, T) but here we will focus on their dependence on the director field $\mathbf{n}(r)$. The field, say $\mathbf{n}_0(r)$, which will be realized in equilibrium will then be the solution of the Euler–Lagrange equation:

$$\frac{\delta \mathfrak{I}_{FO}[\mathbf{n}]}{\delta \mathbf{n}[r]} = 0, \tag{12.3}$$

together with the appropriate boundary conditions describing the anchoring properties of the nematic on the vessel's boundaries.

Using the chain rule for functional derivatives:

$$\frac{\delta \mathfrak{I}\left[\tilde{h}_0\right]}{\delta \mathbf{n}(r)} = \int dr' \int du' \frac{\delta \mathfrak{I}\left[\tilde{h}_0\right]}{\delta \tilde{h}_0(\mathbf{u}' \cdot \mathbf{n}(r'))} \frac{\delta \tilde{h}_0(\mathbf{u}' \cdot \mathbf{n}(r'))}{\delta \mathbf{n}(r)}, \tag{12.4}$$

together with (12.1), (12.3) becomes:

$$\frac{\delta \mathfrak{I}_{El}[\mathbf{n}]}{\delta \mathbf{n}(r)} = 0. \tag{12.5}$$

In the Frank–Oseen (FO) theory the following local approximation for the free energy of deformation is used:

$$\mathfrak{I}_{El}[\mathbf{n}] = \int dr\, f(\nabla \mathbf{n}(r)), \tag{12.6}$$

where the free energy density, $f(\nabla \mathbf{n})$ is an ordinary function (not a functional) of the gradients of $\mathbf{n}(r)$. If the latter are not too strong and $\mathbf{n}(r)$ is a continuous function of r we may tentatively expand this function for small gradients as:

$$f(\nabla \mathbf{n}) = a + b_{ij} \nabla_i \mathbf{n_j} + c_{ijkl}(\nabla_i \mathbf{n}_j)(\nabla_k \mathbf{n}_l) + \mathcal{O}((\nabla \mathbf{n})^3), \tag{12.7}$$

where the Einstein convention for summation over repeated indices has been used for $i = \{x, y, z\}$, etc. The above deformations are termed elastic if the undeformed state is recovered when $(\nabla \mathbf{n}) \to 0$. For (12.7) this is seen to require $a = 0$. For a nematic we must have, $f(\nabla \mathbf{n}) = f(-\nabla \mathbf{n})$, since $\mathbf{n}$ and $-\mathbf{n}$ are equivalent directors and hence $b_{ij} = 0$ in (12.7) and:

$$f(\nabla \mathbf{n}) = c_{ijkl}(\nabla_i \mathbf{n}_j)(\nabla_k \mathbf{n}_l) + \mathcal{O}((\nabla \mathbf{n})^4). \tag{12.8}$$

Finally, $f(\nabla n)$ has also to be a scalar function. There are five scalar functions of second order in ∇ and even in n, which are linearly independent. For these we choose:

$$\left\{(\nabla \cdot n)^2, (n \cdot (\nabla \times n))^2, (n \times (\nabla \times n))^2, \nabla \cdot (n\nabla \cdot n), \nabla \cdot (n \times (\nabla \times n))\right\}, \tag{12.9}$$

because they can be expressed, in terms of the scalar and vectorial product, in a compact form. Other expressions are possible or acceptable but they can be shown not to be independent from those of (12.9), for instance, $((n \times \nabla) \cdot n)^2 = (n \cdot (\nabla \times n))^2$, etc. To dominant order in the gradients of $n(r)$, we find thus that the Frank–Oseen (Helmholtz) free energy of elastic deformation, $\Im_{El}[n]$ of (12.6), can be written as:

$$\Im_{El}[n] = \int dr \left\{ K_1 (\nabla \cdot n)^2 + K_2 (n \cdot (\nabla \times n))^2 + K_3 (n \times (\nabla \times n))^2 \right\}$$

$$+ \int dr \{ K_4 \nabla \cdot (n (\nabla \cdot n)) + K_5 \nabla.(n \times (\nabla \times n)) \}, \tag{12.10}$$

where the coefficients, $K_i = K_i(\rho, T)(i = 1, ..., 5)$ are the Frank–Oseen elastic constants of the nematic. It is seen that the terms of (12.10) containing K_1, K_2, K_3 describe bulk elastic deformations while those containing K_4 and K_5 correspond to surface deformations. The latter can be transformed into surface integrals by using Gauss's theorem and eventually incorporated into the boundary conditions for $n(r)$. In what follows we will thus only be interested in the bulk elastic deformations described by the three elastic constants K_1, K_2, K_3.

12.1.5 Frank–Oseen Elastic Constants

Any bulk elastic deformation of the nematic can thus be represented as the superposition of three elementary deformations whose relative importance is fixed by the numerical value of the elastic constants $K_i (i = 1, 2, 3)$ which in turn fixes their contribution to the following free energy of deformation:

$$\Im_{El}[n] = \int dr \left\{ K_1 (\nabla \cdot n)^2 + K_2 (n \cdot (\nabla \times n))^2 + K_3 (n \times (\nabla \times n))^2 \right\}. \tag{12.11}$$

In order to physically interpret (12.11) we first consider a deformation for which we have:

$$\nabla \times n = 0; \ \nabla \cdot n \neq 0, \tag{12.12}$$

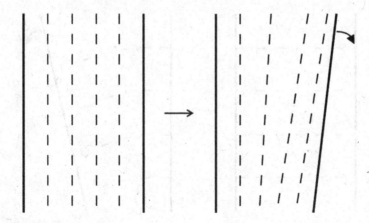

Fig. 12.9 Splay deformation

so that only the term in K_1 will survive in (12.11). To visualize a deformation satisfying (12.12) we consider a slice of nematic contained between two parallel planes. Before the deformation we assume the field lines of $n(r)$ to be everywhere parallel to the two planes, like in a perfectly ordered nematic with planar anchoring on these planes (cf. Fig. 12.8). During the deformation we slightly rotate one of the planes through an axis perpendicular to the original field lines while maintaining a planar anchoring for the rotating plane. This will open the field lines in a fan-like manner (cf. Fig. 12.9). Since all field lines remain straight lines we have $\nabla \times n = 0$, but $\nabla \cdot n \neq 0$ because the volume elements change locally. Such a deformation is called a splay deformation and K_1 is the splay elastic constant.

Next we consider a deformation for which:

$$\nabla \cdot n = 0; n \cdot (\nabla \times n) = 0; n \times (\nabla \times n) \neq 0, \qquad (12.13)$$

so that only the term in K_3 survives in (12.11). To visualize such a deformation we consider again a slice of nematic contained between two parallel planes, as above, but this time we take the two planes to be perpendicular to the undeformed field lines (cf. Fig. 12.10) instead of parallel. We again rotate one of the planes, maintaining a (fictitious) homeotropic anchoring on the planes, so that the field lines remain parallel to each other (cf. Fig. 12.10). Since there is no local volume change we have $\nabla \cdot n = 0$, while the field lines are now curved ($\nabla \times n \neq 0$) in such a way that $\nabla \times n$ is perpendicular to n, so that (12.13) will be satisfied. Such a deformation is called a bending and K_3 is hence the elastic constant of bending.

Consider finally a deformation for which:

$$\nabla \cdot n = 0; n \times (\nabla \times n) = 0; n \cdot (\nabla \times n) \neq 0, \qquad (12.14)$$

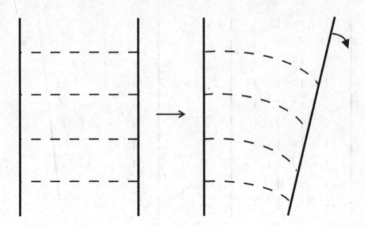

Fig. 12.10 Bending deformation

so that only the term in K_2 will survive in (12.11). Consider again the slice of nematic with a (fictitious) planar anchoring of Fig. 12.9 but this time one of the planes is rotated parallel to itself as shown in Fig. 12.11. The field lines undergo then a twist so that (12.14) will be satisfied and K_2 is the Frank–Oseen twist elastic constant of the nematic.

Any deformation of the nematic can thus be considered to be a linear combination (see (12.11)) of three elementary deformations, a splay, a bending, and a twist.

Although a nematic is an intrinsically anisotropic fluid, in theoretical studies it is often assumed that the nematic is nevertheless elastically isotropic so that, $K_1 = K_2 = K_3 \equiv K$, while (12.11) becomes then:

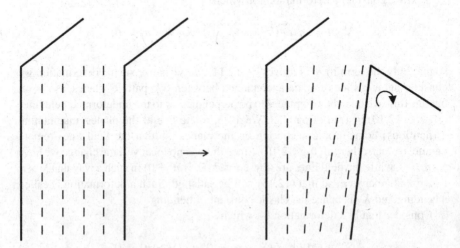

Fig. 12.11 Twist deformation

$$\mathfrak{J}_{El}[n] = K \int dr \{ (\nabla \cdot n)^2 + (\nabla \times n)^2 \}, \tag{12.15}$$

and the Euler–Lagrange equation (12.5) reduces to:

$$\nabla^2 n(r) = 0, \tag{12.16}$$

together with appropriate boundary conditions for $n(r)$. The appearance here of the Laplace equation (12.16) is usually invoked as the reason for the electrostatic analogy encountered above in Sect. 12.1.4. It should be mentioned however that in electrostatics it is the electric potential ($\varphi(r)$), a scalar field, not the electric field $E(r)(E = -\nabla \varphi)$ which satisfies the Laplace equation. Moreover, as mentioned already above, at the level of the vector fields we have $(n(r))^2 = 1$ but $(E(r))^2 \neq 1$. And, last but not least, a real nematic is only rarely elastically isotropic as assumed in (12.15) and (12.16).

12.1.6 Disclinations

The experimental identification of nematics often starts with the study of a thin film of fluid sandwiched between two parallel glass plates. Let us use cylindrical coordinates for $r = \{R, z\}$, with the z-axis perpendicular to the two plates and R a vector parallel to the plates with $R = \{x, y\}$, or in polar coordinates $R = \{R, \theta\}$, with θ being the angle between R and the x-axis. If the two plates are situated at $z = \pm \frac{L}{2}$, then the film thickness will be L (see Fig. 12.12). If we assume planar anchoring of the nematic on these plates, then we must have:

Fig. 12.12 A thin nematic film sandwiched between two parallel planes with planar anchoring

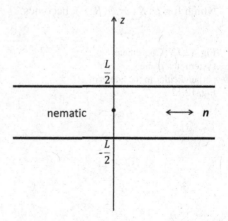

$$n_z\left(\boldsymbol{R}, z = \pm\frac{L}{2}\right) = 0, \tag{12.17}$$

i.e., the component of $\boldsymbol{n}(\boldsymbol{r})$ perpendicular to the plates must vanish in the vicinity of the plates. If moreover, L is small (compared to the gradients of the director), then $n_z(\boldsymbol{R}, z)$ will change little between the two plates, i.e., for $-\frac{L}{2} < z < \frac{L}{2}$, and (12.17) implies then that n_z will be small also between the two planes. This then suggests to use what is called the thin film approximation whereby $\boldsymbol{n}(\boldsymbol{R}, z)$ is supposed to be of the form:

$$\boldsymbol{n}(\boldsymbol{R}, z) = \left\{n_x(\boldsymbol{R}), n_y(\boldsymbol{R}), 0\right\}, \tag{12.18}$$

i.e., $\boldsymbol{n}(\boldsymbol{R}, z)$ is everywhere parallel to the plates and there is no variation with z. Since the condition, $\boldsymbol{n}^2 = 1$, becomes then in view of (12.18), $n_x{}^2 + n_y{}^2 = 1$, we may write:

$$n_x(\boldsymbol{R}) = \cos\varphi(\boldsymbol{R}); \; n_y(\boldsymbol{R}) = \sin\varphi(\boldsymbol{R}), \tag{12.19}$$

where $\varphi(\boldsymbol{R})$ is the angle which $\boldsymbol{n}(\boldsymbol{R})$ makes with the x-axis at the point $\boldsymbol{R}$(see Fig. 12.13). For simplicity (or as a first approximation) we may also consider the nematic to be elastically isotropic (cf. (12.15)) in which case we will have to solve (12.16) with the aid of (12.18) and (12.19). If the deformation around the undeformed director, $\boldsymbol{n}_0 = \{1, 0, 0\}$ is small, so will be $\varphi(\boldsymbol{R})$, and to dominant order (12.16) reduces then to:

$$\nabla^2\varphi(\boldsymbol{R}) = 0, \tag{12.20}$$

which for, $\varphi(\boldsymbol{R}) = \varphi(R, \theta)$, becomes:

Fig. 12.13 Coordinate system in a plane perpendicular to the z-axis of Fig. 12.12

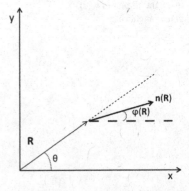

$$\frac{1}{R}\frac{\partial}{\partial R}\left(R\frac{\partial\varphi}{\partial R}\right) + \frac{1}{R^2}\frac{\partial^2\varphi}{\partial\theta^2} = 0, \tag{12.21}$$

while, in order to avoid too specific boundary conditions with respect to R, we henceforth consider only solutions of the form, $\varphi(R,\theta) = \varphi(\theta)$, in which case (12.21) reduces to:

$$\frac{1}{R^2}\frac{\partial^2\varphi}{\partial\theta^2} = 0, \tag{12.22}$$

whose general solution is:

$$\varphi(\theta) = c + k\theta, \tag{12.23}$$

where $c = \varphi(\theta = 0)$ and k is a constant. Returning to (12.18) and (12.19) we can now define a two-dimensional vector field $n(\theta)$ as:

$$n_x(\theta) = \cos(c + k\theta); n_y(\theta) = \sin(c + k\theta), \tag{12.24}$$

since here, $n_z(\theta) = 0$. For polar coordinates we must have that (R, θ) and $(R, \theta + 2\pi)$ represent the same point R and thus here $n(\theta + 2\pi)$ must represent the same director field as $n(\theta)$, or since $n(\theta)$ and $-n(\theta)$ are equivalent directors, we must have:

$$n(\theta + 2\pi) = \pm n(\theta), \tag{12.25}$$

which for (12.24) implies that $2k$ must be either even (for the $+$ sign in (12.25)) or odd [for the—sign in (12.25)]. For a given value of c there will thus be two families of deformations of the type of (12.24), one for k an integer ($k = 0, \pm 1, \pm 2, ...$) and one for k a half-integer $\left(k = \pm\frac{1}{2}, \pm\frac{3}{2}, \pm\frac{5}{2}, ...\right)$.

Let us finally evaluate the free energy cost for deforming the nematic. To this end we substitute (12.23) and (12.24) into (12.15). To dominant order in $\varphi(\theta)$ we obtain then:

$$\mathfrak{I}_{El}[n] = K\int dr\left(\frac{1}{R}\frac{\partial\varphi(\theta)}{\partial\theta}\right)^2 = K\int_{-\frac{L}{2}}^{\frac{L}{2}} dz\int_0^{2\pi} d\theta\int dR\frac{R}{R^2}k^2 = KL2\pi k^2\int_{R_c}^{R_0}\frac{dR}{R}, \tag{12.26}$$

where the last integral is seen to diverge as $R_c \to 0$. Indeed from (12.22) it is seen that our solution (12.24) will not be valid for $R = 0$. In (12.26) we regularized the divergent integral with the aid of a cutoff value R_c such that for $R \geq R_c$ our solution (12.24) remains valid. Moreover in (12.26) R_0 is the radius of the cylinder of height L containing the nematic and hence:

$$\mathfrak{J}_{El}[\boldsymbol{n}] = k^2 2\pi L K \ln\left(\frac{R_0}{R_c}\right), \tag{12.27}$$

i.e., the free energy cost increases with k like k^2. This is the reason why, in practice, one observes only small k deformations, e.g., $k = \pm\frac{1}{2}$ and $= \pm 1$. Figure 12.14 shows the field lines of $\boldsymbol{n}(\theta)$ of (12.24) for a few values of k and c. It is seen that in all cases the director field becomes singular at $R = 0$ for all z-values, with the line of singularities having a "thickness", R_c just as above. This line represents a topological line defect. This defect is called a disclination of axis z and intensity k. It is the type of topological defect most frequently observed in nematics.

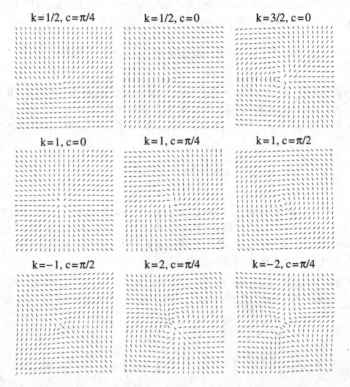

Fig. 12.14 Director field, $n(\theta)$ with $\theta = \text{arctg}(\frac{y}{x})$ for the disclinations of the strength k and constant c of (12.24). For clarity, the double arrows of, say Fig. 12.3, have been reduced here to a short segment ($-$). *Source* Kleman and Lavrentovich [3]; reproduced with permission

12.1.7 Schlieren Texture

In soft matter, in particular for liquid crystals, the Frank–Oseen elastic constants are sufficiently weak so that thermal fluctuations can easily induce deformations. These deformations will usually contain topological defects and the spatial organization of these defects is called the texture of, say, the liquid crystal. In this section we will study such a texture for the case of an uniaxial nematic. To this end we again consider the thin film setup of Fig. 12.12 but assume now that the two glass plates are linear polarizers crossed at $\frac{\pi}{2}$ and lightened from below with white light. Since the nematic is an anisotropic medium it will be optically birefringent so that each incident ray will be divided into an ordinary ray polarized perpendicular to the director n and an extraordinary ray polarized parallel to n. Observing the transmitted light under a microscope an image such as shown in Fig. 12.15 is obtained. It consists of dark points connected by dark lines or branches and surrounded by colored regions. The black regions are regions of total extinction, where no light could come through, while in the colored regions light could come through but with a phase shift which depends on its wavelength (hence the colors). The important parts here are the black regions. The whole set of dark points connected by dark lines forms a spatial structure which is called its texture. The texture shown in Fig. 12.15 is called a Schlieren texture because of the characteristic lines (called "Schlieren" in german). This texture is characteristic of nematics. To show this let us first assume that the fluid in the film would be isotropic. In this case, all the light polarized by the first polarizer will be extinguished by the second polarizer and the image will be all black, which is not the case of Fig. 12.15. Assume next that the fluid in the film is a perfectly ordered nematic with a constant n. If the direction n coincides with the polarization direction of the first polarizer, only the extraordinary rays will pass but will then be stopped by the second polarizer, leading again to a uniformly black image. If, on the contrary, n makes a nonzero angle with the first polarization direction, both ordinary and extraordinary rays will pass and the image will be uniformly white. Since Fig. 12.15 exhibits both black and non-black regions, we must conclude that the nematic is deformed and could contain defects such as the disclinations studied above. Let us assume hence that the black points are the image of singularities, e.g., disclinations of strength k with their axis perpendicular to the plates. As argued above, we only expect small $k$disclinations to be present. Figure 12.16 shows that two different black points are present. Those with two branches and those with four branches. Let us show now that these are related to disclinations of, respectively, strength $k = \pm\frac{1}{2}$ and $k = \pm 1$. Observe first that for an arbitrary point of the image the constant c of (12.24) is arbitrary since $c = \varphi(\theta = 0)$ depends on the arbitrary choice of the x, y-axis. We will thus consider here only disclinations with $c = 0$. Next, consider the director field $n(\theta)$ for $\left(k = \frac{1}{2}, c = 0\right)$ shown in Fig. 12.14. It is seen there that for any θ, the directions $n(\theta)$ and $n(\theta + \pi)$ are orthogonal. Therefore there will always exist a value of θ, say θ_0, such that $n(\theta_0)$ is perpendicular to one of the polarization directions while $n(\theta_0 + \pi)$ is orthogonal to the other polarization direction. The direction $\theta = \theta_0$, will thus be a direction along which there will be

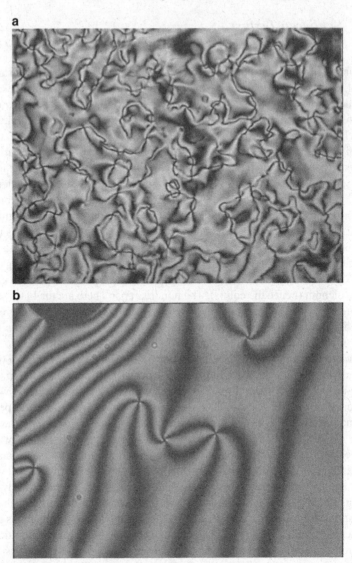

Fig. 12.15 Low (**a**) and higher (**b**) magnification view of a nematic film with planar anchoring conditions and viewed in transmission between crossed polarizers and white light. *Source* Dierking [4]; reproduced with permission

extinction. Indeed, Fig. 12.16a shows the two dark branches leaving the singularity in two diametrically opposed directions. It is however also seen that the two branches are not straight lines but are curved. This is due to the fact that away from the singularity $\varphi(\boldsymbol{R})$ is no longer small, as assumed here, and the terms nonlinear in $\varphi(\boldsymbol{R})$ will then curve the lines as shown. Next, consider the case$(k = 1, c = 0)$, Fig. 12.14 shows that the situation is similar to the $\left(k = \frac{1}{2}, c = 0\right)$ case except that now there exist two

Fig. 12.16 **a** A disclination of strength $k = \pm\frac{1}{2}$, showing two branches or "brushes" leaving the singular point in opposite directions. **b** A disclination of strength $k = \pm 1$, showing fourfold brushes leaving the singular point in two different opposite directions. **c** A region showing simultaneously a twofold and a fourfold brush. *Source* Dierking [4]; reproduced with permission

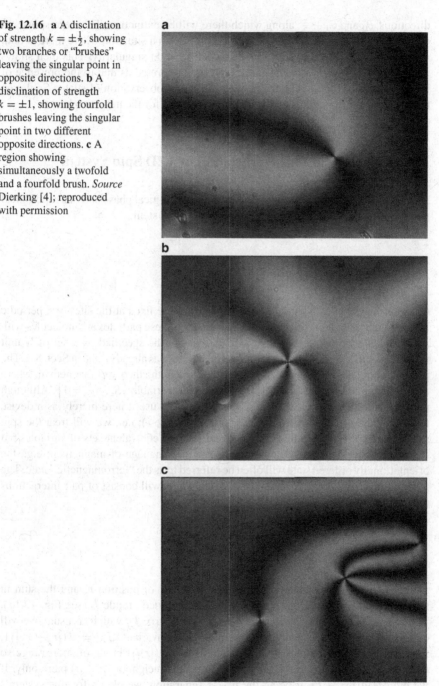

directions, θ_0 and $\theta_0 + \frac{\pi}{2}$, along which there will be extinction. This corresponds to what is seen in Fig. 12.16b, while Fig. 12.16c shows a situation where there is both a two-branched $\left(k = \frac{1}{2}\right)$ and a four-branched $(k = 1)$ singularity. This then shows that pictures like shown in Fig. 12.15 can be interpreted as due to the presence of low-strength disclinations in the nematic film. The observation of such a Schlieren texture is currently used as an identification method for the nematic phase.

12.2 Topological Phase Transition in a 2D Spin System

In order to illustrate the particular features of a topological phase and of a topological phase transition we first introduce a simple model system.

12.2.1 A Classical Spin System

Consider a system of Nparticles whose positions are fixed at the sites of a periodic lattice so that only their orientations can change. If these particles are molecules with a cylindrical symmetry, then their orientations can be specified by a set of N unit vectors $\left\{u_i; u_i^2 = 1\right\}$ along their symmetry axis, as was already done in Sect. 8.4. The present model has however been extensively used in the theory of magnetism, where the particle's orientation is determined by a spin variable $\left\{S_i; S_i^2 = 1\right\}$. Although the spin is a quantum mechanical concept, we will use it here merely as a devise to specify the orientation (S_i) of the particle (at site i); i.e., we will treat the spin as a classical variable. In this case $\{S_i\}$ and $\{u_i\}$ are equivalent sets of variables but it will be convenient here to continue to use the language of magnetism; e.g., the orientationally ordered state will often be referred to as the "ferromagnetic" state. The potential energy V of our system of classical spins will consist of pair interactions represented by:

$$V = \sum_{i,j} S_i . J_{ij} . S_j, \tag{12.28}$$

where S_i is the spin of the particle sitting on site i of position r_i and the sum in (12.28) extends over all the N sites of a square lattice of side L (see Fig. 12.17). The spin–spin interaction is characterized by the matrix J_{ij} which, as usual, we will consider to be translationally and orientationally invariant, $J_{ij} = J\left(\left|r_i - r_j\right|\right)1$, there being no self-interactions $(J(|0|) = 0)$ and $J\left(\left|r_{ij}\right|\right)$ being of short range so that the sum in (12.28) can be restricted to nearest neighbor $(n - n)$ pairs only. If we put,$(J(|a|) = -J)$, a being the $n - n$ separation, we obtain for this system's Hamiltonian:

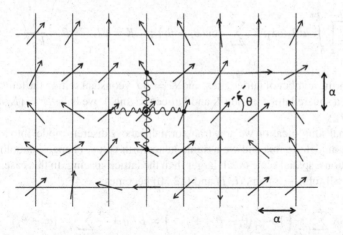

Fig. 12.17 A small piece of a $L \times L$ square lattice of lattice spacing a. A spin of arbitrary direction is placed at every lattice site. The spin's direction can be determined by the angle θ as indicated and also shown is the interaction (wavy line) of each spin (open dot) with its four nearest neighbors (black dots)

$$H = -J \sum_{\langle i,j \rangle} S_i \cdot S_j, \tag{12.29}$$

since the particles have no kinetic energy here ($H \equiv V$) while the restriction to $n - n$ pairs is indicated as $\langle i, j \rangle$ in (12.29). This is the Heisenberg model, first used to study spin–exchange interactions in the context of a quantum mechanical study of magnetism, J being called then the exchange integral. With $J > 0$, the state of lowest energy of (12.29) will correspond to the "ferromagnetic" state, $S_i \cdot S_j = 1$ for all (i, j). Note also that the same type of interaction has already been used in our study of the Heisenberg fluid (cf. (9.26) with $l = 1$). In the above, we did not specify the dimensionality of S. For a 2D-lattice the vectors, S_i, could still be 3D, but here we will consider them to be 2D, $S = (S_x, S_y) = (\cos\theta, \sin\theta)$ where θ is the polar angle of S w.r.t. the x-axis. For this 2D system we have thus, $S_i \cdot S_j = \cos(\theta_i - \theta_j)$, and (12.29) becomes:

$$H = -J \sum_{\langle i,j \rangle} \cos(\theta_i - \theta_j), \tag{12.30}$$

and this model is often referred to as the $x - y$ or planar model, because both the lattice and the spins are in the $x - y$ plane. Since here only the θ_i can change ($0 < \theta_i < 2\pi$) this system is also often referred to as a system of "rotors." The Helmholtz free energy, $F = -k_B T \ln Z$, of this system can then be obtained from, e.g., the following canonical partition function:

$$Z = \prod_{k=1}^{N} \int_{0}^{2\pi} d\theta_k \, exp \, K \sum_{<i,j>} \cos(\theta_i - \theta_j) \; ; \quad K = \beta J, \quad \beta = \frac{1}{k_B T} \quad (12.31)$$

where N is the number of lattice sites, and $K = \beta J$ a constant setting the temperature scale. For a square lattice of side L and lattice spacing a, we have $N = (L/a)^2$ (see Fig. 12.17).

As a final simplification we now transform the above discrete model into a continuous field model. To this end we assume henceforth that θ_i changes smoothly with r_i, at least on a spatial scale much larger than the lattice spacing. In this case, $\theta_i - \theta_j$ will be small for $n - n$ pairs (i, j) and (12.30) becomes:

$$\beta H = -K \sum_{\langle i,j \rangle} \left(1 - \frac{1}{2}(\theta_i - \theta_j) + \dots \right) = \beta E_0 + \frac{K}{2} \sum_{\langle i,j \rangle} (\theta_i - \theta_j)^2 + \dots ;$$
$$(12.32)$$

where $E_0 = -JnN$ with n the number of $n - n$ pairs ($n = 4$ for a square lattice) is an immaterial constant, H and $H - E_o$ leading to the same physics. Let us write, $\theta_i = \theta(r_i)$, then $\theta_i - \theta_j = \theta(r_i) - \theta(r_j) = \theta(r_i) - \theta(r_i + a) = -a.\frac{\partial}{\partial r_i}\theta(r_i) + \cdots$ since for $n - n$ pairs we have $r_j = r_i + a$, a being an elementary lattice vector. Dropping henceforth E_o we obtain for (12.32):

$$\beta H = \frac{K}{2} \sum_{i=1}^{N} \left(a.\frac{\partial\theta(r_i)}{\partial r_i} \right)^2 + \cdots , \quad (12.33)$$

or in the limit of a continuous lattice:

$$\beta H = \frac{K}{2} \int_{D} dr (\nabla\theta(r))^2, \quad (12.34)$$

where $r = (x, y)$, $\nabla = \frac{\partial}{\partial r}$ and it is understood that $\theta(r)$ is now a continuous function of r; i.e., $\theta(r)$ is a (scalar) field. Note that in (12.34), $\int_D dr = \int_a^L dx \int_a^L dy$, since the lattice spacing a operates here as a lower cutoff. The partition function (12.31) corresponding to (12.34) becomes now:

$$Z = \int D[\theta] \, exp - \frac{K}{2} \int dr \left(\nabla\theta(r) \right)^2, \quad (12.35)$$

where $\int D[\theta]$ represents a functional integration (cf. Appendix C) over all possible scalar fields $\theta(r)$ such that $0 \leq \theta(r) \leq 2\pi$. Equations (12.34) and (12.35) determine our classical spin model with $\theta(r)$ representing the polar angle of the spins located in the vicinity of r. Any field, $\theta(r)$, realizing a local minimum of H, i.e., an equilibrium state, must satisfy:

$$\frac{\delta H}{\delta \theta(r)} = 0 \quad \text{or} \quad \nabla^2 \theta(r) = 0. \tag{12.36}$$

In what follows it will often be convenient to rewrite the equations in terms of the distortion field $v(r)$:

$$v(r) = \nabla \theta(r), \tag{12.37}$$

hence

$$\beta H = \frac{K}{2} \int dr (v(r))^2, \tag{12.38}$$

$$Z = \int D[\theta] \exp -\frac{K}{2} \int dr (v(r))^2, \tag{12.39}$$

$$\nabla \cdot v(r) = 0, \tag{12.40}$$

Corresponding, respectively, to (12.34), (12.35) and (12.36). Note finally that $v(r)$, being the gradient of the scalar field $\theta(r)$, is a vector field.

12.2.2 Absence of Long-Range Order in 2D Systems

In analogy with the 3D systems studied in Chap. 8 we expect our 2D system to order at low temperatures. Indeed, $\theta(r) = \theta_0$ is a solution of (12.36) corresponding to a ferromagnetic state, all spins pointing in the same direction. This is a perfectly ordered state but following an argument by Landau and Peierls, we now show that the thermal fluctuations will destroy this long-range order for $d < 3$. Hence, $\theta(r) = \theta_0$, represents an unstable equilibrium state of our 2D spin system.

Consider a fluctuation from, $\theta(r) = \theta_0$ to $\theta(r) = \theta_0 + \varphi(r)$ and generalize (12.34) to d-dimensions as:

$$\beta H_d = \frac{K_d}{2} \int dr (\nabla(\theta_0 + \varphi(r)))^2; \; K_d = Ka^{2-d}, \tag{12.41}$$

where r is now a d-dimensional vector and K_d makes βH_d dimensionless. We first develop the thermal fluctuation $\varphi(r)$ in a Fourier series:

$$\varphi(r) = \sum_k \varphi_k e^{ik \cdot r}, \tag{12.42}$$

$$\nabla \varphi(\boldsymbol{r}) = \sum_{k} i \boldsymbol{k} \varphi_{k} e^{i\boldsymbol{k}.\boldsymbol{r}}, \tag{12.43}$$

$$\int d\boldsymbol{r} (\nabla \varphi(\boldsymbol{r}))^2 = V \sum_{k} k^2 \varphi_{k} \varphi_{-k}, \tag{12.44}$$

where $V = L^d$, and the Fourier components φ_k satisfy, $\varphi_{-k} = (\varphi_k)^*$, since $\varphi(\boldsymbol{r})$ is a real function. In these equations the components $\{k_1, \ldots, k_d\}$ of $\boldsymbol{k}$ are given by $k_i = \frac{2\pi}{L} n_i$ with $n_i = \pm 1, \pm 2, \ldots$, and the sums in (12.42)–(12.44) are limited to $\frac{2\pi}{L} \leq |k_i| \leq \frac{2\pi}{a}$. The change in energy produced by the fluctuations can then be written:

$$\beta H_d = \frac{K_d}{2} L^d \sum_{k} k^2 \varphi_{k} \varphi_{-k}, \tag{12.45}$$

and these fluctuations will occur with a probability density $P(\{\varphi_k\})$ proportional to the Boltzmann factor, $\exp -\beta H_d$, or:

$$P(\{\varphi_k\}) \sim \exp -\frac{K_d}{2} L^d \sum_{k} k^2 \varphi_{\boldsymbol{k}} \varphi_{-\boldsymbol{k}}, \tag{12.46}$$

showing that the amplitudes $\{\varphi_k\}$ are Gaussian random variables. Using the results of Appendix B.4 we have thus:

$$\langle \varphi_k \varphi_{k'} \rangle = \delta_{k+k'} \frac{2}{k^2 K_d L^d}, \tag{12.47}$$

and also (see (B.57)):

$$\left\langle \exp \sum_{k} a_k \varphi_k \right\rangle \begin{cases} = \exp \frac{1}{2} \sum_{k,k'} a_k a_{k'} \langle \varphi_k \varphi_{k'} \rangle \\ = \exp \frac{1}{K_d L^d} \sum_{k} \frac{1}{k^2} a_k a_{-k} \end{cases}, \tag{12.48}$$

where we have used (12.47) in the first line of (12.48).

We now consider two physical properties of our spin system. First, the magnetization at the origin:

$$M_d = \langle \cos(\theta(0)) \rangle = \text{Re} \langle e^{i\theta(0)} \rangle, \tag{12.49}$$

and next, the spin–spin correlation function:

$$g_d(\boldsymbol{r}) = \langle \cos(\theta(\boldsymbol{r}) - \theta(0)) \rangle = \text{Re} \langle e^{i(\theta(\boldsymbol{r}) - \theta(0))} \rangle, \tag{12.50}$$

which for the fluctuation, $\theta(r) \equiv \varphi(r)$, since θ_0 can always be taken to be zero by an appropriate choice of axes, yields using (12.42):

$$M_d = \mathrm{Re}\langle \exp i \sum_k \varphi_k \rangle, \tag{12.51}$$

$$g_d(r) = \mathrm{Re}\langle \exp i \sum_k \varphi_k (e^{ik.r} - 1) \rangle, \tag{12.52}$$

or for Gaussian fluctuations, using (12.48) will $a_k = i$:

$$M_d = \exp -\frac{I_d}{K_d}; \quad I_d = \frac{1}{L^d} \sum_k \frac{1}{k^2}, \tag{12.53}$$

for (12.51), while (12.52) becomes using (12.48) with $a_k = i(e^{ik.r} - 1)$:

$$g_d(r) = \exp -\frac{G_d(r)}{K_d}; \quad G_d(r) = \frac{2}{L^d} \sum_k \frac{1}{k^2}(1 - \cos(k.r)). \tag{12.54}$$

For large enough L values ($L \gg a$) the discrete sums over k can be replaced by integrals:

$$I_d = \int dk \frac{1}{k^2} = \Omega_d \int\limits_{k_-}^{k_+} dk k^{d-3}, \tag{12.55a}$$

$$G_d(r) = 2 \int dk \frac{1}{k^2}(1 - \cos k.r)$$

$$= 2\Omega_d \int\limits_{k_-}^{k_+} dk k^{d-3}(1 - J_0(k.r)), \tag{12.55b}$$

where $k_- = 2\pi/L$, and $k_+ = 2\pi/a$, as discussed above, $\Omega_d = 2\pi^{d/2}/\Gamma(d/2)$ is the surface area of a d-dimensional sphere of unit radius (cf. (4.12)), and, $J_0(kr) = \frac{1}{2\pi}\int_0^{2\pi} d\varphi \cos(kr \cos\varphi)$, the Bessel function of zeroth order. We then obtain for $d = 3, 2$ and 1 (cf. $\Omega_3 = 4\pi$, $\Omega_2 = 2\pi$, $\Omega_1 = 2$) from (12.55a):

$$I_3 = \frac{8\pi^2}{a}\left(1 - \frac{a}{L}\right), \tag{12.56}$$

$$I_2 = 2\pi \ln\left(\frac{L}{a}\right), \tag{12.57}$$

$$I_1 = \frac{a}{\pi}\left(\frac{L}{a} - 1\right), \tag{12.58}$$

yielding for (12.53):

$$M_3 = \exp -\frac{8\pi^2}{K}\left(1 - \frac{a}{L}\right), \tag{12.59}$$

$$M_2 = \left(\frac{a}{L}\right)^{2\pi/K}, \tag{12.60}$$

$$M_1 = \exp -\frac{1}{\pi K}\left(\frac{L}{a} - 1\right). \tag{12.61}$$

Let us now also evaluate (12.55b). We write:

$$\frac{G_d(r)}{K_d} = \frac{2}{K}\Omega_d\left(\frac{a}{r}\right)^{d-2}\int_{u_-}^{u_+} du\, u^{d-3}(1 - J_0(u)), \tag{12.62}$$

where $u = kr$ and $u_\pm = rk_\pm$. We now split this integral into three contributions:

$$\frac{G_d(r)}{K_d} = \frac{2}{K}\Omega_d(A_d(r) + B_d(r) + C_d(r)), \tag{12.63}$$

with:

$$A_d(r) = \left(\frac{a}{r}\right)^{d-2}\int_{u_-}^{1} du\, u^{d-3}(1 - J_0(u)), \tag{12.64}$$

$$B_d(r) = \left(\frac{a}{r}\right)^{d-2}\int_{1}^{u_+} du\, u^{d-3}, \tag{12.65}$$

$$C_d(r) = -\left(\frac{a}{r}\right)^{d-2}\int_{1}^{u_+} du\, u^{d-3} J_0(u). \tag{12.66}$$

For the evaluation of (12.64) we can use, $1 - J_0(u) = \frac{u^2}{4} + \cdots$ for $u \ll 1$, to obtain:

$$A_d(r) \simeq \left(\frac{a}{r}\right)^{d-2} \int\limits_{u_-}^{1} du \frac{u^{d-1}}{4} = \left(\frac{a}{r}\right)^{d-2} \left(\frac{1 - u_-^d}{4d}\right)$$

$$A_d(r) = \frac{1}{4d}\left(\frac{a}{r}\right)^{d-2} - \frac{(2\pi)^d}{4d}\left(\frac{r}{L}\right)^d \left(\frac{a}{r}\right)^{d-2} \tag{12.67}$$

while (12.65) yields:

$$B_3(r) = \frac{a}{r}(u_+ - 1) = 2\pi - \frac{a}{r}, \tag{12.68}$$

$$B_2(r) = \ln u_+ = \ln\left(2\pi\frac{r}{a}\right), \tag{12.69}$$

$$B_1(r) = \frac{r}{a}\left(1 - \frac{1}{u_+}\right) = \frac{r}{a} - \frac{1}{2\pi}. \tag{12.70}$$

Finally, for the evaluation of (12.66) we use the property, $J_o(u) = \sqrt{\frac{2}{\pi u}}\cos\left(u - \frac{\pi}{4}\right) + \cdots$ for $u \gg 1$, to find that the integral tends to a constant, say c_d, when u_+ tends to infinity.

Hence

$$C_d(r) = -c_d\left(\frac{a}{r}\right)^{d-2}. \tag{12.71}$$

Putting these results together we obtain:

$$g_3(r) = \exp -\frac{8\pi}{K}\left\{\frac{1}{12}\left(\frac{a}{r}\right) - \frac{(2\pi)^3}{12}\left(\frac{r}{L}\right)^3\left(\frac{a}{r}\right) + 2\pi - \frac{a}{2} - c_3\left(\frac{a}{r}\right)\right\}, \tag{12.72}$$

$$g_2(r) = \exp -\frac{4\pi}{K}\left\{\frac{1}{8} - \frac{(2\pi)^2}{8}\left(\frac{r}{L}\right)^2 + \ln\left(2\pi\frac{r}{a}\right) - c_2\right\}, \tag{12.73}$$

$$g_1(r) = \exp -\frac{2}{K}\left\{\frac{1}{4}\left(\frac{r}{a}\right) - \frac{\pi}{2}\left(\frac{r}{L}\right)\left(\frac{r}{a}\right) + \frac{r}{a} - \frac{1}{2\pi} - c_1\frac{r}{a}\right\}. \tag{12.74}$$

Let us now consider the physical interpretation of the above results.

For a 3D system we see from (12.59) that in the thermodynamic limit ($L \gg a$) the magnetization becomes:

$$\lim_{L=\infty} M_3 = \exp -8\pi^2\frac{k_B T}{J}, \tag{12.75}$$

whereas the correlation function (12.72) tends asymptotically ($r \gg a$) to:

$$\lim_{r=\infty}\lim_{L=\infty} g_3(r) = \exp -16\pi^2 \frac{k_B T}{J},$$
(12.76)

This is the expected behavior for a 3D ferromagnetic system. Indeed, (12.75) shows that the magnetization has a finite thermodynamic limit whose value increases when T decreases or J increases. From (12.76) we see that $g_3(r)$ has an infinite range, with a limiting value equal to $(M_3)^2$ as expected for a system exhibiting "long-range order" (LRO).

For a 2D system we obtain from (12.60);

$$\lim_{L=\infty} M_2 = \begin{cases} 0 & T \neq 0 \\ 1 & T = 0 \end{cases},$$
(12.77a)

while (12.73) yields:

$$\lim_{r=\infty}\lim_{L=\infty} g_2(r) = \begin{cases} 0 & T \neq 0 \\ 1 & T = 0 \end{cases}$$
(12.77b)

and hence for $T \neq 0$ our 2D spin system can have no macroscopic magnetization nor any LRO. Since the larger the system, the smaller the magnetization ($M_2 = \left(\frac{a}{L}\right)^{\frac{2\pi}{K}}$), it is said that the thermal fluctuations ($T \neq 0$) do destroy the ferromagnetic order for $d = 2$ (but not for $d = 3$). Note that the asymptotic decay of $g_2(r) \sim \left(\frac{a}{r}\right)^{\frac{4\pi}{K}}$ is algebraic. Such a behavior is also encountered for 3D systems, but only at the critical temperature (cf. (10.30)) whereas here the decay of $g_2(r)$ is algebraic for all $T \neq 0$. Because of the algebraically slow decay of $g_2(r)$ our system is said to exhibit quasi-LRO. Such a system can exhibit order over a large region but this region is always of finite extend.

For a 1D system we obtain from (12.61) and (12.74):

$$\lim_{L=\infty} M_1 = \begin{cases} 0 & T \neq 0 \\ 1 & T = 0 \end{cases},$$
(12.78)

$$\lim_{r=\infty}\lim_{L=\infty} g_1(r) = \begin{cases} 0 & T \neq 0 \\ 1 & T = 0 \end{cases},$$
(12.79)

a behavior similar to the $d = 2$ case, except that here the decay of $M_1 \sim \exp -\frac{L}{\pi K a}$ and $g_1(r) \sim \exp -c_1\left(\frac{r}{a}\right)$ is exponential, not algebraic. Exponential decay is usually associated with disordered systems.

In summary, for 3D systems the LRO can resist the thermal fluctuations, in 2D systems they transform the LRO into quasi-LRO while in 1D systems the thermal fluctuations disorder the system. In other words, the influence of the thermal fluctuations increases as we decrease the spatial dimensionality.

Note finally that the destruction of the LRO by the thermal fluctuations, which was shown here for a 2D spin system by the Landau–Peierls argument, is in fact a very general property of any 2D system, as was later shown by Mermin and Wagner.

12.2.3 Topological Phases

As we have seen in the previous section, the well-known ferromagnetic equilibrium phase of 3D spin systems becomes unstable in 2D and hence cannot provide an equilibrium state for our 2D spin system. We will now show that the role played by the "order" in 3D systems is replaced in 2D systems by their "topology." Indeed, the set of all possible solutions to (12.36) can be divided into regular and singular solutions. A solution being called singular if $\theta(r)$ cannot be defined at certain points $\{r_j\}$, $r = r_j$ being then a point of singularity of $\theta(r)$. Obviously the ferromagnetic state, $\theta(r) = \theta_0$ is a regular solution. On the contrary, the following expression $(r = (x, y))$:

$$\theta(x, y) = \text{arctg}\left(\frac{y}{x}\right), \tag{12.80}$$

is a singular solution of (12.36) since (12.80) is undefined at $r = 0$. Other singular solutions of (12.36) have already been encountered before (cf. (12.23)–(12.26)). Each singularity is said to belong to a topological defect. A phase containing topological defects is called a topological phase.

What is needed now is an analytic criterium to decide whether a given $\theta(r)$, solution of (12.36), is singular or not. To this end we use the fact that $\theta(r)$ is an angle whose physical definition is restrained to $0 \le \theta(r) \le 2\pi$. In the presence of a singularity $\theta(r)$ will however become multivalued. In other words, the contour integral of $\theta(r)$ around a closed contour γ will restore the value of $\theta(r)$, when $\theta(r)$ is regular inside γ, but will increase by an integer (n) multiple of 2π, when γ contains a singularity of $\theta(r)$:

$$\int_\gamma d\theta = \oint_\gamma dl.\nabla\theta(r) = \begin{cases} 2\pi n & \text{if } \theta(r) \text{ is singular inside } \gamma \\ 0 & \text{if } \theta(r) \text{ is regular inside } \gamma \end{cases}. \tag{12.81}$$

Note that θ and $\theta + 2\pi n$ correspond to the same direction for the spin. Using this property for a topological phase we obtain:

$$\oint_{\gamma_L} dl.\nabla\theta(r) = \sum_j 2\pi n_j \left(n_j = \pm 1, \pm 2, \dots\right), \tag{12.82}$$

where γ_L is now a contour enclosing the whole system and the sum in (12.82) extends now over all the singularities contained within γ_L, i.e., all the singularities of $\theta(r)$. Let us rewrite (12.82) as:

$$\oint_{\gamma_L} dl.\nabla\theta(r) = 2\pi \int_D dr\rho(r),\qquad(12.83)$$

where $\rho(r)$ is the local density of singularities:

$$\rho(r) = \sum_{j=1}^{N} n_j \delta(r - r_j),\qquad(12.84)$$

and:

$$\int_D dr\rho(r) = \sum_{j=1}^{N} n_j,\qquad(12.85)$$

since all singularities are inside D. In (12.84) it is assumed that there are N-singularities (note that N should not be confused here with the number of lattice sites), each singularity being represented as a point defect located at, say, $r = r_j$ and having a strength n_j, which is a positive or negative integer. For example, (12.80) represents a point defect located at $r = 0$ and of strength $n = 1$. In conclusion, a topological phase can be characterized by its density of point defects, $\rho(r)$, of (12.84).

12.2.4   Relation to the Vorticity

Let us now rewrite (12.66) in terms of the distortion field, $v(r) = \nabla\theta(r)$ (cf. (12.37)):

$$2\pi \sum_j n_j = \int_{\gamma_L} dl.v(r) = \int_{S_{\gamma_L}} dS.(\nabla \times v(r)),\qquad(12.86)$$

where the second equality results from Stokes's theorem (E. 13).

Since here, $dS = \hat{z}dr$, where $\hat{z}$ is the direction normal to the x-y plane, we have comparing with (12.85):

$$2\pi\rho(r) = \hat{z}.(\nabla \times v(r)),\qquad(12.87)$$

Fig. 12.18 An isolated vortex in the x-y model. See also case ($k = 1, c = \frac{\pi}{2}$) of Fig. 12.14. *Source* Kosterlitz and Thouless [5]; reproduced with permission

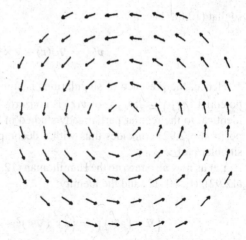

and thus a nonzero $\rho(\mathbf{r})$ implies a nonzero $\nabla \times \mathbf{v}(\mathbf{r})$. In the theory of fluid flow, if $\mathbf{v}(\mathbf{r})$ represents the velocity field of the flow, then $\nabla \times \mathbf{v}(\mathbf{r})$ represents its vorticity. Such a vortex is shown in Fig. 12.18. It is obvious that the arrows in Fig. 12.18 can represent either the flow direction or the spin direction; i.e., Fig. 12.18 can also represent a point defect of our spin system. Equation (12.84) can thus be said to represent a density of vortices of centers $\{r_j\}$ and strength $\{n_j\}$. The point defects and vortex languages being equivalent. As explained in Appendix F, a vector field like $\mathbf{v}(\mathbf{r})$ in (12.87) can always be decomposed in two independent parts:

$$\mathbf{v}(\mathbf{r}) = \nabla \varphi(\mathbf{r}) + \nabla \times \mathbf{A}(\mathbf{r}), \tag{12.88}$$

where the scalar potential $\varphi(\mathbf{r})$ does not contribute to (12.87) while the contribution of the vector potential reads here:

$$\hat{\mathbf{z}} \cdot (\nabla \times \mathbf{v}(\mathbf{r})) = \hat{\mathbf{z}} \cdot \left(\nabla \times (\nabla \times \mathbf{A}(\mathbf{r}))\right) =$$
$$\hat{\mathbf{z}} \cdot \nabla (\nabla \cdot \mathbf{A}(\mathbf{r})) - \nabla^2 \hat{\mathbf{z}} \cdot \mathbf{A}(\mathbf{r}), \tag{12.89}$$

but since, $\hat{\mathbf{z}} \cdot \nabla \equiv 0$, $\left(cf. \nabla = \frac{\partial}{\partial \mathbf{r}}, \mathbf{r} = (x, y)\right)$ we can rewrite (12.87) as:

$$-\nabla^2 \Psi(\mathbf{r}) = 2\pi \rho(\mathbf{r}), \tag{12.90}$$

where we introduced:

$$\Psi(\mathbf{r}) = \hat{\mathbf{z}} \cdot \mathbf{A}(\mathbf{r}), \tag{12.91}$$

so that (12.88) becomes:

$$v(r) = \nabla\varphi(r) + \nabla \times \left(\hat{z}\,\Psi(r)\right). \tag{12.92}$$

Hence, the presence of singularities is signaled here by the fact that the vector potential, $A(r) = \hat{z}\Psi(r)$, of $v(r)$ is nonzero. The scalar potential, $\varphi(r)$, of $v(r)$ is identical to the regular part already studied in Sect. 12.2.2. The Helmholtz decomposition (12.92) coincides thus with a decomposition of $v(r)$ into a regular and a singular part.

Let us now also rewrite the Hamiltonian (12.38) in terms of $\varphi(r)$ and $\Psi(r)$. Using (12.92), (E. 10–11) and the identity:

$$\left(\nabla \times \left(\hat{z}\Psi\right)\right)^2 = \left(\hat{z}\right)^2 (\nabla\Psi)^2 - \left(\hat{z}\cdot\,\nabla\Psi\right)^2 = (\nabla\Psi)^2, \tag{12.93}$$

(cf. $\hat{z}\cdot\nabla \equiv 0$), we obtain:

$$\beta H = \frac{K}{2} \int dr\{(\nabla\varphi(r))^2 + (\nabla\Psi(r))^2\}, \tag{12.94}$$

showing that the regular ($\varphi(r)$) and singular ($\Psi(r)$) contributions to H are additive. Finally, since:

$$\int dr(\nabla\Psi)^2 = \int dr(\nabla.(\Psi\nabla\Psi) - \Psi\nabla^2\Psi), \tag{12.95}$$

we can rewrite (12.94) as:

$$\beta H = \frac{K}{2} \int dr\{(\nabla\varphi(r))^2 - \left(\Psi(r)\nabla^2\Psi(r)\right)\}, \tag{12.96}$$

provided the surface term:

$$\int dr\nabla \cdot \{\Psi(r)\nabla\Psi(r)\} = \int dS.\{\Psi(r)\nabla\Psi(r)\}, \tag{12.97}$$

vanishes. If this is the case we can use (12.90) to rewrite (12.96) as:

$$\beta H = \frac{K}{2} \int dr\{(\nabla\varphi(r))^2 + 2\pi\rho(r)\Psi(r)\}. \tag{12.98}$$

If we formally solve (12.90) as:

$$\Psi(r) = \int dr' G(r - r')\rho(r'),$$ (12.99)

where $G(r)$ is Green's function satisfying:

$$-\nabla^2 G(r) = 2\pi\delta(r),$$ (12.100)

together with some boundary conditions [indeed, operating with ∇^2 on both sides of (12.99) and using (12.100), we recover (12.90)], we finally obtain for (12.98):

$$\beta H = \frac{K}{2} \int dr (\nabla\varphi(r))^2 + \pi K \int dr \int dr' \rho(r) G(r - r')\rho(r'),$$ (12.101)

showing explicitly the contribution of the point defects $\rho(r)$ to the Hamiltonian.

12.2.5 Relation to the 2D Coulomb Plasma

We immediately recognize that (12.90) is the 2D Poisson equation while (12.100) is the equation defining the 2D Coulomb potential (see Appendix E). There is thus also a close relation between the singular part of H and electrostatics. Indeed, $\Psi(r)$ can be viewed (cf. (12.99)) as the (electrostatic) potential due to the (charge) distribution $\rho(r)$. It will thus be said that $\Psi(r)$ is due to some "topological charges" of density $\rho(r)$. From (12.101) one also recognizes the typical expression for the charge–charge interactions in a Coulomb system. The point defects of $\theta(r)$ can thus be interpreted either as a gas of vortices with vortex–vortex interactions (by viewing $v(r)$ as a velocity field) or as a gas of topological charges with Coulombic charge–charge interactions (by viewing $\Psi(r)$ as an electrostatic potential). These interpretations are of course equivalent and from now on we will use them interchangeably.

As shown in Appendix E the infinite space (i.e., $0 < r < \infty, r = |r|$) solution of (12.100) is the 2D Coulomb potential:

$$V_2(r) = \ln\left(\frac{a}{r}\right) = -\ln\left(\frac{r}{a}\right),$$ (12.102)

where a is an arbitrary scale factor necessary in order to make the argument of the logarithm dimensionless. In the present case our spin system is of finite dimension. After having taken the continuous limit of the original $L \times L$ lattice, the lattice spacing a loses its meaning as a distance and becomes a lower cutoff. Since the square character of the integration domain D in (12.34) becomes irrelevant here we henceforth will replace it by a circular domain ($\int_D dr = \int_a^L dr r \int_0^{2\pi} d\varphi \cdots$) so that we need a solution of (12.100) for $a < r < L$. We can thus identify the arbitrary scale factor of (12.102) with the lower cutoff a and write:

$$G(|r|) = \begin{cases} -\ln\left(\frac{|r|}{a}\right) & |r| \geq a \\ 0 & |r| \leq a \end{cases},$$ (12.103)

since nothing can be said here about the $r < a$ region. Let us now consider the other boundary condition, $r \to L$. Using (12.103) in (12.99) we obtain:

$$\Psi(r) = 2\pi \sum_j n_j G(|r - \bar{r}_j|),$$ (12.104)

and, $\Psi(r) \to 2\pi G(L) \sum_j n_j$ when $r \to L$. Hence the surface term (12.97) will vanish provided we impose:

$$\sum_j n_j = \int dr\, \rho(r) = 0,$$ (12.105)

which is the condition of electroneutrality well known for a plasma of real charges. Having satisfied all boundary conditions we now turn to the Hamiltonian (12.101). We write:

$$(\beta H)_{\text{reg}} = \frac{K}{2} \int dr (\nabla \varphi(r))^2,$$ (12.106)

for the regular part already studied in Sect. 12.2.2, and we concentrate now on the contribution to H of the topological charges (cf. (12.84)):

$$(\beta H)_t = \pi K \int dr \int dr' \rho(r) G(|r - r'|) \rho(r') = \pi K \sum_{i,j} n_i n_j G(|r_{ij}|)$$ (12.107)

where we note that because of (12.105), $G(|r_{ij}|)$ and $G(|r_{ij}|) +$ constant, lead to the same Hamiltonian. We now use this freedom in order to fix the ground state of our Hamiltonian. Taking this constant to be $-G(d)$ with $d > a$, we obtain finally:

$$(\beta H)_t = \pi K \sum_{i,j} n_i n_j \{G(|r_{ij}|) - G(d)\}$$

$$= \pi K \sum_{i,j}' n_i n_j \{G(|r_{ij}|) - G(d)\} + \pi K \sum_j (n_j)^2 \{G(0) - G(d)\},$$ (12.108)

or using (12.103)

$$(\beta H)_t = -2\pi K \sum_{i<j} n_i n_j \ln\left(\frac{|r_{ij}|}{d}\right) + \pi K \left(\sum_j n_j^2\right) \ln\frac{d}{a}, \quad (12.109)$$

where we have separated the double sum of (12.108) into a sum over different charges $\left(\sum_{i,j}' = \sum_{\substack{i,j \\ i \neq j}}\right)$ and a self-energy contribution $(\sum_{\substack{i,j \\ i=j}})$ whereas in (12.109) we replaced, $\sum_{i<j}$ by $2\sum_{i<j}$ since each pair (i, j) is counted twice. Note that the first term of (12.109) is identical to the Hamiltonian of a Coulomb gas of charges (cf. $K = \beta J$):

$$q_i = n_i \sqrt{2\pi J}, \quad (12.110)$$

while the second term of (12.109) is a self-energy term. As, $\ln\left(\frac{|r_{ij}|}{d}\right) \to 0$ when $|r_{i,j}| \to d$, we may interpret d as a lower bound to the distance between two charges. Hence d represents the "diameter" of the charges or of the singularity. We must have, $d > a$, because a is a microscopic scale here (the lattice spacing), void of any physics. The last term of (12.109)

$$\beta E_{\text{core}} = \left(\sum_j n_j^2\right) \pi K \ln\left(\frac{d}{a}\right), \quad (12.111)$$

is usually referred to as the core energy, related to what happens inside the core of the singularity. It can also be viewed as an excluded volume term since no physical description is available here for the region $a < r < d$.

12.2.6 A Topological Phase Transition

As already seen in Chap. 9, when two phases can compete with each other, a system may undergo a phase transition when one of its thermodynamic parameters (e.g., its temperature) is modified. When these two phases are topological phases the resulting phase transition is called a topological phase transition. As seen in the previous section our spin system is such a topological phase and we now inquire whether it can undergo a phase transition. As seen before, this system may be viewed either as a gas of vortices with logarithmic interactions or as a Coulomb plasma of topological charges. From the present point of view we must thus first gain some insight into the energetic and thermodynamic aspects of this system. Let us first consider the case where $\theta(r)$ represents a dilute gas of vortices. Assume that γ is a closed contour, say a circle of radius r around the center of a vortex located say at $r = 0$. If the gas is dilute the interior of γ will not feel the presence of the other vortices and we may assume that inside γ we have, $\theta(r) = \theta(|r|)$. In this case (12.81) yields:

$$2\pi r \frac{\partial \theta(r)}{\partial r} = 2\pi n, \tag{12.112}$$

where $r = |r|$, and:

$$\frac{\partial \theta(r)}{\partial r} = \frac{n}{r} \qquad r > a, \tag{12.113}$$

which on substitution into (12.34) yields:

$$(\beta H)_{1v} = \frac{K}{2} 2\pi \int_a^L dr\, r \left(\frac{\partial \theta(r)}{\partial r} \right)^2$$

$$= \pi K \int_a^L dr\, r \left(\frac{n}{r} \right)^2 = n^2 \pi K \ln\left(\frac{L}{a} \right), \tag{12.114}$$

for this one vortex (1v) contribution to H. A remarkable feature of this result is that the energy of a vortex increases when the system size L increases. This is a direct consequence of (12.113) which shows that $\frac{\partial \theta(r)}{\partial r}$ decays very slowly (as $\frac{1}{r}$). In some sense a single vortex is thus already a macroscopic object. The creation of a vortex requires hence always a large energy, and this could easily become an unovercomable thermal barrier except maybe at high temperatures (or small K-values). Note also that, because of the n^2 prefactor in (12.114), the creation of a vortex of higher strength $(n^2 > 1)$ is much less probable than for a vortex of unit strength $(n = \pm 1)$. We therefore disregard the higher strength vortices and henceforth focus only on the unit strength $(n = \pm 1)$ vortices. Note however that even in this case, the electroneutrality condition (12.105) requires an even number of vortices, i.e., an equal number of vortices $(n = +1)$ and of anti-vortices $(n = -1)$. A topological phase should thus contain at least one vortex and one anti-vortex. As an example of a vortex of unit strength (but now without the assumption of spherical symmetry $\theta(r) = \theta(r)$) we can take (12.80). Direct evaluation yields now, $\frac{\partial \theta}{\partial x} = -\frac{y}{r^2}$ and $\frac{\partial \theta}{\partial y} = \frac{x}{r^2}$, so that (12.34) becomes:

$$(\beta H)_{1v} = \frac{K}{2} \int dr \left(\frac{y^2}{r^4} + \frac{x^2}{r^4} \right) = \pi K \int_a^L \frac{dr}{r} = \pi K \ln\left(\frac{L}{a} \right), \tag{12.115}$$

a result identical to (12.114) for $n^2 = 1$.

For a system at constant temperature, the above energetic considerations must be supplemented by entropic ones, since what really counts here is the Helmholtz free energy, $F = E - TS$:

$$\beta F = \beta E - \ln W, \qquad (12.116)$$

where we have used the Boltzmann expression for the entropy, $S = k_B \ln W$, W being the "free" volume available to the system. For one vortex we obtain then from (12.115) for (12.116):

$$(\beta F)_{1v} = (\beta E)_{1v} - \ln (W)_{1v}$$

$$= \pi K \ln\left(\frac{L}{a}\right) - \ln\left(\frac{L}{a}\right)^2$$

$$= (\pi K - 2) \ln\left(\frac{L}{a}\right), \qquad (12.117)$$

where we have used, $(W)_{1v} = \left(\frac{L}{a}\right)^2$ since the vortex can be everywhere in the system of 2D volume L^2.

From our estimate (12.117) it is seen that for:

$$\pi K_- < 2, \qquad (12.118)$$

we have $(\beta F)_{1v} < 0$, and hence it becomes thermodynamically favorable for the system to form a vortex, since this will lower its free energy. Note that this argument is possible only because in (12.117) we could factor out the large factor, $\ln\left(\frac{L}{a}\right)$, between E and S. Indeed, what follows is possible only in systems where the dominant interactions (E) are logarithmic (hence the overwhelming presence of the logarithmic function in all our equations). For high temperatures or small K-values, namely $\pi K < 2$, we expect our system to form vortices. We can thus assume that the high-temperature phase of our 2D spin system (i.e., the phase corresponding to the disordered phase of 3D systems) will be a topological phase consisting of a dilute gas of vortices (or a plasma of free charges).

For low temperatures, or large K-values, the formation of free vortices is thermodynamically unfavorable. In order to circumvent this difficulty we assume now that the positive and negative charges form dipoles, or equivalently, that the vortices ($n = +1$) and anti-vortices ($n = -1$) form bound states. Indeed, because of the electroneutrality condition, our system will consist of, say N, positive charges and N negative charges which could form a dilute gas of bound states [note that the number of singularities is here equal to $2N$, not N as in the previous section. As an example of

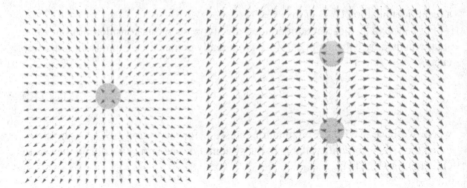

Fig. 12.19 To the left a single vortex configuration and to the right a vortex–anti-vortex pair. The angle θ is shown as the direction of the arrows, and the cores of the vortex and anti-vortex are shaded in red and blue, respectively. Note how the arrows rotate as you follow a contour around a vortex. (Figure by Thomas Kvorning.) *Source* Scientific Background on the Nobel Prize in Physics 2016 [6]; reproduced with permission

such a topological dipole ($n = \pm 1$), located along the x-axis at $\left(\pm \frac{R}{2}, 0 \right)$, we consider (see also Fig. 12.19)]:

$$\theta(x, y) = \operatorname{arctg}\left(\frac{yR}{x^2 + y^2 - \left(\frac{R}{2}\right)^2} \right), \tag{12.119}$$

where R is the length of the dipole, or the center-to-center distance of the vortex–anti-vortex pair. Note that, if γ encloses now the two singularities of (12.119), (12.81) becomes:

$$\oint_\gamma d\boldsymbol{l} . \nabla \theta(\boldsymbol{r}) = 2\pi (1 - 1) = 0, \tag{12.120}$$

the difference with (12.112) being that now $\theta(\boldsymbol{r})$ can no longer be symmetric (cf. $\theta(|\boldsymbol{r}|)$) inside γ, since γ contains two centers now. Equation (12.120) implies then that the contributions of $\nabla \theta(\boldsymbol{r})$ on γ around the two centers have to compensate each other. This compensation will make $\nabla \theta(\boldsymbol{r})$ of short range, and this will in turn strongly affect our previous thermodynamic considerations. Indeed, substituting (12.119) into (12.34), or using (12.109) for a single vortex pair $n_1 = -n_2 = 1$ yields for this two-vortex ($2v$) contribution:

$$(\beta H)_{2v} = 2\pi K \ln\left(\frac{R}{d} \right). \tag{12.121}$$

Comparing (12.121) with (12.115), we see that since, $R \ll L$ and $d > a$, we have, $(\beta H)_{2v} \ll 2(\beta H)_{1v}$, and it will hence be easier for the system to form a bound state of two vortices than to form two free vortices. Using the same argument for the $2v$ case as for the $1v$ case of (12.117) we see that $(\beta F)_{2v} < 0$, provided $L \gg R$. In order to find out whether such a R value, $d < R < L$, exists we proceed as follows. The energy of a dipole of length R is given by (12.121), its probability of occurrence will be proportional to the Boltzmann factor, $\sim \exp -(\beta H)_{2v}$, and the average value of R^2 will then be given by:

$$\langle R^2 \rangle = \int_a^L dr r^3 \exp -2\pi K \ln\left(\frac{r}{d}\right) / \int_a^L dr r \exp -2\pi K \ln\left(\frac{r}{d}\right). \tag{12.122}$$

Using, $x = \frac{r}{a}$, and letting $L \to \infty$ in (12.122) yield:

$$\langle \frac{R^2}{a^2} \rangle = \frac{\int_1^\infty dx x^{3-2\pi K}}{\int_1^\infty dx x^{1-2\pi K}} = \frac{2\pi K - 1}{2\pi K - 3}, \tag{12.123}$$

where we have used: $\int_1^\infty \frac{dx}{x^n} = \frac{1}{n-1}$.

Since (12.123) must be positive, we see that the dipole of length R is well-defined for:

$$\pi K > \frac{3}{2}, \tag{12.124}$$

i.e., in the low temperature or large K-regime. We will thus assume that in this regime our 2D spin system will exhibit a topological phase (corresponding to the ordered phase of the 3D systems) which will consist of a dilute gas of dipoles or vortex–anti-vortex bound states. We thus have two topological phases, one for small K and one for large K, which for some intermediate K-value could compete and lead to a phase transition whereby the dipoles break up into free charges or the vortex–anti-vortex bound states unbind. This phase transition should occur at some intermediate K-value for which both phases are still stable. According to our previous estimates (12.118) and (12.124) this region is:

$$\frac{3}{2} < \pi K < 2. \tag{12.125}$$

This topological phase transition is called the Kosterlitz–Thouless (KT) transition, and the precise value of K for which it occurs, say $\frac{J}{k_B T_c}$, is called the critical temperature (T_c) of this KT-transition. In what follows we will use the renormalisation group (RG) technique explained in Sect. 10.7 in order to obtain a more precise value of its critical temperature. The KT-transition is a phase transition of a different type than those of Chap. 9 because it implies a change in the topology of the singularities of our 2D spin system instead of a change in the order, as was the case for the 3D systems of Chap. 9.

12.2.7   Absence of Screening in 2D Coulomb Systems

As we have seen in Sect. 8.3.4, in a 3D Coulomb plasma the interaction between two external charges is modified by the internal charges. This modification of the original long-ranged Coulomb potential into a short-ranged Debye–Hückel potential is called screening of the external charges by the internal charges. We now show that in the 2D Coulomb plasma there is no screening; i.e., the interaction between the external charges remains logarithmic, even in the presence of medium or internal charges. To show this, we modify the system's charge density $\rho(r)$ (12.84) by introducing two external charges, say one of charge $n = +1$ at $r = R$ and one of charge $n = -1$ at $r = R'$. This corresponds to an external charge density:

$$\rho_{ext}(r) = \delta(r - R) - \delta(r - R'), \tag{12.126}$$

which together with $\rho(r)$ will form then the total (T) charge density according to:

$$\rho_T(r) = \rho(r) + \rho_{ext}(r). \tag{12.127}$$

When changing $\rho(r)$ into $\rho_T(r)$, the system's Hamiltonian will change from H (12.101) into H_T:

$$\beta H_T = \pi K \int dr \int dr' \, \rho_T(r) G\left(\left|r - r'\right|\right) \rho_T\left(r'\right), \tag{12.128}$$

where we have omitted the regular part of H, i.e., put $\varphi(r) = 0$, since the latter plays no role in what follows. Using (12.127) we can decompose (12.128) as:

$$\beta H_T = \beta H_{ext} + \beta H + \beta H_{int}, \tag{12.129}$$

where βH_{ext} is the contribution of (12.126)–(12.128):

$$\beta H_{ext} = \pi K \int dr \int dr' \, \rho_{ext}(r) G\left(\left|r - r'\right|\right) \rho_{ext}\left(r'\right), \tag{12.130}$$

βH the contribution of $\rho(r)$, and βH_{int} the interaction term between $\rho(r)$ and $\rho_{ext}(r)$:

$$\beta H_{int} = 2\pi K \int dr \int dr' \, \rho_{ext}(r) G\left(\left|r - r'\right|\right) \rho\left(r'\right). \tag{12.131}$$

Using $G(|0|) = 0$ (cf. (12.103)), we have:

$$\beta H_{\text{ext}} = -2\pi K G\left(\left|\mathbf{R} - \mathbf{R}'\right|\right), \tag{12.132}$$

and:

$$\beta H_{\text{int}} = 2\pi K \sum_i n_i \left\{ G(|\mathbf{r}_i - \mathbf{R}|) - G\left(\left|\mathbf{r}_i - \mathbf{R}'\right|\right) \right\}, \tag{12.133}$$

while:

$$\beta H = 2\pi K \sum_{i<j} n_i n_j G\left(\left|\mathbf{r}_{ij}\right|\right). \tag{12.134}$$

In order to study the system's response to the external charges we will compare, $\exp -\beta H_{\text{ext}}$ with $\langle \exp -\beta H_T \rangle$, where the average is taken over the internal charges. From (12.129) and (12.130) it follows that:

$$\langle e^{-\beta H_T} \rangle = e^{2\pi K G(|\mathbf{R}-\mathbf{R}'|)} \langle e^{-\beta(H+H_{\text{int}})} \rangle. \tag{12.135}$$

Here, the internal charges consist of N positive and N negative charges described by the Hamiltonian (12.134). Since the exact value of N is not known a priori, we will evaluate the average in (12.135) with the aid of the grand canonical ensemble (cf. Chap. 6). The corresponding partition function Z will read then:

$$Z = \sum_{N=0}^{\infty} \frac{(Z_+)^N}{N!} \frac{(Z_-)^N}{N!} \prod_{k=1}^{2N} \frac{dr_k}{a^2} exp 2\pi K \sum_{i<j} n_i n_j G\left(|r_{ij}|\right), \tag{12.136}$$

where $Z_\pm = exp - \beta E_\pm$ is the fugacity of the $\pm$ charges whose core energy is $E_\pm$. If, as in (12.109), $E_\pm$ is taken to be independent of the sign of the charges ($E_+ = E_- = E_0$), we have $Z_+ = Z_-$ and we can put, $Z_+ Z_- = y^2$:

$$y^2 = \exp -2\pi K \ln\left(\frac{d}{a}\right), \tag{12.137}$$

and (12.136) becomes:

$$Z = \sum_{N=0}^{\infty} \frac{y^{2N}}{(N!)^2} \prod_{k=1}^{2N} dx_k \, exp 2\pi K \sum_{i<j} n_i n_j G(|x_{ij}|) \ . \tag{12.138}$$

where n_i and n_j are unit charges (± 1), $x_k = \frac{r_k}{a}$ is dimensionless with:

$$\int dx_k \cdots = \int_1^{\frac{L}{a}} dx_k x_k \int_0^{2\pi} d\varphi_k \cdots . \tag{12.139}$$

Note also that in (12.136) and (12.138) we have considered the charges of the same sign to be undistinguishable (hence the $(N!)^2$ factor). From (12.137) it follows that y^2 will be small when K is large and hence (12.138) can be viewed as a perturbation series in power of y^2, viz.

$$\mathsf{Z} = 1 + y^2 I_0 + \mathcal{O}(y^4), \tag{12.140}$$

with:

$$I_0 = \int dx \int dx' \exp 2\pi K G\big(|x - x'|\big), \tag{12.141}$$

and we write similarly for (12.135):

$$\langle e^{-\beta(H+H_{\text{int}})} \rangle = \frac{1 + y^2 I(R, R') + \mathcal{O}(y^4)}{1 + y^2 I_0 + \mathcal{O}(y^4)}$$

$$= 1 + y^2 \big\{ I(R, R') - I_0 \big\} + \mathcal{O}(y^4)$$

$$= \exp\big[y^2 \big\{ I(R, R') - I_0 \big\} + \mathcal{O}(y^4) \big], \tag{12.142}$$

or returning to (12.135):

$$\langle e^{-\beta H_T} \rangle = \exp\big\{ 2\pi K G\big(|R - R'|\big) + y^2 \big\{ I(R, R') - I_0 \big\} \big\}, \tag{12.143}$$

to dominant order in y^2. Here:

$$I(R, R') - I_0 = \int dx \int dx' e^{2\pi K G(|x-x'|)} \big\{ e^{2\pi K D(x,x':R,R')} - 1 \big\}, \tag{12.144}$$

where (putting $a = 1$ in order to simplify the notation):

$$D(x, x' : R, R') = G\big(|x - R'|\big) + G\big(|x' - R|\big) - G\big(|x - R|\big) - G\big(|x' - R'|\big), \tag{12.145}$$

and it is understood that the primed variables (x', R') correspond to negative charges and the unprimed ones (x, R) to positive charges.

To proceed with the evaluation of (12.142) we first introduce some new variables:

$$\begin{cases} s = x - x' \\ S = \frac{(x+x')}{2} \end{cases} \begin{cases} x = S + \frac{s}{2} \\ x' = S - \frac{s}{2} \end{cases}, \tag{12.146}$$

which are such that, $dx dx' = ds dS$, whereas (12.144) becomes:

$$I(R, R') - I_0 = \int ds e^{2\pi K G(s)} \int dS \left\{ e^{2\pi K A(s,S;R,R')} - 1 \right\}, \tag{12.147}$$

with:

$$A(s, S; R, R') = G\left(\left| S - R' + \frac{s}{2} \right| \right) + G\left(\left| S - R - \frac{s}{2} \right| \right)$$
$$- G\left(\left| S - R + \frac{s}{2} \right| \right) - G\left(\left| S - R' - \frac{s}{2} \right| \right)$$
$$= \left\{ e^{\frac{1}{2}s.\nabla} - e^{-\frac{1}{2}s.\nabla} \right\} \left\{ G(|S - R'|) - G(|S - R|) \right\}. \tag{12.148}$$

where here $\nabla = \frac{\partial}{\partial S}$ and we have used the property, $e^{\alpha.\nabla} f(S) = f(S + \alpha)$.

Note now that the prefactor, $\exp 2\pi G(s) = \exp -2\pi K \ln s$, in (12.147) will be small except when s is small. We may thus expand $A(s, S; R, R')$ for small s, yielding to dominant order:

$$A(s, S; R, R') = s.\nabla \left\{ G(|S - R|) - G(|S - R'|) \right\} + \mathcal{O}(s^3), \tag{12.149}$$

whereas (12.147) becomes (simplifying the notation somewhat):

$$I(R, R') - I_0 = \int_1^{\frac{L}{a}} ds s e^{2\pi KG(s)} \int d\hat{s} \int dS \left\{ 2\pi K A + \frac{1}{2}(2\pi K A)^2 + \mathcal{O}(s^3) \right\}, \tag{12.150}$$

where $\int d\hat{s} = \int_0^{2\pi} d\varphi$, φ being here the polar angle of s. Consider first the term linear in A of (12.150). We have (cf. (12.149)):

$$\int d\hat{s}\, s \cdot \nabla \{\cdots\} = 0, \tag{12.151}$$

by symmetry. For the quadratic term we have instead:

$$\int dS A^2 = \int dS[s \cdot \nabla[\{\cdots\}s \cdot \nabla\{\cdots\}] - \{\cdots\}(s \cdot \nabla)^2\{\cdots\}], \qquad (12.152)$$

after integration by parts and $\{\cdots\} = \{G(|S - R|) - G(|S - R'|)\}$ (cf. (12.149). For large enough L the surface term of (12.152) will vanish (cf.$\{\cdots\} \to 0$) while the second term of (12.152) can be simplified using first $\int \frac{d\hat{s}}{2\pi}\left(\hat{s} \cdot \nabla\right)^2 = \frac{1}{2}\nabla^2$, and then, $-\nabla^2 G(|S - R|) = 2\pi\delta(S - R)$ and $G(|0|) = 0$ (cf. (12.100) and (12.103)), to yield finally for (12.150):

$$I(R, R') - I_0 = \int\limits_{1}^{\frac{L}{a}} ds s^3 e^{2\pi KG(s)}(2\pi)\frac{1}{2}(2\pi K)^2\frac{1}{2}(+2\pi)(-2)G(|R - R'|)$$

$$= -G(|R - R'|)8\pi^4 K^2 \int\limits_{1}^{\frac{L}{a}} ds s^3 e^{2\pi KG(s)}, \qquad (12.153)$$

which on substituting into (12.143) yields:

$$\langle e^{-\beta H_T}\rangle = e^{2\pi K_R G(|R-R'|)}, \qquad (12.154)$$

where

$$K_R = K - 4\pi^3 K^2 y^2 \int\limits_{1}^{\frac{L}{a}} ds s^{3-2\pi K} + \mathcal{O}(y^4), \qquad (12.155)$$

From (12.154) we see that the medium (internal charges) did not modify the form of the potential $(G(R - R'))$ between the external charges at R and R'. Hence, in a 2D Coulomb plasma there is no screening of the Coulomb potential. From (12.155) we also see that the whole effect of the medium is contained in a reduction $(K_R < K)$ of the strength of the Coulomb potential, without modifying its range (as would be the case in 3D). Since (cf. (12.100)) K determines the charge $(\beta q^2 = 2\pi K)$, (12.155) shows how the plasma medium diminishes the effective charge on the particles. Hence, K_R is called the renormalized value of K. We now study (12.155) in more detail.

12.2.8 RG Equations of the Kosterlitz–Thouless Transition

Let us rewrite the charge renormalization (12.155) as:

$$\frac{1}{K_R} = \frac{1}{K} + 4\pi^3 y^2 \int_1^{\frac{L}{a}} ds\, s^{3-2\pi K} + \mathcal{O}(y^4). \tag{12.156}$$

where, for concreteness, we may use $\frac{d}{a} = 10$ in (12.137) or:

$$y^2 = \exp(-2\pi K \ln 10), \tag{12.157}$$

so that (12.156) depends only on K and $\frac{L}{a}$. When, $\pi K > 2$, the integral in (12.156) is well behaved as $L/a \to \infty$ and the correction $(K_R - K)$ is finite. Note however that, $\pi K > 2$, is just outside the region of interest (12.125) for the study of the KT-transition. In order to extract some useful information from (12.156) for $\pi K \simeq 2$ we now apply the RG ideas of Sect. 10.7 to (12.156). From what was said there it is clear that L is here the diverging length scale (i.e., $L \to \infty$ is responsible for the divergence of (12.156) for $\pi K \simeq 2$) whereas a is an irrelevant microscopic length scale which we will now try to "decimate." To this end we proceed iteratively. In a first step we separate the small length scales, say from a to $a_1 = a.b$ with $b > 1$, by splitting the integration domain in (12.156) as follows:

$$\int_a^L dr = \int_a^{a_1} dr + \int_{a_1}^L dr \quad \text{or} \quad \int_1^{\frac{L}{a}} ds = \int_1^b ds + \int_b^{\frac{L}{a}} ds, \tag{12.158}$$

and absorb the short distances into a renormalization of K:

$$\frac{1}{K_1} = \frac{1}{K} + 4\pi^3 y^2 \int_1^b ds\, s^{3-2\pi K}, \tag{12.159}$$

while the second integral is rewritten as $(s = bu)$:

$$y^2 \int_b^{\frac{L}{a}} ds\, s^{3-2\pi K} = y^2 b^{4-2\pi K} \int_1^{\frac{L}{a_1}} du\, u^{3-2\pi K}$$

so that (12.156) becomes:

$$\frac{1}{K_R} = \frac{1}{K_1} + 4\pi^3 y_1{}^2 \int\limits_1^{\frac{L}{a_1}} ds\, s^{3-2\pi K_1},\tag{12.160}$$

where

$$y_1{}^2 = y^2 b^{4-2\pi K} \text{ or } y_1 = y b^{2-\pi K},\tag{12.161}$$

y_1 being the renormalized value of y. Note that (12.160) is identical in form to (12.156). Repeating now the above decimation process on (12.159) we obtain:

$$\frac{1}{K_R} = \frac{1}{K_2} + 4\pi^3 y_2{}^2 \int\limits_1^b ds\, s^{3-3\pi K},\tag{12.162}$$

generating a RG:

$$\frac{1}{K_n} = \frac{1}{K_{n-1}} + 4\pi^3 y_{n-1}^2 \int\limits_1^b ds\, s^{3-2\pi K_{n-1}},\tag{12.163}$$

$$y_n = y_{n-1} b^{2-\pi K_{n-1}},\tag{12.164}$$

where $a_{n+1} = a_n b = a b^{n+1}$ with $b > 1$ and $K_0 = K,\, y_0 = y,\, n = 1, 2, \ldots$.

The transformations (12.163) and (12.164) leave (12.156) invariant:

$$\frac{1}{K_R} = \frac{1}{K_n} + 4\pi^3 y_n^2 \int\limits_1^{\frac{L}{a_n}} ds\, s^{3-2\pi K_n}.\tag{12.165}$$

The differential equations (cf. Sect. 10.7) characterizing this RG (12.163) and (12.164) can be obtained by writing, $b = 1 + \frac{da}{a}$, and expanding to dominant order in $\frac{da}{a}$. For instance, for (12.164) we have:

$$y_n = y_{n-1}\left(1 + (2 - \pi K)\frac{da}{a} + \cdots\right); \quad y_n - y_{n-1} = dy + \cdots$$

or:

$$a\frac{dy}{da} = (2 - \pi K)y,\tag{12.166}$$

a linear equation for y. From (12.163) we obtain:

$$K_{n+1}^{-1} = K_n^{-1} + 4\pi^3 y_n^2 \int\limits_1^{1+\frac{da}{a}} ds\, s^{3-2\pi K_n}$$

$$K_{n+1}^{-1} - K_n^{-1} = 4\pi^3 y_n^2 \frac{da}{a} 1^{3-2\pi K} + \cdots = dK^{-1} + \cdots$$

or:

$$a\frac{dK^{-1}}{da} = 4\pi^3 y^2, \tag{12.167}$$

a nonlinear equation, which together with (12.166) constitute the RG differential equations corresponding to the RG-transformations (12.163) and (12.164). During the RG-transformations we coarse grain the small length scales, looking at larger and larger a-values. Equation (12.167) shows that during this process K^{-1} increases with a. Therefore, in the RG-language, K^{-1} is called a relevant variable. From (12.166) it is seen that for $\pi K < 2$, y is a relevant variable whereas for $\pi K > 2$, y decreases with a and becomes hence an irrelevant variable. The point,

$$\left(K^{-1} = \frac{\pi}{2}, y = 0\right), \tag{12.168}$$

of the plane (K^{-1}, y), corresponds to the fixed point of the RG-equations (12.166) and (12.167). Let us introduce the new variable,

$$x = K^{-1} - \frac{\pi}{2}, \tag{12.169}$$

and rewrite the RG-equations:

$$a\frac{dx}{da} = 4\pi^3 y^2, \tag{12.170}$$

$$a\frac{dy}{da} = \frac{4}{\pi}xy, \tag{12.171}$$

where, in the vicinity of the fixed point (12.168), we have approximated, $2 - \pi K = \frac{4x}{(\pi+2x)} \simeq \frac{4x}{\pi}$ in (12.171), since x is small there. Observe now that the quantity c:

$$c = x^2 - \pi^4 y^2, \tag{12.172}$$

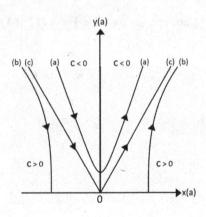

is conserved by (12.170) and (12.171), viz. $a\frac{dc}{da} = 0$. The solutions to the RG-equations (12.170) and (12.171), which are curves in the x-y plane, can therefore be labeled by the value of c. Three examples of such curves (hyperbola) are shown in Fig. 12.20. When c is negative, $c = -|c|$, we have $\pi^4 y^2 = x^2 + |c|$, corresponding to curve (a). Indeed, since $y > 0$, curve (a) represents the upper part of a hyperbola around the y-axis. The flow direction for increasing a is indicated by the arrow. When, $c = |c| > 0$, we have $x^2 = |c| + \pi^4 y^2$, corresponding to an hyperbola around the x-axis (cf. $x = \pm\sqrt{|c| + \pi^4 y^2}$) but since y must be positive, only the upper parts of the hyperbola are physical, as shown by curve (b). Note that curve (b) ends (when $x < 0$) or starts (when $x > 0$) at the line of fixed points $y = 0$, these being stable for $x < 0$ and unstable for $x > 0$. For $c = 0$ we have, $\pi^2 y = \pm x > 0$, i.e., $\pi^2 y = x$ when $x > 0$ and $\pi^2 y = -x$ when $x < 0$ as shown in curve (c).

Note also that this curve (c) is the separatrix of the curves of the type (a) and of the type (b); i.e., curve (c) provides the asymptotes for all the hyperbola. For large K (low T) or $x < 0$, we are in the insulating phase and y should decrease as shown in curve (b). For curve (a), on the contrary, y proceeds to unbounded values, where our perturbation analysis will certainly break down, as expected for the high-T conducting phase. The curve (c), corresponding to the separatrix between the high-T and low-T behavior, corresponds thus to the critical trajectory ($T = T_c$). The critical temperature T_c of the KT-transition $\left(K_c = \frac{J}{k_B T_c}\right)$ can thus be found from, $y = \frac{x}{\pi^2}$ the low-T critical trajectory along which our perturbation theory should remain valid. The critical value, K_c, of our model for the KT-transition must hence be a solution of $y_c = \frac{x_c}{\pi^2}$, or in terms of the original variables:

$$K_c^{-1} = \frac{\pi}{2} + \pi^2 \exp\left(-\pi K_c \ln\frac{d}{a}\right), \qquad (12.173)$$

or to within dominant order of our perturbation analysis:

$$\pi K_c = 2 - 4\pi \, \exp\left(-\pi K_c \ln \frac{d}{a}\right) + \cdots . \qquad (12.174)$$

Since the correction term must be small, the solution of (12.174) will satisfy (12.125). Of course, the precise value of K_c will still depend on the value chosen for $\frac{d}{a}$. Equation (12.125) shows that, as assumed in (12.157), $\frac{d}{a} \geq 10$ is a realistic choice within the present model.

In conclusion, we have seen that our simple 2D spin model predicts a topological phase transition between a high-temperature phase of free vortices and a low-temperature phase of vortex–anti-vortex bound states, or using the Coulomb language between a plasma of free charges and a plasma of dipoles, all charges being here topological charges, i.e., topological point defects. This phase transition is different from what we have seen in Chap. 9. In the Landau classification it is a transition without order parameter, while in Ehrenfest's classification it appears as a transition of infinite order. It has a critical behavior with algebraic decay but the latter is not universal.

The above KT-transition has been observed experimentally in a series of 2D systems, e.g., surface layers or 2D solids, films of superfluid or superconducting material, etc. Topological phases are also encountered in 1D systems (wires) but their study always requires a quantum mechanical treatment and will not be considered here.

Further Reading

1. S. Chandrasekhar, *Liquid Crystals*, 2nd edn. (Cambridge University Press, Cambridge, 1992). *Discusses topological defects in a variety of liquid crystalline phases.*
2. P.M. Chaikin, T.C. Lubensky, *Principles of Condensed Matter Physics* (Cambridge University Press, Cambridge, 1995). *Provides a discussion of topological defects in a more general context.*
3. M. Kleman, O.D. Lavrentovich, *Soft Matter Physics: An Introduction* (Springer, New York, 2003). *Contains a detailed discussion of disclinations and other topological defects.*
4. I. Dierking, *Textures of Liquid Crystals* (Wiley-VCH, Weinheim, 2003). *Provides beautiful color illustrations of liquid crystal textures.*
5. J.M. Kosterlitz, D.J. Thouless, J. Phys. C: Solid State Phys. **6**, (1973). https://doi.org/10.1088/0022-3719/6/7/010; J.M. Kosterlitz, J. Phys. C: Solid State Phys. **7** (1974). *Historic papers for the Kosterlitz-Thouless transition.*
6. Topological Phase Transitions and Topological Phases of Matter. Scientific Background on the Nobel Prize in Physics 2016 (www.nobelprize.org). *Provides an overview of all the topological phases discovered till now.*

Chapter 13
Phase Transformation Kinetics

Abstract In the above, a phase transition was viewed as resulting from changing a control parameter, such as T or p, across a phase boundary, say a coexistence line, of the system's phase diagram. In practice, such a transformation, of say a liquid phase into a crystal phase, will however require a considerable amount of molecular scale reorganization, and this will always take a finite amount of time. The precise way in which the old (liquid) phase is transformed into the new (crystal) phase will hence involve some dynamic (time-dependent) phenomena which cannot be described within the equilibrium statistical mechanical framework used until now. In the present chapter some of the kinematic aspects related to the physical transformation of one phase into another will thus be considered. Needless to say that a good understanding of these kinematic aspects is essential for the correct interpretation of what is seen in the laboratory during a phase transition. For instance, in some cases the transformation is so slow that the new phase cannot be formed, during the finite time of observation, without some help from the experimentalist. In other cases, the amount of reorganization required is so important that the system gets arrested in a glass-like configuration somewhat intermediate between the two equilibrium phases involved in the transition. Unfortunately, several of these aspects are, even today, not well understood. Here, only the classical theories of nucleation and of spinodal decomposition will be considered.

13.1 Homogeneous Nucleation

The most general mechanism of phase transformation is provided by the process of nucleation. If one starts from an initial phase, say phase 1, without any impurities, one qualifies the nucleation as being homogeneous (the case of heterogeneous nucleation will be analyzed in Sect. 13.2). To initiate a homogeneous nucleation process, one transforms the state (p, T) of phase 1, from an initial equilibrium state (p_i, T_i) into a final state (p_f, T_f), where phase 1 is thermodynamically metastable relative to, say, phase 2. The transformation $(p_i, T_i) \rightarrow (p_f, T_f)$ may proceed via any path (cf. Fig. 13.1) of the (T, p)-plane, as long as this path crosses the 1–2 coexistence line. It may proceed in a slow quasistatic or in a rapid quench-like manner. The only

Fig. 13.1 A thermodynamic path ($i \to b \to$ f) transforming the initial equilibrium state (i) of phase 1 into a final state (f) where phase 1 is metastable relative to phase 2. This path crosses the 1–2 phase boundary at (b)

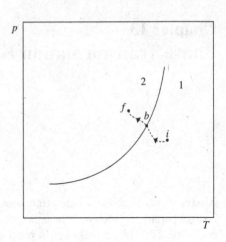

thing which counts is that in the final state, (p_f, T_f), phase 1 is metastable relative to phase 2, which is the equilibrium state corresponding to the values (p_f, T_f) of (p, T). Note that in the case where phase 1 would be unstable in the final state, the process of spinodal decomposition would be involved, instead of the nucleation process considered here (see Sect. 13.3). Although, during the transformation, (p_i, T_i) $\to$ (p_f, T_f), the system crosses the 1–2 phase boundary, nothing particular happens during the crossing. Leaving the system in its final metastable state, an evolution will eventually set in after some time. In the classical theory of homogeneous nucleation, this evolution is described as occurring in three successive steps. In the first step, the thermal fluctuations will produce in phase 1, an embryo or "nucleus" of phase 2; i.e., a small amount of matter where the phase transition 1 $\to$ 2 took place, as a result of the random exploration of the system's free energy surface by the thermal fluctuations. Once the system contains several nuclei, it is observed that the smaller nuclei of phase 2 disappear by retransforming into phase 1, while the larger nuclei eventually grow in size at the expense of the remainder of phase 1. This then corresponds to the second step of the nucleation process. When the larger nuclei grow they eventually coalesce to form still larger ones and progressively transform all of the remainder of phase 1 into a bulk phase of phase 2. This third step represents then the end of the evolution, the system having been, meanwhile, transformed into its new equilibrium state corresponding to the values (p_f, T_f) of (p, T), which is phase 2.

13.1.1   Becker–Döring Theory

An attempt to describe the first step of the nucleation process is provided by the theory of Becker and Doring. Since the first step occurs on the timescale of the thermal fluctuations, no explicit time-dependent considerations will be introduced.

Instead, an attempt will be made to continue to describe the situation using only thermodynamic considerations, although the system is not in a true equilibrium state.

First, consider a system containing one nucleus of phase 2 surrounded by phase 1 and describe it in the context of the so-called capillary approximation of Gibbs. It is thereby assumed that the nucleus of phase 2 can be described in the same way as an ordinary bulk phase, while the nucleus phase 1 interface can be assimilated to a geometric surface. If the system consists of N molecules enclosed in a volume V at the temperature T, and the nucleus contains N_2 molecules and has a volume V_2, the system's Helmholtz free energy, $F(T, V, N)$, will be moreover written as being equal to

$$F(T, V, N) = F_1(T, V_1, N_1) + F_2(T, V_2, N_2) + \sigma_{12}(T)A_{12}(V_2), \qquad (13.1)$$

where $F_1(T, V, N_1)$ is the Helmholtz free energy of a bulk phase-1 containing N_1 $= N - N_2$ molecules occupying a volume $V_1 = V - V_2$, while $F_2(T, V_2, N_2)$ is the corresponding free energy of a bulk phase-2 (representing the nucleus), and $A_{12}(V_2)$ is the surface area of the nucleus on which a surface tension $\sigma_{12}(T)$ is supposed to act. Note that (13.1) is identical to the separation used to describe the equilibrium coexistence between two bulk phases separated by an interface (cf. Chap. 11). In what follows, one will assume the nucleus to be spherical; i.e., it is assumed that F is minimal w.r.t. A_{12} for a given V_2. In practice, this is not always the case, in particular when phase-2 is anisotropic. If $V_2 = 4\pi R^3/3$ and $A_{12}(V_2) = 4\pi R^2$, then a_{12} of (13.1) could moreover still depend on R but, henceforth, this dependence will be neglected and a_{12} will be considered to be a function of T only, as indicated in (13.1). Finally, if the nucleus is treated as a bulk phase, one has $N_2 = \rho_2 V_2$, where ρ_2 is the (constant) density of phase 2.

Since F of (13.1) depends parametrically on V_2 and N_2, i.e., on R, but the value of R is not controlled experimentally, one can let these values fluctuate and impose instead the pressure p acting on V. The thermodynamic potential appropriate to these experimental conditions will hence be the Gibbs free energy $G(T, p, N)$ obtained from $F(T, V, N)$ through the Legendre transform $G(T, p, N) = F(T, V, N) + pV$, with $p = -\partial F/\partial V$, or in the present case one obtains from (13.1)

$$\begin{aligned} G(T, p, N) = (F_1(T, V_1, N_1) + pV_1) + (F_2(T, V_2, N_2) + pV_2) \\ + \sigma_{12}(T)A_{12}(V_2), \end{aligned} \qquad (13.2)$$

where N_2, V_2, N_1, and V_1 are fluctuating but p and N are not. If the nucleus is in thermodynamic equilibrium with phase 1, then the partition of the system into two phases, at fixed T, p, and N, must be such that

$$(dG)_{T,p,N} = 0, \left(\frac{\partial G}{\partial V_i}\right)_{T,p,N} = 0, \left(\frac{\partial G}{\partial N_i}\right)_{T,p,N} = 0, (i = 1, 2). \qquad (13.3)$$

From (13.2) and (13.3) one obtains:

$$\left(\frac{\partial G}{\partial V_1}\right)_{T,p,N} = 0 = \frac{\partial F_1}{\partial V_1} + p, \tag{13.4}$$

and

$$\left(\frac{\partial G}{\partial V_2}\right)_{T,p,N} = 0 = \frac{\partial F_2}{\partial V_2} + p + \sigma_{12}(T)\frac{dA_{12}(V_2)}{dV_2}, \tag{13.5}$$

where

$$\frac{dA_{12}(V_2)}{dV_2} = \frac{d\left(4\pi R^2\right)}{d\left(4\pi R^3/3\right)} = \frac{2}{R}. \tag{13.6}$$

Writing $p_i = -\partial F_i/\partial V_i$ ($i = 1, 2$), (13.4) becomes $p_1 = p$; i.e., the pressure in phase 1 equals the externally imposed pressure p, while (13.5) can then be rewritten as:

$$p_2 = p_1 + \frac{2}{R}\sigma_{12}(T), \tag{13.7}$$

i.e., the pressure inside the nucleus p_2 exceeds ($\sigma_{12}(T) > 0$) the exterior pressure by an amount $2\sigma_{12}(T)/R$, which depends on the radius of curvature R of the nucleus. Equation (13.7) is known as the Laplace law. Finally, from $N = N_1 + N_2$, it follows that

$$\left(\frac{\partial N_1}{\partial N_2}\right)_{T,p,N} = -1, \tag{13.8}$$

and hence

$$\left(\frac{\partial G}{\partial N_1}\right)_{T,p,N} = -\left(\frac{\partial G}{\partial N_2}\right)_{T,p,N}, \tag{13.9}$$

while (13.2) yields

$$\left(\frac{\partial G}{\partial N_1}\right)_{T,p,N} = \frac{\partial F_1}{\partial N_1} + \left[\frac{\partial F_2}{\partial N_2} + \sigma_{12}(T)\frac{\partial A_{12}(V_2)}{\partial N_2}\right]\left(\frac{\partial N_2}{\partial N_1}\right)_{T,p,N}$$
$$= 0 \tag{13.10}$$

with

$$\frac{\partial A_{12}(V_2)}{\partial N_2} = \frac{\partial A_{12}(V_2)}{\partial V_2}\frac{\partial V_2}{\partial N_2} = \frac{2}{R\rho_2}, \tag{13.11}$$

where (13.6) has been used and $N_2\,(V_2) = \rho_2 V_2$. Then (13.10) reads

$$\mu_1(T, p) = \mu_2(T, p) + \frac{2v_2}{R}\sigma_{12}(T), \tag{13.12}$$

where $v_2 = 1/\rho_2, \mu_i = \partial F_i/\partial N_i (i = 1, 2)$. With $p = p_1$ and (13.7) one obtains from (13.12)

$$\mu_1(T, p_1) = \mu_2(T, p_1) + v_2(p_2 - p_1), \tag{13.13}$$

but since $\mu_2(T, p_1) - p_1 v_2 = f_2(T, v_2)$ (see (2.29)), the r.h.s. of (13.13) can be rewritten as $f_2(T, v_2) + p_2 v_2 = \mu_2(T, p_2)$, so that (13.13) becomes

$$\mu_1(T, p_1) = \mu_2(T, p_2), \tag{13.14}$$

i.e., the usual condition of equality of the chemical potentials of the two phases, provided these chemical potentials are referred to the pressure of each phase, these pressures being different on account of (13.7). As a result of (13.14), the equilibrium coexistence curve, say $p = p(T)$, between phase 1 and a nucleus of phase 2 will correspond to a solution of the implicit equation,

$$\mu_1(T, p) = \mu_2\left(T, p + \frac{2}{R}\sigma_{12}(T)\right), \tag{13.15}$$

and this curve will in general be different from the coexistence curve, say $p_0 = p_0(T_0)$, for two bulk phases separated by a planar interface, corresponding to a solution of

$$\mu_1(T_0, p_0) = \mu_2(T_0, p_0). \tag{13.16}$$

To quantify the influence of the curvature of the nucleus on the coexistence curve, one may write $p = p_0 + \Delta p$ and $T = T_0 + \Delta T$, and when the corrections are small one can use the approximation

$$\mu(T_0 + \Delta T, p_0 + \Delta p) \simeq \mu(T_0, p_0) - s_0 \Delta T + v_0 \Delta p, \tag{13.17}$$

where (2.35) has been taken into account, s_0 being the entropy per particle and v_0 the volume per particle. If the temperature is fixed at $T = T_0$, one may use (13.17) with $\Delta T = 0$, together with (13.15) and (13.16) to obtain

$$v_1 \Delta p = v_2\left(\Delta p + \frac{2}{R}\sigma_{12}(T_0)\right), \tag{13.18}$$

showing that the equilibrium pressure will be shifted by an amount

$$\Delta p = \left(\frac{v_2}{v_1 - v_2}\right)\frac{2}{R}\sigma_{12}(T_0) \tag{13.19}$$

having the same sign as $v_1 - v_2 \equiv v_1(T_0) - v_2(T_0)$. Similarly, if the pressure is fixed at $p = p_0$ then one obtains from (13.15–13.17)

$$-s_1\Delta T = -s_2\Delta T + v_2\frac{2}{R}\sigma_{12}(T_0), \tag{13.20}$$

where the correction $\sim (\partial\sigma_{12}(T)/\partial T)\Delta T$ has been neglected since it is usually very small. Equation (13.20) shows then that the equilibrium temperature T_0 will be shifted by an amount

$$\Delta T = -\left(\frac{v_2}{s_1 - s_2}\right)\frac{2}{R}\sigma_{12}(T_0) \tag{13.21}$$

having the same sign as $-(s_1 - s_2) \equiv -(s_1(T_0) - s_2(T_0))$. Since (see (2.75)) $dp_0/dT_0 = (s_1 - s_2)/(v_1 - v_2)$, the signs of Δp and ΔT will be opposite whenever dp_0/dT_0 is positive. These shifts in the coexistence curve, resulting from the curvature of the interface, are usually referred to as the Gibbs–Thomson effect. They imply that the values of (p, T) for which the nucleus of phase 2 will be in equilibrium with phase 1 will in general belong to the metastable regions of the 1–2 bulk phase diagram, i.e., $p(T) \neq p_0(T)$.

Assume now that one starts from an initial situation consisting solely of phase 1. Its Gibbs free energy will thus be $G_i(T, p, N) = N\mu_1(T, p)$. After some time, the thermal fluctuation may be able to create a nucleus of phase 2, which when in equilibrium with phase 1 will lead then to a final situation whose Gibbs free energy will be $G_f(T, p, N) = N_1\mu_1(T, p) + N_2\mu_2(T, p) + \sigma_{12}(T)A_{12}(V_2)$. The corresponding change in free energy $\Delta G = G_f - G_i$ can then be written as

$$\begin{aligned}\Delta G &= (N_1 - N)\mu_1(T, p) + N_2\mu_2(T, p) + \sigma_{12}(T)A_{12}(V_2)\\ &= -N_2(\mu_1(T, p) - \mu_2(T, p)) + \sigma_{12}(T)A_{12}(V_2),\end{aligned} \tag{13.22}$$

i.e.,

$$\begin{aligned}\Delta G &= -\frac{4\pi R^3}{3}\rho_2\Delta\mu + 4\pi R^2\sigma_{12}(T)\\ &= 4\pi R^2\sigma_{12}(T)\left(1 - \frac{R}{R_0}\frac{\Delta\mu}{|\Delta\mu|}\right),\end{aligned} \tag{13.23}$$

where $\Delta\mu = \mu_1(T, p) - \mu_2(T, p)$, and

$$R_0 = \frac{3v_2\sigma_{12}(T)}{|\Delta\mu|}, \tag{13.24}$$

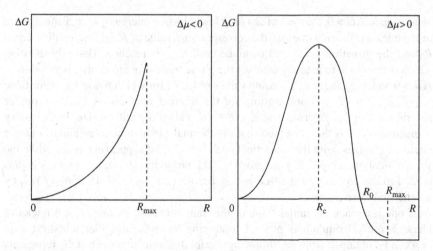

Fig. 13.2 Qualitative behavior of ΔG of (13.23) as a function of R, according to the sign of $\Delta \mu = \mu_1 - \mu_2$

is a characteristic radius. For the transformation $i \to f$ to be thermodynamically favorable one must have $\Delta G < 0$, which according to (13.23) is possible only when $\Delta \mu > 0$, i.e., $\mu_1(T, p) > \mu_2(T, p)$ or when phase 1 is metastable relative to phase 2 for the given (p, T)-values. Indeed, only when $\Delta \mu > 0$ can the volume term ($\sim R^3$) of (13.23) compete with the surface term ($\sim R^2$), the latter being essential to the stability of the nucleus and being always positive. Note also from (13.23) that ΔG, when viewed as a function of R, has a qualitatively different behavior according to the sign of $\Delta \mu$ (cf. Fig. 13.2).

Since N_2 is limited by N, one expects R to be limited by R_{max}, where $4\pi R_{max}^3/3$ is of the order of Nv_2. From (13.23) or Fig. 13.2 it is seen that, when $\Delta \mu < 0$, ΔG is positive for $0 < R < R_{max}$ and increases with R. If, on the contrary, $\Delta \mu > 0$ then ΔG is positive for $0 < R < R_0$ with a maximum at $R = R_c$, and negative for $R_0 < R < R_{max}$. From (13.23) it is easily seen that $R_c = 2R_0/3$. When $\Delta G < 0$, i.e., for $\Delta \mu > 0$ and $R_0 \lesssim R < R_{max}$, the transition $i \to f$ is favorable. The system can moreover further decrease its free energy by increasing the size R of the nucleus till at $R \simeq R_{max}$ the phase transition $1 \to 2$ is completed. Creating such a large nucleus ($R_0 < R < R_{max}$) by a thermal fluctuation is however a process of very small probability, and this process can, therefore, not explain why a phase transformation does occur. When $\Delta G > 0$, i.e., for $0 < R < R_{max}$ when $\Delta \mu < 0$ or $0 < R < R_0$ when $\Delta \mu > 0$, the transition $i \to f$ will involve a cost in free energy. In this case ΔG is moreover an increasing function of R, for all R when $\Delta \mu < 0$ and for $0 < R < R_c$ when $\Delta \mu > 0$. The system can thus lower its free energy by decreasing R. One expects thus that even if a small nucleus (for instance, $0 < R < R_c$) is created by a thermal fluctuation it will disappear again after some time. There remains however the case, $\Delta \mu > 0$ and $R_c < R < R_0$. Here, although ΔG is positive, it is now a decreasing function of R. Hence, once such a nucleus is created by a thermal fluctuation, which will involve some cost in free energy since $\Delta G > 0$, the system can only lower its free energy by increasing the size of the nucleus. When it does so, the corresponding value of R will eventually exceed R_0 and the growth process will continue until R_{max} is reached. The only situation which is favorable to actually creating the phase transformation corresponds thus to $\Delta \mu > 0$ and $R \gtrsim R_c t$, i.e., choosing values of (p, T) for which phase 1 is metastable $(\mu_1(T, p) > \mu_2(T, p))$ and waiting for the thermal fluctuations to create one or several nuclei of an intermediate size $R \simeq R_c$ called the critical size. In the theory of nucleation, it is thus assumed that the thermal fluctuations continuously create nuclei of all sizes, with the subcritical nuclei $(R < R_c)$ disappearing again, while the supercritical nuclei $(R > R_c)$ are growing. The phase transformation process is thus related to the crossing of the free energy barrier $(\Delta G)_{R=R_c} = 4\pi R_c^2 \sigma_{12}(T)/3$, by the thermal fluctuations.

If one imagines the initial stage of the transformation to consist of n nuclei of phase 2 embedded in a bulk phase 1 containing N_1 molecules, then a first estimate of n can be obtained from the following thermodynamic argument. If G_n represents the free energy of such a state, then the value of n which will be realized when the nuclei are in equilibrium with phase 1, say n_{eq}, must be such as to minimize G_n. If one writes $G_n = G_1 + \Delta G_n - T\Delta S_n$, where G_1 represents the free energy of phase 1, ΔG_n represents the cost in free energy to create the n nuclei and ΔS_n the gain in entropy of mixing the n nuclei with the N_1 molecules of phase 1, then the equilibrium value of n will be the solution of $\partial G_n/\partial n = 0$, i.e.,

$$\left(\frac{\partial \Delta G_n}{\partial n}\right)_{n=n_{eq}} = \left(T\frac{\partial \Delta S_n}{\partial n}\right)_{n=n_{eq}}. \tag{13.25}$$

To simplify (13.25), assume that all the n nuclei have the same radius R so that $\Delta G_n = n\Delta G(R)$, where $\Delta G(R)$ is given by (13.23), and that ΔS_n can be approximated by the following entropy of mixing of an ideal mixture containing N_1 objects of type 1 and n objects of type 2:

$$\Delta S_n = k_B \ln \frac{(N_1 + n)!}{N_1! n!}. \tag{13.26}$$

Since one expects n to increase with N_1 and $N_1 \gg 1$, one may use Stirling's approximation $\ln N! = N (\ln N - 1)$ for $N \gg 1$, to rewrite (13.26) as

$$\Delta S_n \simeq k_B(N_1 + n)(\ln(N_1 + n) - 1)$$

$$-k_B[N_1(\ln N_1 - 1) + n(\ln n - 1)], \tag{13.27}$$

so that (13.C5) yields

$$\Delta G(R) = k_B T \ln\left(\frac{N_1 + n_{eq}}{n_{eq}}\right) \simeq k_B T \ln\left(\frac{N_1}{n_{eq}}\right), \tag{13.28}$$

for $n_{eq} \ll N_1$, and hence

$$n_{eq}(R) = N_1 e^{-\beta \Delta G(R)}, \tag{13.29}$$

i.e., a Boltzmann distribution relative to the free energy barrier $\Delta G(R)$.

13.1.2 Zeldovich–Frenkel Theory

Although the initial creation of the nuclei could be described in thermodynamic terms, their subsequent growth or decay must involve genuine kinetic aspects. The latter will be described here in the context of the theory of Zeldovich and Frenkel. In this theory, the elementary step involved in the growth or decay of the nuclei is considered to be the attachment or detachment of a molecule of phase 1 to an existing nuclei. This process can then be represented in the form of the following "polymerization" reaction:

$$M_1 + M_p \rightleftarrows M_{p+1} \tag{13.30}$$

where M_p represents a nucleus containing p molecules, M_1 being the molecule of phase 1 which is being attached to the nucleus M_p with a frequency v_p^+, or being detached from the nucleus M_{p+1} with a frequency v_p^-. If $N(p, t)$ represents the number of nuclei of p molecules present at time t, the net flux $I(p, t)$ resulting from (13.30) will be

$$I(p, t) = v_p^- N(p+1, t) - v_p^+ N(p, t). \tag{13.31}$$

When the reaction (13.30) reaches "chemical" equilibrium, the net flux in (13.31) is zero and the corresponding equilibrium distributions $N^{eq}(p)$ will satisfy

$$v_p^- N^{eq}(p+1) = v_p^+ N^{eq}(p), \tag{13.32}$$

which allows us to eliminate v_p^- or v_p^+ in favor of the $N^{eq}(p)$, and to rewrite (13.31) as

$$I(p, t) = v_p^+ N^{eq}(p) \left[\frac{N(p+1, t)}{N^{eq}(p+1)} - \frac{N(p, t)}{N^{eq}(p)} \right]. \tag{13.33}$$

If the reaction (13.30) is viewed as a stochastic process, the time evolution of $N(p, t)$ will be given by the master equation (7.105), which can be rewritten here as

$$\frac{\partial N(p, t)}{\partial t} = \sum_{p'} {}^* \left[T(p'|p) N(p', t) - T(p|p') N(p, t) \right], \tag{13.34}$$

where $N(p, t)/N_T$ is the probability to observe a nucleus containing p nuclei, and

$$N_T = \sum_p N(p, t) \tag{13.35}$$

is the total number of nuclei. Note that since the latter is here a conserved quantity, N_T is independent of time and (13.34) is equivalent to (7.108). In (13.34), the $T(p'|p)$ are the transition probabilities per unit time associated with the frequency with which the process $p' \rightarrow p$ occurs. Since, according to the one-step process described by (13.30), one must have $p' - p = \pm 1$, and

$$v_p^+ = T(p|p+1), \quad v_p^- = T(p+1|p), \tag{13.36}$$

it is seen that (13.32) is equivalent to the condition of detailed balance (7.109), while (13.34) can be rewritten here as

$$\frac{\partial N(p, t)}{\partial t} = T(p+1|p)N(p+1, t) + T(p-1|p)N(p-1, t)$$
$$- T(p|p+1)N(p, t) - T(p|p-1)N(p, t), \tag{13.37}$$

or

$$\frac{\partial N(p, t)}{\partial t} = \left[v_p^- N(p+1, t) - v_p^+ N(p, t) \right]$$
$$- \left[v_{p-1}^- N(p, t) - v_{p-1}^+ N(p-1, t) \right]$$
$$= I(p, t) - I(p-1, t). \tag{13.38}$$

Now, if p is large enough ($p \gg 1$), one may treat p as a continuous variable and use the result

$$\frac{f(p + \Delta p) - f(p)}{\Delta p} \simeq \frac{\partial f(p)}{\partial p}, \tag{13.39}$$

valid for $|\Delta p/p| \ll 1$, and any smooth function $f(p)$, to rewrite (13.38) as

$$\frac{\partial N(p, t)}{\partial t} \simeq \frac{\partial I(p, t)}{\partial p}, \tag{13.40}$$

where $|\Delta p/p| = 1/p \ll 1$ has been used. When p is considered as a continuous variable, (13.40) is nothing but the continuity equation, which expresses the conservation of $N_T = \int dp N(p, t)$, with $I(p, t)$ being the flux conjugated to $N(p, t)$. In this case (13.33) becomes:

$$I(p, t) = v_p^+ N^{eq}(p) \frac{\partial}{\partial p} \left[\frac{N(p, t)}{N^{eq}(p)} \right]. \tag{13.41}$$

so that, according to (13.40), $N(p, t)$ must be a solution of the following Fokker–Planck equation:

$$\frac{\partial N(p, t)}{\partial t} = \frac{\partial}{\partial p}\left\{v_p^+ N^{eq}(p)\frac{\partial}{\partial p}\left[\frac{N(p, t)}{N^{eq}(p)}\right]\right\}, \tag{13.42}$$

i.e., a diffusion equation in p-space,

$$\frac{\partial N(p, t)}{\partial t} = \frac{\partial}{\partial p}\left[D_p\frac{\partial N(p, t)}{\partial p}\right] - \frac{\partial}{\partial p}[A_p N(p, t)], \tag{13.43}$$

with a diffusion coefficient $D_p \equiv v_p^+$ and a friction coefficient A_p:

$$A_p = -v_p^+ N^{eq}(p)\frac{\partial}{\partial p}\left[\frac{1}{N^{eq}(p)}\right]. \tag{13.44}$$

To evaluate (13.44), one uses (13.29), $N^{eq}(p) = n_{eq}(R)$, where the change of variables from R to p is given by $p = 4\pi R^3 \rho_2/3$, so that $A_p = -v_p^+[\partial \beta \Delta G(p)]/\partial p$, with $\Delta G(p)$ given by (13.23) expressed in terms of p instead of R. Equation (13.43) then becomes

$$\frac{\partial N(p, t)}{\partial t} = \frac{\partial}{\partial p}\left[D_p\frac{\partial N(p, t)}{\partial p}\right] + \frac{\partial}{\partial p}\left[D_p N(p, t)\beta\frac{\partial \Delta G(p)}{\partial p}\right], \tag{13.45}$$

i.e., the Zeldovich–Frenkel kinetic equation for $N(p, t)$. This equation is difficult to solve, but clearly shows that the evolution of $N(p, t)$ is governed by the competition between a diffusion process which is purely kinetic, $D_p = v_p^+$, and a friction process which is controlled by the thermodynamic force, $-\partial \Delta G(p)/\partial p$, which changes sign (cf. Fig. 13.2) at $p = p_c$, with $p_c = 4\pi R_c^3 \rho_2/3$.

In order to avoid having to solve (13.45), the method of absorbing boundary conditions will be applied to the present nucleation problem. One, therefore, assumes that once a nucleus has reached the characteristic size $p \geq p_0$, or $R > R_0$ (cf. (13.24)), where $\Delta G(p) \leq 0$, it will evolve spontaneously and uncouple from the polymerization reaction (13.30). If this is the case then (13.30) will be unable to reach equilibrium, but will reach eventually a stationary state characterized by the fact that the larger nuclei $(p > p_0)$ are constantly being withdrawn from the reaction (13.30). Such a stationary state will be characterized by a distribution, $\partial N(p, t)/\partial t = 0$ or $N(p, t) = N^0(p)$, but $N^0(p) \neq N^{eq}(p)$, and according to (13.40) by a flux, $I(p, t) = I^0(p)$, such that $\partial I^0(p)/\partial p = 0$ or $I^0(p) = I^0 \neq 0$; i.e., the stationary flux must be independent of p. This flux can be identified with the nucleation rate, i.e., the rate at which the nuclei cross the $p = p_c < p_0$ barrier. According to (13.41) this nucleation rate will be given by

$$I^0 = -D_p N^{eq}(p)\frac{\partial}{\partial p}\left[\frac{N^0(p)}{N^{eq}(p)}\right], \tag{13.46}$$

or

$$\frac{I^0}{D_p N^{eq}(p)} = -\frac{\partial}{\partial p}\left[\frac{N^0(p)}{N^{eq}(p)}\right]. \tag{13.47}$$

Integrating (13.47) up to a value p, formally $p = \infty$ but physically sufficiently large to apply the boundary condition, one obtains

$$I^0 \int_p^\infty dp' \frac{1}{D_{p'} N^{eq}(p')} = \frac{N^0(p)}{N^{eq}(p)} - \left(\frac{N^0(p)}{N^{eq}(p)}\right)_{p=\infty}, \tag{13.48}$$

where $\lim_{p\to\infty}\left(N^0(p)/N^{eq}(p)\right) = 0$. Indeed, since the nuclei with $p > p_0$ are absorbed (withdrawn), one expects $N^0(p) < N^{eq}(p)$ for $p > p_0$, and since $\Delta G(p)$ is negative for $p > p_0$, one expects $N^{eq}(p) \sim e^{-\beta\Delta G(p)}$ to be much larger than $N^0(p)$ for $p > p_0$. On the contrary, when $p \to 0$, one expects (13.30) to be only weakly perturbed so that $N^0(p) \simeq N^{eq}(p)$ for $p < p_c$, or $\lim_{p\to 0}\left(N^0(p)/N^{eq}(p)\right) = 1$. Using these boundary conditions, (13.48) implies

$$I^0 = \left[\int_0^\infty dp' \frac{1}{D_{p'} N^{eq}(p')}\right]^{-1}, \tag{13.49}$$

and

$$\frac{N^0(p)}{N^{eq}(p)} = \left[\int_p^\infty dp' \frac{1}{D_{p'} N^{eq}(p')}\right]\left[\int_p^\infty dp' \frac{1}{D_{p'} N^{eq}(p')}\right]^{-1}, \tag{13.50}$$

i.e., an expression for $N^0(p)$ and I^0 in terms of the kinematic factor D_p and of the equilibrium distribution $N^{eq}(p)$ containing the Boltzmann factor $e^{-\beta\Delta G(p)}$. For instance, if the nucleation barrier $\Delta G(p)$ is very pronounced around $p = p_c$ (cf. Fig. 13.2), one may evaluate the nucleation rate from (13.49) as

$$\frac{1}{I^0} = \int_0^\infty dp' \frac{1}{D_{p'} N^{eq}(p')} \simeq \frac{e^{\beta\Delta G(p_c)}}{D_{p_c} N_1}\int_0^\infty dp\, e^{-\beta[\Delta G(p_c)-\Delta G(p)]}, \tag{13.51}$$

or, expanding $\Delta G(p)$ around $\Delta G(p_c)$ up to second order in $p - p_c$, one obtains

$$I^0 \simeq D_{p_c} N^{eq}(p_c)\sqrt{\frac{\beta a}{\pi}} \tag{13.52}$$

Fig. 13.3 Schematic behavior of the nucleation rate I^0 of (13.52) as a function of $1/T$

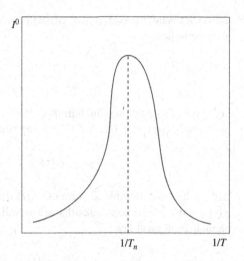

where $2a = -\left(\partial^2 \Delta G(p)/\partial p^2\right)_{p_c} > 0$. Hence, from (13.52) the nucleation rate I^0 is seen to be the product of a purely kinematic factor, D_{pc}, and a purely thermodynamic factor, $N^{eq}(p_c)$. To obtain a substantial nucleation rate it is hence not sufficient to be in a thermodynamically favorable situation, one must also have a kinematically favorable case. For instance, when (13.52) is analyzed as a function of the temperature T, it is found to be a sharply peaked function of the type shown in Fig. 13.3, where T_n is different from the bulk phase coexistence temperature at which $\Delta\mu$ of (13.23) vanishes. Hence, for most values of T, the nucleation process appears to be blocked $\left(I^0 \simeq 0\right)$, while for $T \simeq T_n$ it almost explodes. Therefore, the nucleation process is usually described as a process of "all or nothing."

13.1.3 Avrami Theory

In the late stage of the nucleation process, the respective growth processes of the supercritical nuclei ($p > p_c$) hinder each other, the nuclei coalesce and phase 2 gradually fills the available volume by a complicated hydrodynamic process which depends on the details of the experimental setup. Avoiding the latter aspects, one usually describes the late stage of the nucleation process in terms of the theory of Avrami, which only takes into account the steric aspects of this hindrance. Let $V_2(t)$ be the volume occupied by phase 2 at time t, and $\phi(t) = V_2(t)/V$ the corresponding volume fraction. Moreover, let $\overline{V}_2(t)$ be the volume which phase 2 would occupy at time t if each nucleus could continue to grow without being hindered by the other nuclei, and let $\overline{\phi}(t) = \overline{V}_2(t)/V$ be the corresponding fictitious volume fraction. The theory of Avrami is based on the assumption that, during the late stage of the evolution of the nucleation process, $d\phi(t)/dt$ and $d\overline{\phi}(t)/dt$ will remain proportional to each other. Since, however, $d\phi(t)/dt$ must vanish when, ultimately, $V_2(t) \rightarrow V$, whereas

$d\overline{\phi}|(t)/dt$ is unbounded, one also assumes $d\phi(t)/dt$ to be proportional to $1 - \phi(t)$, and one writes

$$\frac{d\phi(t)}{dt} = (1 - \phi(t))\frac{d\overline{\phi}(t)}{dt}, \tag{13.53}$$

i.e., Avrami's equation, which implies $\ln(1 - \phi(t)) + \overline{\phi}(t) = C$, but since $\phi(t) = 0$ must imply $\overline{\phi}(t) = 0$, one has $C = 0$ and hence

$$\phi(t) = 1 - e^{-\overline{\phi}(t)}. \tag{13.54}$$

Although some attempts to compute $\overline{\phi}(t)$ do exist (which is easier than computing $\phi(t)$), most experimental results are usually fitted to (13.54) by using Avrami's phenomenological law

$$\overline{\phi}(t) = Kt^n, \tag{13.55}$$

where K is a constant and n the, so-called, Avrami exponent of the nucleation process. One then usually finds $1 < n < 4$, with often a non-integer n-value.

Summarizing, the theory of homogeneous nucleation considers the $1 \to 2$ phase transition to proceed in three successive stages:

(1) Creation of nuclei of phase 2 in the midst of a metastable phase 1. This process is expected to occur on a timescale characteristic of the thermal fluctuations, the latter having to cross a free energy barrier (13.29) related to the creation of the nucleus phase 1 interface.

(2) Growth of the nuclei whose size is supercritical ($R > R_c$). This process is occurring on a kinetic timescale ($1/v_p^+$) and leads to a nucleation rate I^0 (13.52) which is a complicated, non-exponential, function of the temperature T.

(3) Coalescence of the growing nuclei, which invade the whole of the available volume V. This process is occurring on a hydrodynamic timescale and leads to a volume fraction $\phi(t)$ (13.54 and 13.55) which is a non-exponential function of the time t.

In practice, these three stages are often seen to overlap; i.e., one stage begins before the previous stage has ended.

13.2 Heterogeneous Nucleation

From the preceding theory of homogeneous nucleation one gets the (correct) impression that it is difficult to produce a phase transition via this process. This conclusion, however, is in sharp contrast with the widespread occurrence of phase transitions in nature, the explanation being that the latter usually proceed via the process of heterogeneous nucleation. Indeed, the theory of homogeneous nucleation assumes that one starts from a system containing only phase 1, whereas most systems will contain moreover impurities, the system being then heterogeneous with respect to its composition. These "impurities" can be dust particles, foreign ions, boundary surfaces made of a different material, etc. Since to initiate a nucleation process one has to first form a nucleus of a supercritical size, one easily realizes that when this nucleus is created around a dust particle, which has already a considerable size, this will greatly facilitate the formation of large nuclei, i.e., nuclei whose interfaces have a small curvature. This difference between homogeneous and heterogeneous nucleation is even more spectacular when the dust particle is charged. Consider hence that one introduces into a clean system a foreign ion of charge e (with e a positive number) and radius a. The only modification to the Becker–Döring expression (13.23) of $\Delta G = G_f - G_i$ will amount then to adding to the system's free energy G an electrostatic contribution G^e. In the initial state where only phase 1 is present, the electrostatic contribution of the ion will be

$$G_i^e = \frac{\varepsilon_1}{8\pi} \int_V d\mathbf{r}(\mathbf{E}_1(\mathbf{r}))^2 = \frac{e^2}{2\varepsilon_1} \int_a^L dr \frac{1}{r^2} = \frac{e^2}{2\varepsilon_1}\left(\frac{1}{a} - \frac{1}{L}\right), \tag{13.56}$$

where ε_1 is the dielectric constant of phase 1, the ion has been placed at the origin so that $|\mathbf{E}_1(\mathbf{r})| = e/\varepsilon_1 r^2$ is the amplitude of the electric field in phase 1, and for facility it has been assumed that V is a sphere of radius L. In the final state, phase 1 contains a nucleus of radius R which contains the ion. It is possible to show that the corresponding electrostatic contribution,

$$G_f^e = \frac{\varepsilon_1}{8\pi} \int_{V_1} d\mathbf{r}(\mathbf{E}_1(\mathbf{r}))^2 + \frac{\varepsilon_2}{8\pi} \int_{V_2} d\mathbf{r}(\mathbf{E}_2(\mathbf{r}))^2, \tag{13.57}$$

will be minimal when the ion is located at the center of the nucleus, in which case the electric field inside the nucleus will have an amplitude $|\mathbf{E}_2(\mathbf{r})| = e/\varepsilon_2 r^2$ with ε_2 being the dielectric constant of phase 2, and (13.57) becomes

$$\begin{aligned} G_f^e &= \frac{e^2}{2\varepsilon_1} \int_R^L dr \frac{1}{r^2} + \frac{e^2}{2\varepsilon_2} \int_a^R dr \frac{1}{r^2} \\ &= \frac{e^2}{2\varepsilon_1}\left(\frac{1}{R} - \frac{1}{L}\right) + \frac{e^2}{2\varepsilon_2}\left(\frac{1}{a} - \frac{1}{R}\right), \end{aligned} \tag{13.58}$$

where $L > R > a$ since the nucleus surrounds the ion. The electrostatic contribution $\Delta G^e = G_f^e - G_i^e$ to ΔG will then read

$$\Delta G^e = -\frac{e^2}{2}\left(\frac{1}{\varepsilon_1} - \frac{1}{\varepsilon_2}\right)\left(\frac{1}{a} - \frac{1}{R}\right), \tag{13.59}$$

leading to a total ΔG (cf. (13.23)),

$$
\Delta G = -\frac{4\pi R^3}{3}\rho_2\Delta\mu + 4\pi R^2\sigma_{12}(T)
$$
$$
-\frac{e^2}{2}\left(\frac{1}{\varepsilon_1}-\frac{1}{\varepsilon_2}\right)\left(\frac{1}{a}-\frac{1}{R}\right), \tag{13.60}
$$

and, therefore, the sign of the electrostatic term will be determined by the relative magnitude of the dielectric constants of the two phases. In particular, if phase 2 is the more condensed phase one will have $\varepsilon_2 > \varepsilon_1$ and the electrostatic term of (13.60) will be negative, opposing the effect of the surface tension term, hereby greatly facilitating the nucleation process. Note that one can now have $\Delta G < 0$ even when $\Delta\mu < 0$, provided the electrostatic term is large enough. This effect is exploited in Wilson's cloud chamber which is filled with a weakly undersaturated vapor. Hence, since $\Delta\mu < 0$ homogeneous nucleation is excluded but as soon as an ion, for instance a charged elementary particle, enters the chamber, the electrostatic term of (13.60) makes ΔG negative allowing nucleation to proceed spontaneously. The particle's trajectory becomes then visible as a track formed by the nuclei of liquid droplets. Putting the chamber in external electric and magnetic fields, a study of this track (trajectory) allows then a determination of the particle's charge and mass, i.e., an identification of this elementary particle. A similar process also explains why it often rains after a thunderstorm, the water vapor condensing on the many ions formed by the thunderstorm. In conclusion, the nucleation process, which is a rare event in the homogeneous case, becomes a frequent event in the presence of natural or artificial impurities.

13.3   Spinodal Decomposition

In the preceding sections it has been seen how the new phase 2 nucleates in the midst of the old phase 1 when the latter is in a metastable state. An alternative scenario is possible for phase transitions which involve unstable states. As already noticed in Chap. 2, an isostructural phase transition can give rise to a van der Waals loop, in which case the coexistence region will contain an unstable region, delimited by two spinodal lines. In such a case, one can first transform the old phase 1 from a stable initial state into a final state inside the spinodal region. Since in the spinodal region the thermodynamic state of phase 1 is unstable, it will immediately undergo a transformation, which is called a "spinodal decomposition," of the old phase 1 and, which will result in the creation of a new stable state, the new phase 2. This scenario is different from a nucleation but in order to realize it the transformation from initial to final state must proceed sufficiently rapidly. Indeed, between the spinodal and the binodal there is always a region where phase 1 is metastable and where a nucleation could start. To avoid this, one has to transform the initial stable state of phase 1 rapidly into an unstable state, so that along the intermediate metastable states the nucleation has no time to proceed. Such a rapid transformation of the initial into the final state is called a "quench." Although the spinodal decomposition process is more easily observed in mixtures, it will, for simplicity, be illustrated here for the case of the liquid–vapor transition of a one-component system.

Assume then that one starts from an initially uniform fluid state of density ρ_i and temperature T_i and that one subsequently quenches the system into a state (ρ, T) where the fluid is unstable. Suppose, moreover, that during the whole process the system's temperature is kept fixed at the value $T = T_i$. Since at (ρ, T) the system is unstable, its density will immediately start to evolve in time and space $\rho \to \rho_1(\mathbf{r}, t) = \rho + \delta\rho_1(\mathbf{r}, t)$, where $\rho_1(\mathbf{r}, 0) = \rho$ and $\delta\rho(\mathbf{r}, t)$ represents a spatial fluctuation:

$$\frac{1}{V} \int_V d\mathbf{r} \, \delta\rho_1(\mathbf{r}, t) = 0, \tag{13.61}$$

which for $t > 0$ structures the system at constant average density ρ. Although the state $(\rho_1(\mathbf{r}, t), T)$ is a non-uniform non-equilibrium unstable state, one first tentatively estimates its intrinsic Helmholtz free energy functional $\mathcal{F}[\rho_1]$ from the square gradient approximation of Chap. 11. The fluctuation $\rho \to \rho + \delta\rho_1(\mathbf{r}, t)$ will be thermodynamically acceptable if it lowers the system's free energy (11.20), i.e., when

$$\Delta \mathcal{F}[\rho_1] = \int_V d\mathbf{r} \left[\overline{f}(T, \rho_1(\mathbf{r}, t)) - \overline{f}(T, \rho) + \frac{1}{2} a_2 \{\nabla\delta\rho_1(\mathbf{r}, t)\}^2 \right], \tag{13.62}$$

is negative. In the early stage of the evolution process, one may consider $\delta p_1(\mathbf{r}, t)$ to be small and (13.62) can be rewritten as

$$\Delta \mathcal{F}[\rho_1] = \frac{1}{2} \int_V d\mathbf{r} \left[\frac{\partial^2 \overline{f}(T, \rho)}{\partial \rho^2} \{\delta\rho_1(\mathbf{r}, t)\}^2 + a_2 \{\nabla\delta\rho_1(\mathbf{r}, t)\}^2 \right], \tag{13.63}$$

where (13.61) eliminates the linear term and the higher-order terms have been neglected on account of the smallness of $\delta\rho_1(\mathbf{r}, t)$. Since for the unstable state (ρ, T) the free energy density $\overline{f}(T, \rho)$ is no longer convex, one has $\partial^2 \overline{f}(T, \rho)/\partial\rho^2 < 0$ and it follows from (13.63) that $\Delta \mathcal{F}[\rho_1] < 0$ provided the non-uniformity $\nabla\delta\rho_1(\mathbf{r}, t)$ is not too strong (remember that $a_2 > 0$). For instance, if $\delta\rho_1(\mathbf{r}, t) \sim e^{i\mathbf{k}\cdot\mathbf{r}}\delta\rho_\mathbf{k}(t)$, then (13.63) implies that $\Delta \mathcal{F}[\rho_1] < 0$ provided $k < k_c$, where the critical wavenumber k_c is given by

$$k_c^2 = -\frac{1}{a_2} \frac{\partial^2 \overline{f}(T, \rho)}{\partial\rho^2} > 0, \tag{13.64}$$

and it has been taken into account that $\delta\rho_1$ $(\mathbf{r}, t)$ must be real. Hence, the system will accept any long-wavelength spatial modulation ($k < k_c$) of its density. The uniform fluid phase will thus transform into regions of high (liquid-like) density and low (vapor-like) density separated by an interface. The form of these regions still depends on the applied boundary conditions but their characteristic size will always be given by $\lambda_c = 2\pi/k_c$, which is an intrinsic wavelength. Once formed, these regions will moreover evolve in time as a result of the system's instability. The latter evolution of $\delta\rho_1$ $(\mathbf{r}, t)$ is usually described in terms of the following generalized diffusion equation:

$$\frac{\partial \delta\rho_1(\mathbf{r}, t)}{\partial t} = M \nabla^2 \frac{\delta\mathcal{F}[\rho_1]}{\delta\rho_1(\mathbf{r}, t)}, \tag{13.65}$$

where M is a phenomenological coefficient describing the mobility (diffusion) of the non-uniform domains of high and low densities. In conjunction with the above square gradient approximation (13.62) for $\mathcal{F}[\rho_1]$, (13.65) reduces to

$$\frac{\partial \delta\rho_1(\mathbf{r}, t)}{\partial t} = M \nabla^2 \left(\frac{\partial \overline{f}(T, \rho_1(\mathbf{r}, t))}{\partial \rho_1(\mathbf{r}, t)} - a_2 \nabla^2 \delta\rho_1(\mathbf{r}, t) \right), \tag{13.66}$$

which is usually referred as the Cahn–Hilliard equation. When $\delta\rho_1$ $(\mathbf{r}, t)$ is small (13.66) reduces to

$$\frac{\partial \delta\rho_1(\mathbf{r}, t)}{\partial t} = M \left(\frac{\partial^2 \overline{f}(T, \rho)}{\partial \rho^2} - a_2 \nabla^2 \right) \nabla^2 \delta\rho_1(\mathbf{r}, t), \tag{13.67}$$

so that for $\delta\rho_1(\mathbf{r}, t) \sim e^{i\mathbf{k}\cdot\mathbf{r} + \omega_k t} \delta\rho_\mathbf{k}(0)$, one obtains

$$\omega_\mathbf{k} = a_2 M \, k^2 \left(k_c^2 - k^2 \right), \tag{13.68}$$

showing that the stable domains ($k < k_c$) will grow exponentially in time ($\omega_\mathbf{k} > 0$). Once this initial growth has reached a stage where $\delta\rho_1$ $(\mathbf{r}, t)$ is no longer small, one has to return to the full nonlinear equations, and the later stage of the spinodal decomposition will then evolve according to an hydrodynamic process, not unlike the one encountered in the late stage of the nucleation process. One then finds that the size of the domains grows (a process called coarsening) like $t^{1/3}$, a result known as the Lifshitz–Slyozov law. In practice, one also often finds it necessary to include higher-order gradients terms into $\mathcal{F}[\rho_1]$, as a result of the steepening of the interfaces between the domains. In the end, the high-density domains will merge and evolve toward a single domain, separated by a single interface from the low-density region, when the phase transition is completed and the system has reached a stable two-phase coexistence equilibrium state.

13.4 Glass Transition

As it has been seen above, the process of homogeneous nucleation is difficult to initiate. This then raises the question of what will happen to the system when one tries to bypass the very possibility of nucleation. Consider, therefore, a third scenario where one starts from a stable initial state of phase 1 and rapidly quenches it into a new state where phase 1 is metastable. In other words, one now applies the rapid quench method, which was used in the spinodal decomposition scenario to avoid the nucleation process, to a situation where nucleation is the only process available to produce a phase transition. To reinforce the absence of nucleation, assume that the target phase 2 has a broken symmetry and, to fix the ideas, one may consider phase 1 to be a liquid and phase 2 a crystal. In this case, the creation of a nucleus, in particular of a small anisotropic crystal, will require a considerable amount of molecular reorganization. To further prevent this, assume moreover that one has performed a deep - quench, for instance to a temperature which is so low that the molecular motions are considerably inhibited. After such a rapid and deep quench of, say, a liquid phase, the latter will hence be kinematically arrested; i.e., its molecules will only be able to execute small oscillations (like in a crystal) around the positions these molecules did have initially, i.e., positions characteristic of a liquid configuration. Although the final state of the system is thermodynamically metastable, any further evolution, in a finite time, has now become impossible, the system being in a non-equilibrium stationary state. Assume now that the corresponding temperature quench did proceed at a constant speed b, i.e., $T(t) = T_i - bt$ with $T_i = T(0)$, and that one measures the specific volume $v = 1/\rho$ of this metastable liquid as a function of the temperature, like one would do for an equilibrium situation. Compare the results for three different speeds b_0, b_1, and b_2, and assume that b_0 is so small that nucleation can proceed, while b_1 corresponds to a rapid quench for which nucleation is excluded, and b_2 to an even more rapid quench. From the results, schematically shown in Fig. 13.4,

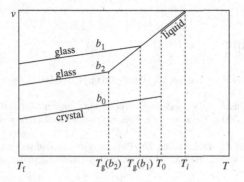

Fig. 13.4 Schematic representation of the evolution of the specific volume v with the temperature $T(t) = T_i - bt$, for three values of b, with $b_0 \ll b_1 < b_2$. Here T_0 corresponds to the equilibrium phase transition temperature and T_g, with $T_g(b_2) < T_g(b_1)$, to the thermodynamic glass transition temperature

one may conclude the following with respect to v as a function of T. For the case $b = b_0$, there are two equilibrium phases with the liquid being the stable phase for $T_0 < T < T_i$ and the crystal being the stable phase for $T_f < T < T_0$, while at the liquid–crystal coexistence temperature (for the given pressure) T_0 there is a discontinuity in v signaling the phase transition.

For the case $b = b_1$, v is continuous everywhere but dv/dT exhibits a discontinuity at $T = T_g$, T_g being called the thermodynamic glass transition temperature, because for $T_f < T < T_g$ the system is said to be in a glass phase. Note that nothing particular has happened at $T = T_0$ indicating hereby that the nucleation process was indeed inhibited. The liquid phase is still the stable phase for $T_0 < T < T_i$ but is called supercooled in the region $T_g < T < T_0$. Note also that the values of dv/dT are very similar for the two cases. Finally, the case $b = b_2$ is similar to the case $b = b_1$, except for the fact that T_g is a decreasing function of b. To avoid this sensitivity of T_g to b, one often introduces a conventional glass transition temperature T_g' defined as the temperature for which the shear viscosity reaches the (conventional) value of 10^3 poise. One then finds that T_g' is much less sensitive to the value of b than is the case for the thermodynamic glass transition temperature T_g. One should, however, not be confused by the terminology being used here: The glass transition is a kinematic transition, not a true equilibrium phase transition. The latter being always a transition between two equilibrium states, whereas the glass phase is a metastable, hence, non-equilibrium state. Therefore, although the glass phase will remain stable on any human-like timescale, some of its properties may still very slowly evolve. One then says that the glass phase is aging. Nowadays, most of the new materials (plastics, ceramics, metallic glasses, etc.) are in fact glass phases, but the nomenclature used refers, of course, to the way in which window glass is being produced. The glass transition is hence a central topic in what is called, at present, materials science. It is a difficult topic which mixes in a subtle way the equilibrium and non-equilibrium aspects of the general theory of phase transitions. Needless to say that a detailed study of the glass transition is beyond the scope of the present book.

Further Reading

1. R. Becker, *Theory of Heat*, 2nd edn. (Springer-Verlag, Berlin, 1967). *Provides a good discussion of the theory of nucleation.*
2. J. L. Barrat, J.P. Hansen, *Basic concepts for Simple and Complex Liquids* (Cambridge University Press, Cambridge, 2003). *Contains an introduction to spinodal decomposition and to the glass transition..*
3. P. Papon, J. Leblond, P.H.sE. Meijer, *The Physics of Phase Transitions: Concepts and Applications* (Springer-Verlag, Berlin, 2002). *Describes many industrial applications of phase transitions.*

Appendix A
Legendre Transformations

A.1 Case of One Variable

In order to introduce the concept of Legendre transformation, consider an analytic function $y = y(x)$, whose derivative is $p \equiv y'(x)$ (in what follows, a prime denotes a derivative with respect to the argument). If one eliminates x from this equation, $x = x(p)$ and then substitutes in the original function the result is $y = y(x(p)) \equiv y(p)$, which is a first-order differential equation whose general solution contains an arbitrary integration constant. Since only for a certain value of the integration constant one recovers the function $y = y(x)$, the functions $y = y(p)$ and $y = y(x)$ are not equivalent. In spite of this, it is possible to find a transformation of the function $y = y(x)$ into another equivalent function in which p is the independent variable. To that end, note that a curve is defined either by the set of points (x, y) such that $y = y(x)$ or by the envelope of the curve. The latter is formed by the set of points (p, y_0), where y_0 is the ordinate at the origin of the straight line whose slope is $p = y'(x)$, which is the tangent to the curve at the point x. Since the equation of the tangent to the curve that passes through the point (x, y) is given by

$$y = y_0 + px, \tag{A.1}$$

the curve is completely characterized by the function $y_0(p)$, namely

$$y_0(p) = y(p) - px(p), \tag{A.2}$$

where $x(p)$ is obtained when one eliminates x from the equation

$$p = y'(x), \tag{A.3}$$

and $y(p)$ is the function:

$$y = y(x(p)) \equiv y(p). \tag{A.4}$$

© Springer Nature Switzerland AG 2021
M. Baus and C. F. Tejero, *Equilibrium Statistical Physics*,
https://doi.org/10.1007/978-3-030-75432-7

The transformation (A.2) by which the function $y = y(x)$ is replaced by the function $y_0 = y_0(p)$, in which the derivative of the original function is the independent variable, is known as a Legendre transformation, and the function $y_0(p)$ is referred to as the Legendre transform of the function $y(x)$.

Differentiating (A.2) one has

$$dy_0 = dy - pdx - xdp = -xdp, \tag{A.5}$$

so that

$$x = -y_0'(p). \tag{A.6}$$

The inverse Legendre transform of the function $y_0(p)$ is obtained when one eliminates the derivative $p = y'(x)$ from (A.6), and so $y(x)$ may be expressed, according to (A 2). as

$$y(x) = y_0(y'(x)) + xy'(x), \tag{A.7}$$

which is the original function.

Note that from (A.3) and (A.6) it follows that

$$p'(x) = y''(x), \quad x'(p) = -y_0''(p),$$

and since

$$p'(x) = \left[x'(p)\right]^{-1},$$

one finally obtains

$$y''(x)y_0''(p) = -1. \tag{A.8}$$

It may be shown that the Legendre transform exists and is unique when the function $y(x)$ is convex or concave. A function $y(x)$ is convex when $y''(x) > 0$ and concave when $y''(x) < 0$. From (A.8) it follows that the Legendre transform of a convex function is a concave function and vice versa. Some simple examples of convex functions on the real axis are $y(x) = x^2$ and $y(x) = e^x$, whose Legendre transforms are, respectively, $y_0(p) = -p^2/4$ and $y_0(p) = p(1 - \ln p)$. The Legendre transform of the function $y = \ln x$, which is convex on the real positive semi-axis, is $y_0(p) = -(1 + \ln p)$.

It may happen that in some interval $x_1 < x < x_2$, $y''(x)$ does not exist. In such cases one may use the following (more general) definitions of convexity and Legendre transform. If $x_k = kx_1 + (1 - k)x_2(0 < k < 1)$ is an arbitrary point of the interval (x_1, x_2), the function $y(x)$ is convex in the interval when

$$y(x_k) < ky(x_1) + (1 - k)y(x_2), \tag{A.9}$$

Fig. A.1 Convex function $y(x)$ in the interval (x_1, x_2). Note that $y(x)$ lies below the straight line joining the points $(a, y(a))$ and $(b, y(b))$ for any $x_1 \leq a < b \leq x_2$

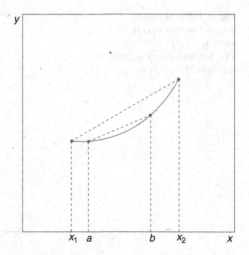

which expresses that $y(x)$ lies below the straight line joining the points $(x_1, y(x_1))$ and $(x_2, y(x_2))$ (Fig. A.1).

If $y'(x)$ exists, (A.9) is equivalent to

$$y(x_k) > y(x_0) + (x_k - x_0)y'(x_0), \tag{A.10}$$

i.e., $y(x)$ lies above the tangent to the curve at x_0, for all $x_k \neq x_0$ (Fig. A.2).

Finally, if $y''(x)$ exists, (A.10) is equivalent to

$$y''(x_k) > 0. \tag{A.11}$$

In the general case, the Legendre transform of $y(x)$ and its inverse $y_0(p)$ are defined as

$$y_0(p) = \min_x(y(x) - xp), \quad y(x) = \max_p(y_0(p) + px), \tag{A.12}$$

where x and p are the conjugate variables of the transformation. Note that $p = y'(x)$ and $x = -y_0'(p)$, when the derivatives do exist.

A.2 Case of n Variables

The concepts developed for one variable may be extended to the case in which $y = y(x_1, \ldots, x_n)$ is a function of n variables. If

$$p_k \equiv \frac{\partial y(x_1, \ldots, x_n)}{\partial x_k} \quad (k = 1, \ldots, n), \tag{A.13}$$

Fig. A.2 Differentiable
convex function $y(x)$ in the
interval (x_1, x_2). Note that
$y(x)$ lies above the tangent tô
the curve at a (or b) for any
$x_1 < a < x_2$

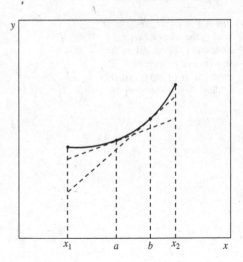

different Legendre transforms may be defined, depending on the partial derivatives
(A.13) that one wants to have as independent variables in the transformation. Let $s \leq$
n be these variables. The partial Legendre transform of the function $y(x_1, \ldots, x_n) \equiv$
$y(\{x_s\}; \{x_{n-s}\})$, in which $\{x_s\}$ denotes the set of variables that are going to be replaced
by their derivatives $\{p_s\}$ in the Legendre transform and $\{x_{n-s}\}$ denotes the set of
variables that are not affected by the transformation, is defined by the following
equation:

$$y_0(\{p_s\}; \{x_{n-s}\}) = y(\{p_s\}; \{x_{n-s}\}) - \sum_{k=1}^{s} p_k x_k(\{p_s\}; \{x_{n-s}\}), \qquad (A.14)$$

where $x_k(\{p_s\}; \{x_{n-s}\})$ is obtained by eliminating the variables $x_k(k = 1, \ldots, s)$
from (A.13) in terms of the derivatives $\{p_s\}$ and of the remaining $n - s$ variables
$\{x_{n-s}\}$, and $y(\{p_s\}; \{x_{n-s}\})$ is the function:

$$y(\{p_s\}; \{x_{n-s}\}) = y(\{x_s(\{p_s\}; \{x_{n-s}\})\}; \{x_{n-s}\}). \qquad (A.15)$$

Differentiating (A.14) one finds

$$dy_0 = dy - \sum_{k=1}^{s} [p_k dx_k + x_k dp_k] = \sum_{i=n-s}^{n} p_i dx_i - \sum_{k=1}^{s} x_k dp_k, \qquad (A.16)$$

so that

$$x_k = -\frac{\partial y_0(\{p_s\}; \{x_{n-s}\})}{\partial p_k}, \quad (k = 1, \ldots, s). \qquad (A.17)$$

The inverse Legendre transform of the function $y_0(\{p_s\}; \{x_{n-s}\})$ is obtained when one eliminates the derivatives $p_k = p_k(\{x_s\}; \{x_{n-s}\})$ from (A.17), in such a way that y may be derived from (A.14) as

$$y(\{x_s\}; \{x_{n-s}\}) = y_0(\{x_s\}; \{x_{n-s}\}) + \sum_{k=1}^{s} x_k p_k(\{x_s\}; \{x_{n-s}\}), \qquad (A.18)$$

which is the original function with

$$y_0(\{x_s\}; \{x_{n-s}\}) = y_0(\{p_s(\{x_s\}; \{x_{n-s}\})\}; \{x_{n-s}\}). \qquad (A.19)$$

A.2.1 Some Examples

The Lagrangian $L_N(q, \dot{q}) \equiv L_N(q_1, \ldots, q_{3N}, \dot{q}_1, \ldots, \dot{q}_{3N})$ of a system of 3 N degrees of freedom is a function of the coordinates $q_1, \ldots, q_{3N}$ and of the generalized velocities $\dot{q}_1, \ldots, \dot{q}_{3N}$ (see Chap. 1). The equations of motion of the system or Lagrange equations are

$$\frac{d}{dt}\left(\frac{\partial L_N(q, \dot{q})}{\partial \dot{q}_i}\right) - \frac{\partial L_N(q, \dot{q})}{\partial q_i} = 0, \quad (i = 1, \ldots, 3N). \qquad (A.20)$$

The generalized momenta are defined as

$$p_i = \frac{\partial L_N(q, \dot{q})}{\partial q_i}, \qquad (A.21)$$

and the Hamiltonian $H_N(q, p) \equiv H_N(q_1, \ldots, q_{3N}, p_1, \ldots, p_{3N})$, with a change of sign, is the Legendre transform of $L_N(q, \dot{q})$ when the generalized velocities are replaced by the generalized momenta, namely

$$-H_N(q, p) = L_N(q, \dot{q}) - \sum_{i=1}^{3N} p_i \dot{q}_i. \qquad (A.22)$$

Differentiating (A.22) one has

$$-dH_N(q, p) = \sum_{i=1}^{3N} (\dot{p}_i dq_i - \dot{q}_i dp_i), \qquad (A.23)$$

where use has been made of (A.20) and (A.21). Therefore,

$$\dot{q}_i = \frac{\partial H_N(q, p)}{\partial p_i}, \quad \dot{p}_i = -\frac{\partial H_N(q, p)}{\partial q_i}, \quad (i = 1, \ldots, 3N), \tag{A.24}$$

which are the Hamilton equations (see Chap. 1).

The fundamental equation of thermodynamics (see Chap. 2) of a one-component system whose only external parameter is the volume is $E = E(S, V, N)$, where E is the internal energy, S is the entropy, V is the volume, and N is the number of particles. The intensive parameters temperature, T, pressure, p, and chemical potential, μ, are the derivatives:

$$T = \frac{\partial E}{\partial S}, \quad p = -\frac{\partial E}{\partial V}, \quad \mu = \frac{\partial E}{\partial N}. \tag{A.25}$$

Throughout the text different partial Legendre transforms of the fundamental equation appear. Three of these transforms or thermodynamic potentials are the Helmholtz free energy

$$F(T, V, N) = E(S, V, N) - TS, \tag{A.26}$$

wherein (A.26) $S = S(T, V, N)$ is obtained from the solution of the first of the implicit equations in (A.25), the Gibbs free energy

$$G(T, p, N) = G(S, V, N) - TS + pV, \tag{A.27}$$

wherein (A.27) $S = S(T, p, N)$ and $V = V(T, p, N)$ are obtained from the solution of the first two implicit equations in (A.25), and the Landau free energy or grand potential

$$\Omega(T, V, \mu) = E(S, V, N) - TS - \mu N, \tag{A.28}$$

wherein (A.28) $S = S(T, V, \mu)$ and $N = N(T, V, \mu)$ are obtained from the solution of the first and the third implicit equations in (A.25).

Further Reading

1.   R. Balian, *From Microphysics to Macrophysics*, vol. 1, (Springer-Verlag, Berlin, 1991). *Provides a detailed discussion of the Legendre transformation method.*

Appendix B
Random Variables

B.1 One-Dimensional Random Variable

One defines a one-dimensional continuous random variable X, which is assumed to take values on the real axis, by its probability density $\rho_X(x)$, which is a nonnegative function that satisfies the normalization condition:

$$\int_{-\infty}^{\infty} dx \rho_X(x) = 1. \tag{B.1}$$

The probability $P[x_1 < x < x_2]$ that the variable takes values in the interval $[x_1, x_2]$ is given by

$$P[x_1 < x < x_2] = \int_{x_1}^{x_2} dx \rho_X(x), \tag{B.2}$$

so that (B.1) expresses that the probability that X takes any value on the real axis is one. Note that the random variable has been denoted by a capital letter and its possible values by the same small letter.

If X is a random variable, one defines the moment of order n, $\langle X^n \rangle$, as

$$\langle X^n \rangle \equiv \int_{-\infty}^{\infty} dx \, x^n \rho_X(x), \tag{B.3}$$

and the average value of a function $f(X)$ by

© Springer Nature Switzerland AG 2021
M. Baus and C. F. Tejero, *Equilibrium Statistical Physics*,
https://doi.org/10.1007/978-3-030-75432-7

$$\langle f(X) \rangle \equiv \int\limits_{-\infty}^{\infty} dx \, f(x)\rho_X(x). \tag{B.4}$$

The first moment is thus the average value of the variable. The dispersion or fluctuation of the random variable, σ_X^2, is defined as

$$\sigma_X^2 \equiv \langle [X - \langle X \rangle]^2 \rangle = \langle X^2 \rangle - \langle X \rangle^2 \geq 0. \tag{B.5}$$

Consider a real function $Y = f(X)$ of the random variable X, whose probability density is $\rho_X(x)$. Y is a random variable whose probability density $\rho_Y(y)$ may be determined as follows. The probability that Y takes values in the interval $[y, y + \Delta y]$ is the probability that X takes values on the real axis, such that $y < f(x) < y + \Delta y$, namely

$$\rho_Y(y)\Delta y = \int\limits_{-\infty}^{\infty} dx [\Theta(y + \Delta y - f(x)) - \Theta(y - f(x))] \rho_X(x), \tag{B.6}$$

and so dividing by Δy and taking the limit $\Delta y \to 0$, one has

$$\rho_Y(y) = \int\limits_{-\infty}^{\infty} dx \, \delta(y - f(x))\rho_X(x). \tag{B.7}$$

When the probability density $\rho_X(x)$ is of the form

$$\rho_X(x) = \sum_n p_n \delta(x - x_n), \quad (p_n \geq 0), \tag{B.8}$$

where the sum extends over a finite or numerable infinite set of points of the real axis, one says that the random variable is discrete. In (B.8) p_n is then the probability that the random variable takes the value x_n and the normalization condition (B.1) is expressed as

$$\sum_n p_n = 1. \tag{B.9}$$

B.1.1 Some Examples

A continuous random variable X is Gaussian when its probability density $\rho_X(x)$ is

$$\rho_X(x) = \frac{1}{\sqrt{2\pi\sigma_X^2}} e^{-(x-\langle X \rangle)^2/2\sigma_X^2}, \tag{B.10}$$

where $\langle X \rangle$ is the average value of the variable and σ_X^2 its dispersion. Observe that in the case of Gaussian random variables all the moments can be expressed in terms of the first two moments. When $\langle X \rangle = 0$ and $\sigma_X = 1$, one says that the random variable is a normal Gaussian random variable in which case the moments of odd order are zero, by symmetry, and the moments of even order are given by

$$\langle X^{2n} \rangle = \frac{1}{\sqrt{2\pi}} \int_{-\infty}^{\infty} dx\, x^{2n} e^{-x^2/2} = \frac{2^n}{\sqrt{\pi}} \Gamma\left(n + \frac{1}{2}\right), \tag{B.11}$$

where $\Gamma(x)$ is the Euler gamma function.

The Fourier transform of the Gaussian distribution (B.10) is

$$\tilde{\rho}_X(k) = \int_{-\infty}^{\infty} dx\, e^{-ikx} \rho_X(x) = e^{-ik\langle X \rangle - k^2\sigma_X^2/2}, \tag{B.12}$$

an expression that will be used later on.

A discrete random variable X is binomial, when the probabilities p_n are given by

$$p_n = \frac{N!}{n!(N-n)!} p^n (1-p)^{N-n}, \quad (n = 0, 1, \ldots, N), \tag{B.13}$$

where N is an integer and p a real number $0 < p \leq 1$. The average value and the dispersion of the binomial distribution (B.13) are $\langle X \rangle = N p$, $\sigma_X^2 = N p(1-p)$.

A discrete random variable X is a Poisson random variable, when the probabilities p_n are given by

$$p_n = \frac{1}{n!} \lambda^n e^{-\lambda}, \quad (n = 0, 1, \ldots), \tag{B.14}$$

where λ is a positive real number. The average value and the dispersion of the Poisson distribution (B.14) are $\langle X \rangle = \lambda$, $\sigma_X^2 = \lambda$.

B.2   Approximation Methods

Throughout the text one finds probability densities of random variables which have the form

$$\rho_X(x) = \frac{1}{Z(N)} e^{N f(x)}, \tag{B.15}$$

where $f(x)$ is an analytic function on the real axis and N an integer, $N \gg 1$. The constant $Z(N)$ in (B.15) is obtained from the normalization condition (B.1), namely

$$Z(N) = \int_{-\infty}^{\infty} dx\, e^{N f(x)}. \tag{B.16}$$

Assume now that $f(x)$ has a single maximum at x_0, i.e., $f'(x_0) = 0$ and $f''(x_0) < 0$. If $N \gg 1$, the function

$$a(x) \equiv e^{N[f(x)-f(x_0)]}, \tag{B.17}$$

has a very pronounced peak in the vicinity of x_0 and tends to zero exponentially as one gets away from it. An approximate expression for $a(x)$ may then be obtained by performing a Taylor expansion of the function $f(x) - f(x_0)$ that appears in the exponent up to second order, leading to

$$a(x) \simeq e^{-N|f''(x_0)|(x-x_0)^2/2}. \tag{B.18}$$

Integrating with respect to x, one has

$$\int_{-\infty}^{\infty} dx\, a(x) \simeq \int_{-\infty}^{\infty} dx\, e^{-N|f''(x_0)|(x-x_0)^2/2} = \sqrt{\frac{2\pi}{N|f''(x_0)|}}, \tag{B.19}$$

and hence

$$Z(N) \simeq \sqrt{\frac{2\pi}{N|f''(x_0)|}} e^{N f(x_0)}. \tag{B.20}$$

The probability density (B.15) is, according to (B.18) and (B.19), given by

$$\rho_X(x) \simeq \sqrt{\frac{N|f''(x_0)|}{2\pi}} e^{-N|f''(x_0)|(x-x_0)^2/2}, \tag{B.21}$$

which implies that X is a Gaussian random variable of average value x_0 and dispersion:

$$\sigma_X^2 = \frac{1}{N|f''(x_0)|}. \tag{B.22}$$

If one considers more terms in the Taylor expansion of the function $f(x) - f(x_0)$ in (B.17), expanding the resulting exponentials, except the Gaussian, one obtains

$$\int_{-\infty}^{\infty} dx a(x) = \sqrt{\frac{2\pi}{N|f''(x_0)|}} \left[1 + O\left(\frac{1}{N^2}\right)\right], \tag{B.23}$$

and hence

$$\lim_{N \to \infty} \frac{1}{N} \ln Z(N) = f(x_0). \tag{B.24}$$

Note that, by the same type of arguments, if $b(x)$ is a smoothly varying function in the neighborhood of x_0, one has

$$\int_{-\infty}^{\infty} dx b(x) a(x) \simeq b(x_0) \int_{-\infty}^{\infty} dx a(x), \tag{B.25}$$

i.e.,

$$\int_{-\infty}^{\infty} dx b(x) e^{Nf(x)} \simeq \sqrt{\frac{2\pi}{N|f''(x_0)|}} b(x_0) e^{Nf(x_0)}. \tag{B.26}$$

B.3 n-Dimensional Random Variable

An n-dimensional random variable $(X_1, \ldots, X_n)$ is defined by the joint probability density $\rho_{X_1,\ldots,X_n}(x_1, \ldots, x_n)$, which is a nonnegative function that satisfies the normalization condition,

$$\int_{-\infty}^{\infty} dx_1 \ldots \int_{-\infty}^{\infty} dx_n \rho_{X_1,\ldots,X_n}(x_1, \ldots, x_n) = 1 \tag{B.27}$$

and which, when integrated over a region R^n, is the probability that the n-dimensional variable takes values in that region. When it may lead to no confusion, the joint probability density is denoted by $\rho_n(x_1, \ldots, x_n)$ and even by $\rho_n(x_1, \ldots, x_n)$. This last notation is the one used in the text.

The marginal probability density $\rho_{X_1,\ldots,X_s}(x_1, \ldots, x_s)$ of a subset of $s < n$ variables is defined as

$$\rho_{X_1,\ldots,X_s}(x_1, \ldots, x_s) = \int\limits_{-\infty}^{\infty} dx_{s+1} \ldots \int\limits_{-\infty}^{\infty} dx_n \rho_{X_1,\ldots,X_n}(x_1, \ldots, x_n), \qquad \text{(B.28)}$$

whose normalization is, according to (B.27),

$$\int\limits_{-\infty}^{\infty} dx_1 \ldots \int\limits_{-\infty}^{\infty} dx_s \rho_{X_1,\ldots,X_s}(x_1, \ldots, x_s) = 1. \qquad \text{(B.29)}$$

The first and second moments of an n-dimensional random variable are defined as

$$\langle X_j \rangle = \int\limits_{-\infty}^{\infty} dx_1 \ldots \int\limits_{-\infty}^{\infty} dx_n x_j \rho_{X_1,\ldots,X_n}(x_1, \ldots, x_n)$$

$$= \int\limits_{-\infty}^{\infty} dx_j x_j \rho_{X_j}(x_j), \qquad \text{(B.30)}$$

and

$$\langle X_i X_j \rangle = \int\limits_{-\infty}^{\infty} dx_1 \ldots \int\limits_{-\infty}^{\infty} dx_n x_i x_j \rho_{X_1,\ldots,X_n}(x_1, \ldots, x_n)$$

$$= \int\limits_{-\infty}^{\infty} dx_i \int\limits_{-\infty}^{\infty} dx_j x_i x_j \rho_{X_i,X_j}(x_i, x_j). \qquad \text{(B.31)}$$

From (B.30) and (B.31) an $n \times n$ matrix may be defined whose elements are

$$C(X_i, X_j) = \langle X_i X_j \rangle - \langle X_i \rangle \langle X_j \rangle, \qquad \text{(B.32)}$$

The diagonal terms are the dispersions of $\sigma_i^2 = \langle X_i^2 \rangle - \langle X_i \rangle^2$, and the off-diagonal terms are called the correlations.

The random variables $X_1, \ldots, X_n$ are called statistically independent when

$$\rho_{X_1,\ldots,X_n}(x_1, \ldots, x_n) = \prod_{i=1}^{n} \rho_{X_i}(x_i), \qquad \text{(B.33)}$$

in which case $\langle X_i X_j \rangle = \langle X_i \rangle \langle X_j \rangle (i \neq j)$ and the correlation between any pair of variables is zero.

Consider two continuous random variables X_1 and X_2 whose joint probability density is $\rho_{X_1, X_2}(x_1, x_2)$. The probability density of the random variable $Y = X_1 + X_2$ may be determined using a similar reasoning to the one considered in (B.6), namely

$$\rho_Y(y) \Delta y = \int_{-\infty}^{\infty} dx_1 \int_{-\infty}^{\infty} dx_2 [\Theta(y + \Delta y - x_1 - x_2) - \Theta(y - x_1 - x_2)]$$

$$\times \rho_{X_1, X_2}(x_1, x_2). \tag{B.34}$$

Dividing (B.34) by Δy in the limit $\Delta y \to 0$, one has

$$\rho_Y(y) = \int_{-\infty}^{\infty} dx_1 \int_{-\infty}^{\infty} dx_2 \delta(y - x_1 - x_2) \rho_{X_1, X_2}(x_1, x_2)$$

$$= \int_{-\infty}^{\infty} dx_1 \rho_{X_1, X_2}(x_1, y - x_1), \tag{B.35}$$

so that if X_1 and X_2 are statistically independent variables,

$$\rho_Y(y) = \int_{-\infty}^{\infty} dx_1 \rho_{X_1}(x_1) \rho_{X_2}(y - x_1), \tag{B.36}$$

i.e., the probability density of the sum of two independent random variables is the convolution of the individual probability densities. Taking the Fourier transform of (B.36), one has

$$\tilde{\rho}_Y(k) = \tilde{\rho}_{X_1}(k) \tilde{\rho}_{X_2}(k). \tag{B.37}$$

Consider N continuous random variables $X_1, \ldots, X_N$ statistically independent and identical (the individual probability density $\rho_X(x)$ is the same for all of them) of average value $\langle x_i \rangle = 0$ and dispersion σ_X^2. Note that the restriction on the average value is not important, since if this is not zero one may take as random variables $X_i - \langle X_i \rangle$. Consider the random variable

$$Y = \frac{1}{\sqrt{N}} (X_1 + \cdots + X_N) \tag{B.38}$$

of zero average value and dispersion of $\sigma_Y^2 = \sigma_X^2$, as may be verified by squaring (B.38) and taking the average value, since $\langle X_i X_j \rangle = 0 \; (i \neq j)$.

Let $\tilde{\rho}_Y(k)$ and $\tilde{\rho}_X(k)$ be the Fourier transforms of the probability densities $\rho_Y(y)$ and $\rho_X(x)$. One then has

$$\tilde{\rho}_Y(k) = \left[\tilde{\rho}_X\left(\frac{k}{\sqrt{N}}\right)\right]^N. \tag{B.39}$$

Since

$$\tilde{\rho}_X\left(\frac{k}{\sqrt{N}}\right) = \int_{-\infty}^{\infty} dx\, e^{ikx/\sqrt{N}} \rho_X(x) = 1 - \frac{k^2\sigma_X^2}{2N} + O\left(\frac{1}{N^{3/2}}\right), \tag{B.40}$$

it follows that

$$\tilde{\rho}_Y(k) = \left[1 - \frac{k^2\sigma_X^2}{2N} + O\left(\frac{1}{N^{3/2}}\right)\right]^N, \tag{B.41}$$

which in the limit $N \to \infty$ reads

$$\lim_{N\to\infty} \tilde{\rho}_Y(k) = \lim_{N\to\infty}\left[1 - \frac{k^2\sigma_X^2}{2N} + O\left(\frac{1}{N^{3/2}}\right)\right]^N = e^{-k^2\sigma_X^2/2}, \tag{B.42}$$

where use has been made of the definition of the number e. According to (B.12), Y is a normal Gaussian random variable. Note that to arrive at this result it has not been necessary to specify the particular form of $\rho_X(x)$ and it has only been assumed that the random variables are identical and independent. This result is known as the central limit theorem.

B.4 n-Dimensional Gaussian Fluctuations

Consider a joint probability density of the form:

$$\rho_{X_1\cdots X_n}(x_1,\ldots x_n) = C_n \exp - \frac{1}{2}\sum_{i=1}^{n}\sum_{j=1}^{n}(x_i - \langle X_i\rangle)B_{ij}(x_j - \langle X_j\rangle) \tag{B.43}$$

where C_n is determined by the normalisation condition (B.27) and $\{B_{ij}\}$ are constants satisfying $B_{ij} = B_{ji}$. Note that (B.43) is the n-dimensional generalisation of (B.10). The random variables $\{X_1,\ldots,X_n\}$ will thus be referred to as Gaussian random variables. When $\langle X_i\rangle = 0$ for all $i = 1,\ldots,n$, these random variables are referred to as fluctuations and the probability density (B.43) becomes for Gaussian fluctuations:

$$\rho_{X_1\cdots X_n}(x_1,\ldots x_n) = C_n \exp - \frac{1}{2}\sum_{i=1}^{n}\sum_{j=1}^{n}x_i B_{ij} x_j, \tag{B.44}$$

with

$$\frac{1}{C_n} = \int\limits_{-\infty}^{\infty} dx_1 \cdots \int\limits_{-\infty}^{\infty} dx_n \exp - \frac{1}{2}\sum_{i,j} x_i B_{ij} x_j. \tag{B.45}$$

In order to evaluate (B.45) we introduce a linear transformation:

$$x_i = \sum_{j=1}^{n} A_{ij} x'_j (i = 1,\ldots,n), \tag{B.46}$$

which diagonalizes the quadratic form:

$$\sum_{i,j} x_i B_{ij} x_j = \sum_{i,j}\sum_{i',j'} x'_{i'} A_{i,i'} B_{ij} A_{jj'} x'_{j'} = \sum_{i',j'} x'_{i'} x'_{j'}\delta_{i',j'} = \sum_i (x'_i)^2. \tag{B.47}$$

Equation (B.47) implies that:

$$\sum_{i,j} A_{ii'} B_{ij} A_{jj'} = \delta_{i'j'}. \tag{B.48}$$

Using the new variables $\{x'_i\}$,(B.45) becomes:

$$\frac{1}{C_n} = |A| \int\limits_{-\infty}^{\infty} dx'_1 \int\limits_{-\infty}^{\infty} dx'_n \exp -\frac{1}{2}\sum_j (x'_{j0})^2$$

$$= |A| \left(\int\limits_{-\infty}^{\infty} x\, e^{-\frac{x^2}{2}}\right)^n = |A|(2\pi)^{\frac{n}{2}} \tag{B.49}$$

where $|A|$ is the determinant of the matrix (A_{ij}). Taking the determinant of (B.48) we obtain $|A||B||A| = |I|$, where I is the unit matrix and hence, $|A|^2|B| = 1$, so that (B.49) yields:

$$C_n = \frac{|B|^{\frac{1}{2}}}{|2\pi|^{\frac{n}{2}}}, \tag{B.50}$$

where $|B|$ is the determinant of the $n \times n$ matrix $\{B_{ij}\}$.

Let us now evaluate $\langle x_i x_j \rangle$. To this end we start from the expression (B.43) with $\langle X_i \rangle = a_i$:

$$\langle x_i \rangle \equiv \frac{|B|^{\frac{1}{2}}}{(2\pi)^{\frac{n}{2}}} \int \prod_k dx_k x_i \exp - \frac{1}{2}\sum_{i,j}(x_i - a_i)B_{ij}(x_j - a_j) = a_i \tag{B.51}$$

Taking the derivative of both sides with respect to a_j one obtains:

$$\frac{|B|^{\frac{1}{2}}}{(2\pi)^{\frac{n}{2}}} \int \prod_k dx_k x_i \sum_j B_{ij}(x_j - a_j)\exp -\frac{1}{2}\sum_{i,j}(x_i - a_i)B_{ij}(x_j - a_j) = \delta_{ij},$$

(B.52)

which, after putting all $a_i = 0$, yields:

$$\sum_j B_{ij}\langle x_i x_j\rangle = \delta_{ij},$$

(B.53)

and hence:

$$\langle x_i x_j\rangle = B_{ij}^{-1},$$

(B.54)

where B_{ij}^{-1} is the element (i, j) of the inverse matrix B^{-1} of B.

Consider now the following average value:

$$\left\langle \exp \sum_i a_i x_i \right\rangle = \frac{|B|^{\frac{1}{2}}}{(2\pi)^{\frac{n}{2}}} \int \prod_{k=1}^n dx_k \exp \left\{ \sum_i a_i x_j - \frac{1}{2}\sum_{i,j} x_i B_{ij} x_j \right\},$$

(B.55)

where $\{a_1, \cdots, a_n\}$ are arbitrary constants. Using (B.46) we obtain:

$$\left\langle \exp \sum_i a_i x_i \right\rangle = \frac{|B|^{\frac{1}{2}}}{(2\pi)^{\frac{n}{2}}}|A| \int \prod_k dx'_k \, \exp\left\{ \sum_{i,j} a_i A_{ij} x'_j - \frac{1}{2}\sum_j (x'_j)^2 \right\}$$

$$= \frac{|B|^{\frac{1}{2}}}{(2\pi)^{\frac{n}{2}}}|A| \int \prod_k dx'_k \, \exp\left\{ -\frac{1}{2}\sum_j \left(x'_j - \sum_i a_i A_{ij} \right)^2 + \frac{1}{2}\sum_{i,j,k} a_i A_{ij} A_{jk} a_k \right\}$$

$$= \exp\left\{ \frac{1}{2}\sum_{i,j,k} a_i A_{ij} A_{jk} a_k \right\} \frac{1}{(2\pi)^{\frac{n}{2}}} \int \prod_k dx'_k \, \exp -\frac{1}{2}\sum_j \left(x'_j - \sum_i a_i A_{ij} \right)^2$$

$$= \exp\left\{ \frac{1}{2}\sum_{i,j,k} a_i \, A_{ij} \, A_{jk} \, a_k \right\},$$

(B.56)

while using (B.48) and (B.54) finally yields:

$$\left\langle \exp \sum_i a_i x_i \right\rangle = \exp \frac{1}{2}\sum_{i,j} a_i a_j \langle x_i x_j\rangle.$$

(B.57)

Hence, Gaussian fluctuations are described by the probability density (B.44), (B.45), with $\{B_{ij}\}$ an arbitrary symmetric matrix, and are fully characterized by (B.54) and (B.57) where the $\{a_i\}$ are arbitrary constants.

Further Reading

1. N.G. van Kampen, *Stochastic Processes in Physics and Chemistry* (North-Holland, Amsterdam, 1981). *A classic introduction to various stochastic methods.*

Appendix C
Functional and Functional Derivative

C.1 Definition of a Functional

Consider first a function of n variables $F(\rho_1, ..., \rho_n)$, where $\rho_i = \rho(x_i)$ $(i = 1, ..., n)$ is the value of the function $\rho(x)$ at a point x_i in the interval in which the function is defined (in what follows it will be assumed that this interval is the whole real axis). The function $F(\rho_1, ..., \rho_n)$ is an application that associates a number $F(\rho_1, ..., \rho_n)$ to any set of values $\rho_1, ..., \rho_n$. As a generalization of the definition of a function of n variables, the concept of a functional of a function arises as the limit of $F(\rho_1, ..., \rho_n)$ when n tends to infinity in such a way that the $\{\rho_1\}$ cover the whole curve $\rho = \rho(x)$. In this way a functional associates a number to a curve. Consider, for instance, the integral

$$F[\rho] = \int_{-\infty}^{\infty} dx \rho(x) \simeq \sum_i \rho_i \Delta x_i, \qquad (C.1)$$

which associates to each function $\rho(x)$ a number $F[\rho]$ (whenever the integral exists). Note that the functional dependence is denoted by a bracket, instead of by a parenthesis as in the case of a function, and hence two observations are pertinent here. The first one is that, since $F[\rho]$ is not a function of x, the notation $F[\rho(x)]$, which may lead to confusion, is not adequate. The second one is that $F[\rho]$ is not a function of a function, which is denoted by a parenthesis. For example, if $\rho(x) = x$ and $g(x) = x^2 + 1$, one has $g(\rho(x)) = (\rho(x))^2 + 1$, which is a function of the function $\rho(x)$, not a functional of $\rho(x)$.

© Springer Nature Switzerland AG 2021
M. Baus and C. F. Tejero, *Equilibrium Statistical Physics*,
https://doi.org/10.1007/978-3-030-75432-7

C.2   The Functional Derivative

The differential of $F(\rho_1, \ldots, \rho_n)$ when the variables $\{p_i\}$ change from p_i to $p_i + dp_i$ is

$$dF = \sum_{i=1}^{n} \frac{\partial F(\rho_1, \ldots, \rho_n)}{\partial \rho_i} d\rho_i, \tag{C.2}$$

which may be written as

$$dF = \sum_{i=1}^{n} \Delta x_i \frac{1}{\Delta x_i} \frac{\partial F(\rho_1, \ldots, \rho_n)}{\partial \rho_i} d\rho_i, \tag{C.3}$$

where $\Delta x_i = x_{i+1} - x_i$. Consider now the limit $n \to \infty$, $\Delta x_i \to 0$ of (C.3). When some mathematical conditions are met, this sum may be written in the aforementioned limit as the integral

$$\delta F[\rho] = \int dx \frac{\delta F[\rho]}{\delta \rho(x)} \delta \rho(x), \tag{C.4}$$

where $\delta \rho(x)$ is the increment of the function $\rho(x)$ (as the limit of $\{d\rho_i\}$), $\delta F[\rho]$ is the increment of the functional (as the limit of dF), and one has defined the functional derivative $\delta F[\rho]/\delta \rho(x)$ as the limit of

$$\frac{1}{\Delta x_i} \frac{\partial F(\rho_1, \ldots, \rho_n)}{\partial \rho_i}.$$

Note that, in much the same way that the functional dependence is denoted by a bracket, the symbol δ is reserved for the functional increments. Observe also that the functional derivative is a functional of $\rho(x)$ (due to the limit $n \to \infty$) and a function of x (due to the dependence on $\Delta x_i \to 0$) and that the dimension of the functional derivative is not the one of the partial derivative, due to the factor $1/\Delta x_i$.

C.3   Some Examples

The next are a few simple applications.

(a)   From the identity

$$\delta \rho(x) = \int dx' \delta(x - x') \delta \rho(x'), \tag{C.5}$$

where $\delta(x)$ is the Dirac delta, it follows that from the comparison of (C.4) with (C.5) one has

$$\frac{\delta\rho(x)}{\delta\rho(x')} = \delta(x - x'). \tag{C.6}$$

(b) Note that the rules of functional derivation are similar to the ones of the derivation of functions. For instance, given the functional

$$F[\rho] = \int dx \ln \rho(x), \tag{C.7}$$

the increment of the functional $\delta F[\rho] = F[\rho + \delta\rho] - F[\rho]$ when the function increases by $\delta\rho(x)$ is

$$\delta F[\rho] = \int dx \ln[\rho(x) + \delta\rho(x)] - \int dx \ln \rho(x)$$
$$= \int dx \ln\left[1 + \frac{\delta\rho(x)}{\rho(x)}\right] = \int dx \left[\frac{\delta\rho(x)}{\rho(x)} + \ldots\right], \tag{C.8}$$

i.e.,

$$\frac{\delta F[\rho]}{\delta\rho(x)} = \frac{1}{\rho(x)}. \tag{C.9}$$

This result may also be obtained by functional derivation of (C.7), using the rules for the derivation of functions and (C.6), namely

$$\frac{\delta F[\rho]}{\delta\rho(x)} = \frac{\delta}{\delta\rho(x)} \int dx' \ln \rho(x') = \int dx' \frac{\delta \ln \rho(x')}{\delta\rho(x)}$$
$$= \int dx' \frac{1}{\rho(x')} \frac{\delta\rho(x')}{\delta\rho(x)} = \int dx' \frac{1}{\rho(x')} \delta(x - x')$$
$$= \frac{1}{\rho(x)}. \tag{C.10}$$

In this way, if $g(x)$ is a function independent of $\rho(x)$, one has

$$\frac{\delta}{\delta\rho(x)} \int dx' g(x')\rho(x') = g(x), \tag{C.11}$$

$$\frac{\delta}{\delta\rho(x)} \int dx' g(x')[\rho(x')]^n = ng(x)[\rho(x)]^{n-1}, \tag{C.12}$$

$$\frac{\delta}{\delta\rho(x)} \int dx' g(x')\rho(x') \ln \rho(x') = g(x)[\ln \rho(x) + 1] \tag{C.13}$$

(c) Consider a system of particles contained in a closed region R of volume V in an external potential $\phi(\mathbf{r})$. The Helmholtz free energy $F[\phi]$ is a functional of $\phi(\mathbf{r})$ (note that this is a generalization of the one-dimensional problem). The functional derivative with respect to the external potential is the local density of particles $\rho_1(\mathbf{r})$ (see Sect. 5.5):

$$\rho_1(\mathbf{r}) = \frac{\delta F[\phi]}{\delta \phi(\mathbf{r})}. \tag{C.14}$$

In the case of functionals, the Legendre transformation has to be modified due to the definition of the functional derivative. The intrinsic Helmholtz free energy functional $\mathcal{F}[\rho_1]$ is defined as

$$\mathcal{F}[\rho_1] = F[\phi] - \int_R d\mathbf{r}\phi(\mathbf{r})\rho_1(\mathbf{r}), \tag{C.15}$$

where $\phi(\mathbf{r})$ in (C.15) is a function of the local density of particles that is obtained from the solution of the implicit equation (C.14).

The variation of (C.15) is

$$\delta\mathcal{F}[\rho_1] = \delta F[\phi] - \int_R d\mathbf{r}[\phi(\mathbf{r})\delta\rho_1(\mathbf{r}) + \rho_1(\mathbf{r})\delta\phi(\mathbf{r})], \tag{C.16}$$

and since

$$\delta F[\phi] = \int_R d\mathbf{r}\frac{\delta F[\phi]}{\delta\phi(\mathbf{r})}\delta\phi(\mathbf{r}), \tag{C.17}$$

from (C.14), (C.16), and (C.17) it follows that

$$\delta\mathcal{F}[\rho_1] = -\int_R d\mathbf{r}\phi(\mathbf{r})\delta\rho_1(\mathbf{r}), \tag{C.18}$$

i.e.,

$$\phi(\mathbf{r}) = -\frac{\delta\mathcal{F}[\rho_1]}{\delta\rho_1(\mathbf{r})}. \tag{C.19}$$

Therefore, the first derivative of the intrinsic Helmholtz free energy functional with respect to the local density of particles is the external potential with a change of sign.

Equations (C.14) and (C.19) show that $\phi(\mathbf{r})$ and $\rho_1(\mathbf{r})$ are the conjugate variables (fields) of the functional Legendre transformation (C.15).

Further Reading

1. R. Balescu, *Equilibrium and Nonequilibrium Statistical Mechanics* (Wiley Interscience, New York, 1975). *Contains a good introduction to the concept of functional.*

2. J. P. Hansen and I. R. McDonald, *Theory of Simple Liquids*, 2nd ed. (Academic Press, London, 1986). *Provides a good summary of the Density Functional Theory.*

Appendix D
A Quasicrystalline Lattice

As has been indicated in Sect. 8.2, in a crystal the particles (atoms or molecules) oscillate around the equilibrium positions of a periodic lattice (a Bravais lattice). In 1912 von Laue suggested that crystals might behave as diffraction lattices for X-rays, so that if the wavelength of the rays was comparable to the average distance between the atoms, it should be possible to determine the symmetries of the lattice through diffraction experiments. Some symmetries are forbidden in a crystal (as shown below), and hence with the observation in 1984 of crystals with symmetries that do not correspond to periodic lattices, but to quasiperiodic lattices, the study of this kind of "crystals," called quasicrystals, was initiated.

D.1 Forbidden Symmetries of Periodic 2D Lattices

Consider a periodic lattice, that for simplicity will be assumed to be two-dimensional, formed by the vertices of regular polygons that produce a tiling covering the plane (infinite). Since not all regular polygons may cover the plane, it is clear that there must exist forbidden symmetries in periodic lattices. This may be demonstrated with the following reasoning. If the regular polygon has n sides ($n > 2$), the angle that forms the segments that join the center of the polygon with the vertices of one side is $2\pi/n$ (Fig. D.1), and the axis perpendicular to the plane that passes through the center of the polygon is a symmetry axis of order n. Assume now that the axis perpendicular to the plane that passes through one vertex has a symmetry of order q, i.e., that the angle formed by the side of the polygon and the segment that joins one vertex with the center is π/q. One then has (see Fig. D.1) $2\pi/n + 2\pi/q = \pi$, namely $1/n + 1/q = 1/2$, or, alternatively, $q = 2n/(n - 2)$. Note that the only interesting solutions to this equation are those for which n and q are integers. Therefore, if $n = 3$ (triangles) then $q = 6$, when $n = 4$ (squares) $q = 4$, if $n = 5$ (pentagons) $q = 10/3$, which is not an integer, and when $n = 6$ (hexagons) $q = 3$. For $n > 6$ there are no solutions with q being an integer, since when $n \to \infty$ one has $q \to 2$, i.e., $3 > q > 2$ if $6 < n < \infty$ and there is no integer between 2 and 3. As a consequence, the only periodic lattices in

© Springer Nature Switzerland AG 2021
M. Baus and C. F. Tejero, *Equilibrium Statistical Physics*,
https://doi.org/10.1007/978-3-030-75432-7

Fig. D.1 Angles formed by
the segments that join the
center of a regular polygon
of n sides with the vertices of
a side of the polygon and by
one of these segments and
one of the sides of the
polygon

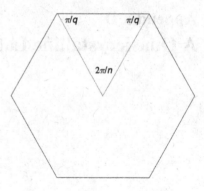

the plane correspond to $n = 3, 4$, and 6, with a symmetry of order $q = 6, 4$, and 3, respectively. The symmetries $q = 5$ and $q > 6$ are forbidden.

To construct a periodic two-dimensional lattice one may adopt the convention by which the vertices (x, y) of the lattice are represented by complex numbers $z = x + iy$.

Consider, for instance, the triangular lattice ($n = 3, q = 6$) and let

$$\omega = e^{2\pi i/6} = \frac{1}{2} + i\frac{\sqrt{3}}{2}, \tag{D.1}$$

be a root of the equation $\omega^6 = 1$. The vertices of the triangular lattice are given by the set of complex numbers

$$\{z_t\} = \{k_1 + k_2\omega\}, \tag{D.2}$$

where k_1 and k_2 are integers ($k_1, k_2 = 0, \pm1, \pm2\ldots$). Note that in this lattice the side of the triangle has been taken equal to one. If the side of the triangle is a, the set of complex numbers defining the lattice is, instead of $\{z_t\}$, $\{az_t\}$.

D.2   A Quasiperiodic 2D Lattice

This convention established for a crystal may now be generalized to a quasicrystal. Thus, consider the set of complex numbers,

$$\{z_d\} = \{z_1 + z_2\xi\}, \tag{D.3}$$

Where

$$\xi = e^{2\pi i/12} = \frac{\sqrt{3}}{2} + i\frac{1}{2}, \tag{D.4}$$

Fig. D.2 Part of the infinite
quasiperiodic lattice
constructed as indicated in
the text

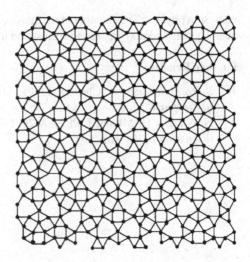

is a root of the equation $\xi^{12} = 1$. Note that ξ is a vector oriented along a forbidden
direction of a two-dimensional lattice. In (D.3) $z_1 = k_1 + k_2\omega$ and $z_2 = k_1' + k_2'\omega$
are two arbitrary elements of the set of complex numbers (D.2), namely

$$\{z_d\} = \left\{k_1 + k_2\omega + k_1'\xi + k_2'\omega\xi\right\} \tag{D.5}$$

where $k_1, k_2, k_1', k_2' = 0, \pm 1, \pm 2, \ldots$ and $\omega\xi = e^{5\pi i/6} = -\sqrt{3}/2 + i/2$ (note that
$\omega = \xi^4$ and $\omega\xi = \xi^{\,5}$).

Observe that while (D.2) is a periodic lattice in real space, (D.5) is a periodic
lattice in complex space, i.e., the number of dimensions of space is duplicated from
two to four. In the first case, k_1 and k_2 are the "coordinates" of each point of space,
which in the second are k_1, k_2, k_1', and k_2'. This interpretation refers, of course, to
abstract mathematical spaces, which does not prevent (D.5) from also representing
points in the space (x, y). For this to be so, one needs to "project" this space of
dimension four onto a two-dimensional lattice. The way to perform this projection
is not unique and, for instance, the quasicrystalline lattice of Fig. D.2 corresponds to
points $\{z_d\}$ that satisfy the following conditions:

$$\left[(\mathrm{Re}z_d)\frac{\cos[(2k+1)\pi/12]}{\cos(\pi/12)} + (\mathrm{Im}z_d)\frac{\sin[(2k+1)\pi/12]}{\sin(\pi/12)}\right]^2 \leq 1, \tag{D.6}$$

where $z_d = \mathrm{Re}\, z_d + i\mathrm{Im}\, z_d$ and $k = 0,1,2,3,4,5$. Note that the lattice is not periodic, but rather a superposition of two periodic lattices, of periods ω and ξ, and
an "interference" term $\omega\xi$. If the values of ω and ξ or the conditions (D.6) are
modified, one obtains different quasiperiodic lattices, although not all of them are of

interest in physics. The example of the quasiperiodic lattice of Fig. D.2 is observed in quasicrystals and may be seen to correspond to a "mixture" of triangles and squares.

Further Reading

1.   Janot, *Quasicrystals* (Clarendon Press, Oxford, 1994). *A good first introduction to the quasi-cristalline solids.*
2.   N. Niizeki, H. Mitani, J. Phys. A **20**, L405 (1987). *Provides the basis for the discussion in the main text.*

Appendix E
The Bi-axial Nematic

The Landau theory of order–disorder transitions (cf. Sectis. 9.3–4) is based on the idea that the thermodynamic potential, e.g. the Gibbs free energy per particle $\mu(p, T)$ (cf. Eq. (2.34)), of the ordered phase can be written in terms of an order parameter q which quantifies the presence of order, i.e. a broken symmetry. If $\mu(p, T; q)$ represents this potential for a given order q then the equilibrium ordered phase will correspond to $\min_q \mu(p, T; q)$ while the equilibrium disordered phase will correspond to $\mu(p, T; q = 0)$. Landau's theory asserts then that when $\mu(p, T; q)$, is an even function of q, i.e. $\mu(p, T; q) = \mu(p, T; -q)$, then the corresponding order–disorder transition will be continuous, whereas otherwise it will be a discontinuous transition.

The search of an appropriate order parameter is not always easy as we will illustrate here for the case of an anisotropic fluid ($\rho(r, u) = \rho h(u)$). If the angular distribution $h(u) \equiv 1$ the fluid will be orientationally disordered or isotropic, whereas when $h(u) \neq 1$ it will be ordered or anisotropic. The quantity, $h(u) - 1$, will provide thus a measure of the amount of order in the system. This quantity however is a function of u, not a parameter as needed for Landau's theory. In order to turn it into a parameter we will consider the following integral:

$$q = \int du \varphi(u)\{h(u) - 1\}, \tag{E.1}$$

where $\varphi(u)$ is still to be specified, and (8.105) was used so that $\int du h(u) = 1$. The simplest possible choice for $\varphi(u)$ is a scalar function of the vector u. However, the only scalar which can be formed with u is $u \cdot u$ but since here $u \cdot u = 1$, $\varphi(u)$ will be a constant and (E.1) yields then $q \equiv 0$. Therefore, no scalar order parameter can be found for an anisotropic fluid. Next we may try a vectorial function for $\varphi(u)$. The only possibility is now $\underline{\varphi}(u) = u$ and (E.1) will yield then:

$$\underline{q} = \int du\, u\, h(u), \tag{E.2}$$

© Springer Nature Switzerland AG 2021
M. Baus and C. F. Tejero, *Equilibrium Statistical Physics*,
https://doi.org/10.1007/978-3-030-75432-7

a vectorial order parameter, provided $h(u) \neq h(-u)$. Such a fluid is said to exhibit ferromagnetic order. Such a ferromagnetic fluid can thus be described by a vectorial order parameter. Note however that since $\mu\left(p, T; \underline{q}\right)$, is a scalar function it can depend on q only through the scalar combination $\underline{q}^2$ and hence $\mu\left(p, T; \underline{q}\right)$, will be an even function of q and the ferromagnetic-isotropic transition will always be a continuous transition. When, $h(u) = h(-u)$, the fluid is said to exhibit nematic order and (E.2) will yield $q = 0$. Hence, a nematic fluid cannot be described by a vectorial order parameter either. Let us try next a second rank tensor $\varphi(u) = \underset{=}{uu}$ leading to a tensorial order parameter for (E.1):

$$\underset{=}{q} = \int d\boldsymbol{u} \; \boldsymbol{uu}\{h(\boldsymbol{u}) - 1\}, \qquad (E.3)$$

with now $\underset{=}{q} \neq 0$ even for a nematic $(h(u) = h(-u))$. Eq. (E.3) represents a symmetric matrix $(q_{ij} = q_{ji})$ of zero trace $\left(\sum_j q_{jj} = 0\right)$ and this is the simplest possible order parameter for a nematic fluid. This symmetric matrix can always be diagonalized:

$$\underset{=}{q} = \sum_{i=1}^{3} \lambda_i \; \boldsymbol{\varepsilon}_i \boldsymbol{\varepsilon}_i; \quad \boldsymbol{\varepsilon}_i \cdot \boldsymbol{\varepsilon}_j = \delta_{ij}, \qquad (E.4)$$

in terms of its eigenvectors $(\boldsymbol{\varepsilon}_i)$ and eigenvalues (λ_i)

$$\underset{=}{q} \cdot \boldsymbol{\varepsilon}_i = \lambda_i \; \boldsymbol{\varepsilon}_i; \quad \lambda_i = \boldsymbol{\varepsilon}_i \; \cdot \underset{=}{q} \cdot \boldsymbol{\varepsilon}_i, \qquad (E.5)$$

with $i = 1, 2, 3$. In order to analyse $\mu\left(p, T; \underset{=}{q}\right)$, we must first consider the rotational invariants which can be constructed with the aid of $\underset{=}{q}$. According to the Caley-Hamilton theorem of matrix theory these are:

$$q_n = Tr \underset{=}{q}^n = \sum_{i=1}^{3} \lambda_i^n \quad n = 1, 2, 3, \qquad (E.6)$$

but since here, $q_1 = \sum_i \lambda_i = 0$, only two of the eigenvalues $\lambda_1, \lambda_2, \lambda_3,$ are independent. One, say λ_1, must be positive, another one, say λ_2, must be negative while the third one (λ_3) can be either positive or negative. Let us write $\lambda_1 = \lambda$, $\lambda_2 = -\frac{1}{2}(\lambda + \lambda')$, $\lambda_3 = \frac{1}{2}(\lambda' - \lambda)$ so that $\lambda_1 + \lambda_2 + \lambda_3 = 0$ and $\lambda = \lambda_1$, and $\lambda' = \lambda_3 - \lambda_2$ are two independent parameters with $\lambda > 0$. In terms of (λ, λ') Eq. (E.4) becomes:

$$\underline{q} = \lambda\varepsilon_1\varepsilon_1 - \frac{1}{2}(\lambda + \lambda')\varepsilon_2\varepsilon_2 + \frac{1}{2}(\lambda' - \lambda)\varepsilon_3\varepsilon_3 = \frac{3}{2}\lambda\left(\varepsilon_1\varepsilon_1 - \frac{1}{3}1\right) + \frac{\lambda'}{2}(\varepsilon_3\varepsilon_3 - \varepsilon_2\varepsilon_2),$$

$$(E.7)$$

where $1 = \sum_{i=1}^{3} \varepsilon_i\varepsilon_i$.

When $\lambda' = 0$ the nematic is called uni-axial. It is seen from Eq. (E.7) that in this case ε_1 is a symmetry axis which is called the director n. From (E.5) we obtain then:

$$\lambda = n \cdot \underline{q} \cdot n$$

$$= \int du(u \cdot n)^2\{h(u) - 1\}$$

$$= \frac{2}{3}\langle P_2(u \cdot n)\rangle \qquad (E.8)$$

and

$$\underline{q} = \langle P_2(u \cdot n)\rangle\left(n\,n - \frac{1}{3}1\right), \qquad (E.9)$$

where $\langle P_2(u \cdot n)\rangle$ is the average taken with $h(u)$ of the second order Legendre polynomial $P_2(u \cdot n)$. Note that this is the same order parameter as used in Sect. 9.3 for the description of a uni-axial nematic. There the above discussion was avoided by assuming from the start that $h(u) = h(u \cdot n)$ in which case both descriptions are equivalent.

When $\lambda' \neq 0$ the nematic is called bi-axial because now (E.7) indicates that the nematic has a second axis of symmetry, namely the diagonal of the $(\varepsilon_2, \varepsilon_3)$ plane. Since, $q_2 = \frac{1}{2}(3\lambda^2 + \lambda'^2)$ and $q_3 = \frac{3}{4}\lambda(\lambda^2 - \lambda'^2)$ are even in λ', the Landau theory of the uni-axial to bi-axial transition of a nematic has to be a continuous phase transition.

Further Reading

1. S. Chandrasekhar, *Liquid Crystals*, 2nd ed. (Cambridge University Press, Cambridge, 1992).

Appendix F
The Helmholtz Decomposition of a Vector Field

Any vector field, $v(r)$, can be decomposed into two independent fields:

$$v(r) = v_\parallel(r) + v_\perp(r) . \tag{F.1}$$

Here $v_\parallel(r)$ is the longitudinal or irrotational part of $v(r)$:

$$\nabla \times v(r) = \nabla \times v_\perp(r); \quad \nabla \times v_\parallel(r) = 0; \quad v_\parallel(r) = \nabla\varphi(r) , \tag{F.2}$$

which can be derived from the scalar potential $\varphi(r)$, whereas $v_\perp(r)$ is the transverse or divergenceless part of $v(r)$:

$$\nabla \cdot v(r) = \nabla \cdot v_\parallel(r); \nabla \cdot v_\perp = 0; v_\perp(r) = \nabla \times A(r), \tag{F.3}$$

which can be derived from the vector potential $A(r)$.

The above Helmholtz decomposition (F.1) of the vector field $v(r)$ is very simple if $v(r)$ admits a Fourier transform v_k:

$$v_k = \int dr e^{-ik.r} v(r) \tag{F.4}$$

$$v(r) = \int \frac{dk}{(2\pi)^d} e^{ik.r} v_k , \tag{F.5}$$

If, $\hat{k} = \frac{k}{|k|}$, denotes the direction of the wave vector k, then we may use the identity, $1 = \hat{k}\hat{k} + \left(1 - \hat{k}\hat{k}\right)$, to write:

$$v_k = \left[\hat{k}\hat{k} = \left(1 + \hat{k}\hat{k}\right)\right].v_k \tag{F.6}$$

© Springer Nature Switzerland AG 2021
M. Baus and C. F. Tejero, *Equilibrium Statistical Physics*,
https://doi.org/10.1007/978-3-030-75432-7

$$= \hat{k}\hat{k} \cdot v_k + \left(\hat{k} \times v_k\right) \times \hat{k} \tag{F.7}$$

$$= v_k^{\parallel} + v_k^{\perp}. \tag{F.8}$$

where (F.8) is the Fourier transform of (F.1) and (F.7) shows that $v_k^{\parallel}$ is the component of v_k along $\hat{k}$, hence the name "longitudinal", while $v_k^{\perp}$ is the component of v_k perpendicular to $\hat{k}, \hat{k}$, hence the name "transverse". Note also that (F.6) is nothing but a projection of v_k onto two orthogonal subspaces:

$$v_k^{\parallel} \cdot v_k^{\perp} = 0 . \tag{F.9}$$

In real space the equivalent of the Fourier space expression (F.9) becomes:

$$\int dr v_{\parallel}(r) \cdot v_{\perp}(r) = \int dr (\nabla \varphi(r)) \cdot (\nabla \times A(r))$$

$$= \int dr \nabla \cdot \{\varphi(r)\nabla \times A(r)\}$$

$$- \int dr \varphi(r) \nabla \cdot (\nabla \times A(r)) , \tag{F.10}$$

and using the identity, $\nabla \cdot (\nabla \times A) \equiv 0$, (F.10) is seen to become a surface term:

$$\int_V dr \, \nabla \cdot \{\varphi(r)\nabla \times A(r)\} = \int_S dS. \{\varphi(r) \, \nabla \times A(r)\} ,, \tag{F.11}$$

which can always be made zero by an appropriate choice of the potentials on the surface S of the volume V, and hence:

$$\int dr v_{\parallel}(r) \cdot v_{\perp}(r) = 0 , \tag{F.12}$$

showing that the two fields, $v_{\parallel}(r)$ and $v_{\perp}(r)$, are independent:

$$\int dr (v(r))^2 = \int dr \left\{ \left(v_{\parallel}(r)\right)^2 + (v_{\perp}(r))^2 \right\} . \tag{F.13}$$

Finally from Gauss's theorem:

$$\int_S dS \cdot v(r) = \int_V dr \, \nabla \cdot v(r) \equiv \int_V dr \, \nabla \cdot v_{\parallel}(r), \tag{F.14}$$

it follows that any surface integral of $v(r)$ is completely determined by $v_{\parallel}(r)$, while from Stokes's theorem:

$$\oint_{\gamma} dr \cdot v(r) = \int_{S} dS \cdot (\nabla \times v) \equiv \int_{S} dS \cdot (\nabla \times v_{\perp}), \qquad (F.15)$$

it follows that any contour integral of $v(r)$ is completely determined by $v_{\perp}(r)$. Here, in (F.15) γ is a closed contour delimiting a surface S, while in (F.14) S is a closed surface delimiting the volume V.

Further Reading

1. G. A. Korn and T. M. Korn, *Mathematical Handbook for Scientists and Engineers* (Mc Graw-Hill, New York, 1961).

Appendix G
The d-Dimensional Coulomb Potential

The d-dimensional Coulomb potential, $V_d(r)$, is the solution in all space $(0 < r < \infty)$ of:

$$-\frac{1}{\Omega_d}\nabla^2 V_d(r) = \delta(\mathbf{r}),\tag{G.1}$$

i.e., the Poisson equation for a unit charge density, $\delta(\mathbf{r})$. Using the properties of the Dirac delta function $\delta(\mathbf{r})$ it follows from (G.1) that:

$$-\frac{1}{\Omega_d}\int d\mathbf{r}\,\nabla^2 V_d(r) = 1\tag{G.2}$$

and

$$\nabla^2 V_d(r) = 0 \quad r \neq 0.\tag{G.3}$$

Because $V_d(r)$ only depends on $r = |\mathbf{r}|$, there is no contribution of the angular part of ∇^2 and (G.2) becomes:

$$\left[r^{d-1}\frac{\partial V_d(r)}{\partial r}\right]_{r=0}^{r=\infty} = -1,\tag{G.4}$$

while (G.3) yields then:

$$r^{d-1}\frac{\partial V_d(r)}{\partial r} = constant \quad r \neq 0.\tag{G.5}$$

Solving, instead of (G.5), the equation:

$$\frac{\partial V_d(r)}{\partial r} = c_1 r^{1-d}\tag{G.6}$$

© Springer Nature Switzerland AG 2021
M. Baus and C. F. Tejero, *Equilibrium Statistical Physics*,
https://doi.org/10.1007/978-3-030-75432-7

yields:

$$V_d(r) = c_1 \frac{r^{2-d}}{2-d} + c_2,$$ (G.7)

which is not well defined at $r = 0$ for $d2$. For $d = 2$, Eq. (G.6) yields, $V_2(r) = c_1 \ln r + c_2$, while if we replace (G.4) by:

$$\lim_{a=0} \left[r^{d-1} \frac{\partial V_d(r+a)}{\partial r} \right]_{r=0}^{r=\infty} = -1,$$ (G.8)

so as to regularize the $r = 0$ singularity for $d = 3$, Eq. (G.8) is seen to imply $c_1 = -1$. We thus have, $V_1(r) = -r + c_2$ or $V_1(r) = -r$ since c_2, is immaterial here. We also have, $V_2(r) = -\ln r + c_2$ or $V_2(r) = -\ln\left(\frac{r}{a}\right)$, with $c_2 = \ln a$ since the argument of the logarithm has to be dimensionless, a being an arbitrary length scale. Finally, $V_3(r) = \frac{1}{r} + c_2$ with $c_2 = 0$ so that $V_3(\infty) = 0$.

Summarizing, we have thus:

$$V_3(r) = \frac{1}{r}$$ (G.9)

$$V_2(r) = -\ln\left(\frac{r}{a}\right)$$ (G.10)

$$V_1(r) = -r.$$ (G.11)

It should be observed that both $V_3(r)$ and $V_2(r)$ are singular for $r = 0$, while $V_1(r)$ is not and that both $V_2(r)$ and $V_1(r)$ diverge as $r \to \infty$, while $V_3(r)$ does not. It can thus be said that $V_2(r)$ operates some sort of transition between $V_3(r)$ and $V_1(r)$. Note finally that $V_d(r)$ is long ranged for all d-values.

Further Reading

1.   G.A. Korn, T.M. Korn, *Mathematical Handbook for Scientists and Engineers* (Mc Graw-Hill, New York, 1961).

Index

© Springer Nature Switzerland AG 2021
M. Baus and C. F. Tejero, *Equilibrium Statistical Physics*,
https://doi.org/10.1007/978-3-030-75432-7

Printed in the United States
by Baker & Taylor Publisher Services